ROWAN UNIVERSITY
CAMPBELL LIBRARY
201 MULLICA HILL RD.
GLASSBORO, NJ 08028-1701

Weeds

HISTORY OF THE URBAN ENVIRONMENT

Martin V. Melosi and Joel A. Tarr, Editors

WEEDS

An Environmental History of Metropolitan America

ZACHARY J. S. FALCK

UNIVERSITY OF PITTSBURGH PRESS

Published by the University of Pittsburgh Press, Pittsburgh, Pa., 15260
Copyright © 2010, University of Pittsburgh Press
All rights reserved
Manufactured in the United States of America
Printed on acid-free paper
10 9 8 7 6 5 4 3 2 1

Library of Congress Cataloging-in-Publication Data

Falck, Zachary J. S.
Weeds : an environmental history of metropolitan America / Zachary J.S. Falck.
p. cm. — (History of the urban environment)
Includes bibliographical references and index.
ISBN 978-0-8229-4405-8 (cloth : alk. paper)
1. Weeds—Social aspects—United States. 2. Urban ecology—United States. 3. Cities and towns—United States—Growth. 4. United States—Environmental conditions. I. Title.
SB612.A2F25 2010
581.6'520973091732—dc22 2010046093

3 3001 00997 7017

For Nicole

CONTENTS

FIGURES

PREFACE

I often wondered when I was researching and writing this book if there was a better word for the plants that I was studying than weeds. Frederic Clements's *vigorous growers*, Edgar Anderson's *peregrinators*, Viktor Muhlenbach's *synanthropes*, Martha Zierden and Bernard Herman's *disturbance plants*, and other scientists' and writers' colorful words and phrases appeared potentially useful. However, the inventors of these words gave them specific meanings that do not apply to all of the plants that appear in this history or describe them accurately. To avoid misusing such terms, I devised the phrases *fortuitous flora*, *happenstance plants*, *metropolitan photosynthesizers*, and *urban herbs*. Occasionally, I have coupled adjectives such as *dense*, *motley*, *ordinary*, *uncultivated*, and *vigorous* with the plain words *plants* and *vegetation*. When and where these plants were especially beneficial or superbly situated, I have employed the phrase *just plants*. Nowhere in this book do the words *weed* and *weeds* express my judgment of plants. *Weed* and *weeds* are used only to convey the attitudes, feelings, and perceptions of past people who were predisposed or committed to maligning some plants and of past people who were unable or unwilling to name the plants around them accurately.

Past Americans' ideas about so-called human weeds posed a similar problem, although I did not develop a parallel solution. Where the phrase *human weeds* appears in this book, it only refers to my understanding of some past people's views of other people. The reader should neither conclude nor infer that I consider any people, past or present, to be human weeds. However, it seems acceptable to me to describe our species generally as human weeds, as in: all people since the dawn of human origins have been human weeds. Whether such thinking is useful in addressing local and global environmental challenges is for the reader to decide.

ACKNOWLEDGMENTS

The advice, assistance, encouragement, expertise, generosity, guidance, instruction, kindness, openness, patience, and spontaneity of dozens of teachers, colleagues, archivists, librarians, professionals, and friends sustained my effort to research and write this book. I am especially grateful to John Anderson, Kenneth Cobb, Edward Curtis IV, Gerald Early, Michael Friedlander, Ty Gourley, William Grand, Richard Grigsby, Andrew Hurley, Douglas Lambert, James Longhurst, Malcolm Nicolson, and Grant Snider for what they taught me and what they made available to me along the way. Caroline Acker, Iver Bernstein, Scott Sandage, John Soluri, and most importantly Joel Tarr were enthusiastic about my endeavors and thoughtfully evaluated my writing; without them, this book would not be. Two anonymous reviewers for the University of Pittsburgh Press scrutinized the manuscript and provided valuable constructive criticisms. The University of Pittsburgh Press's Cynthia Miller, Deborah Meade, Alex Wolfe, and Ann Walston, copy editor Carol Sickman-Garner, and indexer Dave Lujak pushed me to improve the manuscript in many ways and saved me from making many embarrassing mistakes as well.

At different times and in various ways during the past two decades, I received tremendous support from Washington University in St. Louis and Carnegie Mellon University. Friends and family beyond St. Louis and Pittsburgh opened their homes to me for days and sometimes for weeks at a time. The astounding generosity and graciousness of Sean Gorman, Eli and Andrea Varol, Tom Prugh and Evanne Browne, Terra Ishee and Derek Stroup, Fred and Dorothea Reading, Devon Cox and Paulette Reading, Nicole Kleman and Jason Bowers, Lori Bowes and Gavin Haentjens, and John Mullen allowed me to undertake much of my research. Closer to home, even when they lived far away, Noah Falck, Maxine Falck, the late Sara Lloyd, and the late Harold and Freda Sopher supported me before and throughout this project. Long ago, visits to see the late Sallie and Jim Riber and the late Grace and Newman Bryant were filled with opportunities to wander around and wonder about the world. Jane and Russel Falck and Nicole Reading always respected this adventure, and they made sure that I never lacked food, shelter, communication and transportation technologies, open ears, devoted readers, or love. Their gifts and sacrifices allowed me to get deep enough into the weeds to discover as much as I could about the past but also prevented me from getting so deep in the weeds that I lost sight of the good things in the present.

Weeds

INTRODUCTION

Placing Weeds in History

The memorials for George Washington, Thomas Jefferson, and Abraham Lincoln in Washington, D.C., constitute what historian Alan Havig has called "the nation's most significant garden of honor." Abundant granite, limestone, and marble sustain this flowering of American ideals year-round. The ribbons and rings of concrete and asphalt leading to the structures are traveled by millions of people each year. Swaths of grass, small groves of trees, and shrubs decorate this inorganic splendor, but they offer little sense of nature in Washington. However, a mile and a half northwest of the Lincoln Memorial is Lock 3 of the Chesapeake and Ohio Canal. Water whishes through the lock gates, and vehicles' worn struts thump when meeting 30th Street's gentle rise. Chickweed, dandelions, grasses, Indian strawberries, kidneyleaf buttercups, pimpernel, purslanes, white clover, and yellow wood sorrel—plants that have colored the District's landscape for decades—border a bronze marker identifying Georgetown as a Registered National Historic Landmark (RNHL). A few feet away, under the shade of a basswood tree, a bronze bust of Supreme Court Justice William Douglas gazes west.[1]

These modest memorials recognize efforts to shape and regulate Georgetown's environment. Installed in 1977, the bust of Douglas celebrates his leadership in

Figure I.1. Looking west from 30th Street NW, Georgetown, Washington, D.C., 2010. Photograph by Nicole Reading.

protecting the Chesapeake and Ohio Canal, which had become a National Historical Park in 1971. The RNHL marker certifying Georgetown as a site possessing "exceptional value in commemorating and illustrating the history of the United States" dates from 1967. In 1954, Douglas helped turn public opinion against a plan to construct a highway next to the canal by leading naturalists, sportsmen, and newspaper editors on a 170-mile hike along the canal from western Maryland into Georgetown. He imagined the corridor to be like "a wilderness area . . . a place not yet marred by the roar of wheels and the sound of horns," right outside the "Capitol's back door." Congress had earlier passed the Old Georgetown Act in 1950 to establish the "Georgetown Historic District" in order "to preserve and protect" the neighborhood's historic sites, array of architectural styles, and compact urban form. Indeed, efforts to retain Georgetown's historic amenities began before World War II. The National Park Service restored the canal from Georgetown to Seneca to create "a recreational waterway" and opened it in 1940. In 1924, Georgetown civic leaders persuaded the District's zoning commission to limit the height of new structures in Georgetown's residential areas to forty feet. They worried that the construction of large apartment buildings would increase population density and automobile congestion. But while the bronze and brick on this parcel attest to the surrounding environment's historical significance, they leave the lot's environmental history unacknowledged.[2]

Figure I.2. Looking west from C&O Canal Lock 3 in Georgetown, circa 1939. Photograph courtesy of Chesapeake & Ohio Canal National Historical Park.

The grand landscape symbolized by the Douglas bust and the historic charm proclaimed by the RNHL marker rest on ground that for many decades lacked "exceptional value" and seemed detrimental to Georgetown. The parcel on which the RNHL marker sits belatedly realized substantial monetary value in 1957, when the federal government acquired it for the National Capital Parks System. In condemnation proceedings, the government paid $42,500 for 3,449 square feet of land assessed at $2,587. Twelve years beforehand, Georgetown Citizens Association members had claimed that "very unsanitary" housing in the area caused "trouble to the authorities and the community." Houses on 30th Street from M Street south to K Street, where many African Americans lived, lacked electricity and plumbing. Some properties had "yard toilets"; others sent wastewater into the canal. When a proposal was made to turn the canal into a "high-speed roadway" in the mid-1930s, the canal bed was a mix of crumbled canal wall, vegetation, and pools of pollution. Grasses and herbs grew between leaning fence planks and the canal's edge. Ailanthus trees shaded wooden sheds. Automobile owners parked next to the canal, which had ceased operating in 1924. As the canal's economy had deteriorated, older, wealthier residents came to believe a slum was emerging. In the first years of the twentieth century, the land around the RNHL marker was earth perennially home to whatever plants happened to be growing there. Happenstance plants had probably been there for much of the preceding century as well. From the division of John Southgate's property into the canal corridor and a somewhat trapezoidal parcel in 1829, until that parcel's remainder was purchased by the government in 1957, the property's various owners never built on it.[3]

Although the lot seemed to lack enough space to be useful, Washingtonians did make use of it to reject ordinary plants as weeds that did not belong in the nation's capital. Shortly after real estate broker Galen Green acquired the land in May 1906 for ten dollars, his management of it ensnared him in court. Perhaps too busy speculating in land in and around the District and luring would-be home buyers into the sparsely populated Brightwood neighborhood, Green had not tended the small patch, and in September, officials determined the plants on this ground violated the District's weed law. The condition of Green's land, however, was common. Although from 1870 to 1900 Washington's population increased from sixty thousand to three hundred thousand people, and its developed area increased from three to ten square miles, undeveloped land inhabited by happenstance plants such as those on Green's parcel stretched from the center of the city to its edges and beyond. Fleabane, moth mullein, and bull thistle lived on the grounds of Georgetown College. Chicory and ragweed sprouted annually where buildings had not. Carpetweed, false daisies, and yellow clover reached up from unpaved streets. Such fortuitous flora demonstrated that Washington and nature were not so far apart. However, the head of the District's commissioners labeled such plants "a menace to the public comfort and safety," and the *Evening Star* listed weed removal, like smoke prevention and building height regulations, among the local laws passed in 1899 that would promote the city's "grandeur and beauty." After Green's challenge to the law was defeated in March 1907, Washingtonians could hope that weeds would not mar their city in the future.[4]

The District's 1899 weed law arbitrarily prohibited plants "4 or more inches in height" and did not identify particular plants as weeds. This crude definition was a biologically determined aesthetic assessment. However, plants' biologies and people's judgments do not necessarily or neatly intersect. Today, some scientists decide whether plants are weeds on the basis of biological characteristics, such as the abilities to germinate in diverse environments, to grow rapidly, and to produce high volumes of seed continuously. People's complaints about unsightly or useless plants are aesthetic impressions that often form when unsightly or useless plants abound. How a plant functions and how people perceive a plant are potentially conflicting rather than reinforcing aspects of what makes a plant a weed. Plants that grow rapidly are not necessarily useless. Plants that are considered unsightly may not germinate in diverse environments. That any plant can be a weed and that no plant is always a weed mean that plants are only thought to be weeds at particular times and in particular places. For this reason, plants called weeds are best understood with a historical perspective. This book is about such plants in twentieth-century American cities.[5]

Plants maligned as useless and worse by past city dwellers, rather than being organisms that require no study, can be used to understand the evolving relationship of city people and the natural world. Ceaseless changes have made happenstance plants natural vegetation in cities. The details of when and where urban Americans believed that such plants were "out of place"—weeds—present opportunities to examine the nature of urban growth. Since these plants endured the twentieth century, investigating their history helps characterize ecology in cities. As they arrived, thrived, and died, just plants contributed to the development of metropolitan landscapes by contributing to how cities felt, looked, smelled, and sounded. They influenced how people experienced growing, changing, and deteriorating cities. Moreover, the ecological dynamics created by hostility toward them continue to shape urban environments and places beyond.[6]

Ecological processes occurring across municipal borders and throughout metropolitan areas integrated cities with the natural world even as urban life influenced the workings of the natural world. Growing cities were entangled with ecological processes, and they altered the ecologies of organisms living in and around them. Understanding the ecology in and the ecology of American cities requires an encompassing sense of the distant bounds of American urban life. *Urban* is not a mere synonym or euphemism for central business districts or slums. Cities extend beyond the densest creations of asphalt, concrete, glass, and steel. Cities consist of towers and tunnels, gardens and parks, residential neighborhoods and industrial sites, filled-in sinkholes and restored streams, dry streambeds and steep hillsides. Cities are linked by power, communication, and transportation technologies. Happenstance plants do not recognize the borders between cities, suburbs, countryside, and wilderness areas. Ubiquitous just plants sprouting in and moving around cities in ecological time have long blurred the edges of urban places, metropolitan areas, and the environments surrounding them.[7]

Throughout the twentieth century, the relationships of people and plants in cities changed as ecological processes were interrupted, neglected, and encouraged. Evolving environments made plants charming or dangerous, even as plants also made environments desirable or chaotic. Metropolitan photosynthesizers inhabited landscapes of change, from the forgotten pockets of old neighborhoods to the newest fringes. Making cities cleaner, healthier, and safer by combating weeds improved and expanded urban life. Yet progress in eradicating weeds most often resulted from developing an area fully, which altered local economies and landscapes, produced newly vacant land, and promoted additional growth of happenstance plants. In areas perceived to be in decline, weeds seemed to damage buildings, sidewalks, parking lots, and roads. In environments of despair, weeds sometimes intensified poverty and danger. Weed-dominated spaces that deflected investment could reinforce and provoke changes elsewhere in metropolitan areas. Some weeds that created problems for urban Americans grew well beyond where these people lived and where concrete and asphalt sealed much of the ground, and control of such weeds generated tensions with surrounding environs. The perpetual growth of urban herbs and people's frequent attacks on them were a recurring pattern, an ecology of cities that shaped vegetation. Discontent with weeds, like other city dwellers' dissatisfactions, compelled outward urban growth. Failing to accept or appreciate these plants, condemning them instead, alienated people from ecological time.[8]

While cities may seem less natural and be less full of nature than lightly settled places, their disturbance dynamics are important to study because these dynamics rework land across the globe. Urban environmental historians often examine what alternately fascinates some scholars about the lives of people and others about the lives of different creatures in the same spaces. Integrating such knowledge requires working carefully with ecological ideas to detect how patterns in nature shaped the past and where nature is an outcome of history. This work is paralleled by the interests of scientists who research "human-dominated" ecosystems to learn how the ways in which people live within cities alter and respond to changes in ecological systems. When skillfully fused, history and ecology advance our understanding of how past urban landscapes became the places where we now live and where people around the world will someday live.[9]

Since twentieth-century cities contained only an iota of past forests, prairies, croplands, and pastures, and since cities allowed people to move to rhythms other than those of the sun, many scientific and sentimental Americans believed that cities, in the words of geographer Yi-Fu Tuan, had a great "distance from nature." However, cities remained embedded in the natural world; ecological time flowed through them with the operation of biochemical cycles and the repetition of the seasons. Over ecological time, the order imposed by destruction—whether a tornado ripping up the ground or a city block being razed—faded as plant and animal populations appeared, disappeared, and sustained an ever-changing rhythm of life. The building, destruction, reconstruction, neglect, refinement, and preservation of cities frequently interrupted and impeded this time, but did not terminate it. Ecological time quietly advanced, especially in spaces that home owners, real estate agents, and municipal officials labeled abandoned, empty, neglected, undeveloped, or vacant. Where this time was halted or undone, nature may have seemed absent. People's "distance from nature" in cities was not just spatial, then, but temporal, and this temporal distance intensified and increased as city dwellers interfered with and tried to undo the flow of ecological time. The frequent collisions of nature, culture, ecology, economy, and society in urban environments scrambled time. Happenstance plants from myriad places and via many movements shaped spaces and influenced how city dwellers experienced the world around them.[10]

The persistence of happenstance plants, which makes them useful for studying the nature of cities, derives from their long coexistence with people. In 1952, botanist Edgar Anderson wrote that "the history of weeds is the history of man." Although Anderson was skeptical of using historical periodizations such as the Classical Era to comprehend weeds, peoples' relationships with the plants do provide insights into different periods. Historian Alfred Crosby's statement that "weeds thrive on radical changes, not stability," conveys ecologists' concept of ruderal plants—plants adapted to settle or survive in disturbed places and drastically transformed spaces. Taken together, these remarks suggest that the plants are recurring markers of and participants in times and places of change. The plants inhabited land during the emergence and transformation of the United States, from Europeans' colonization of the Americas, through Americans' creation of farms across the continent, to the ongoing city-building process. For more than three centuries, people living in America have commented on the weeds around them. Weeds seemed to make environments inhospitable to colonizers and settlers; they have occupied war-torn landscapes and complicated agriculture. Americans' familiarity with weeds has also enabled them to use the ubiquitous plants to talk about their times and their social and cultural worlds, as when some frustrated and alarmed Americans have compared people with weeds.[11]

The colonizers of America did not find or create a weed-free land. Virginia Company leaders considered "idle and wicked" immigrants among "the weedes and ranknesse of this land." In 1621, members of Parliament debating the importation of tobacco referred to the colony's key export as a "vile weed." *Datura stramonium*, which contains the alkaloids hyoscyamine and scopolamine, sickened British troops who ate the plant during Bacon's rebellion. After rebels burned Jamestown, the plant also supposedly grew from the ruins. This landscape change, folklorist Charles Skinner later wrote, was why a plant Europeans called thorn-apple became known to Americans as Jamestown or jimson weed. In New England, Thomas Hooker sowed biblical imagery by comparing sinners to "wildernesses overgrown with weeds." John Cotton, who emigrated from England to Massachusetts in 1633, wrote in 1620 that those who seized "vacant soyle" and cultivated it won the "Right" to it. Colonists eager to cultivate "vacant" land may have been eager to identify it by finding weed-covered ground. Such an outlook may explain in part why they observed weeds among the Native American women's polyculture plantings. Francis Pastorius, who emigrated from Frankfurt in the 1680s and established Germantown outside of Philadelphia, compared the area to other places in the world where "Thorns, Thistles, Tares and noisom Weeds abound." By the end of the seventeenth century, colonists' fields and roads were home to ordinary European plants such as dandelion, henbane, wormwood, and mayweed.[12]

While revolutionaries likened traditional English politics to weeds in a field, after the Revolutionary War, Americans employed the word *weeds* to criticize democratic culture and politics as well as social trends. In July 1776, congressional delegate Benjamin Rush wrote, "The republican soil is broke up—but we have still many monarchical & aristocratical weeds to pluck up from it." In 1800, Federalist Joseph Dennie, writing as "an Enemy of Innovation," dismissed the American neologisms that Noah Webster intended to include in his dictionary as "noxious weeds." Dennie's *Gazette of the United States* compared Webster to "a Maniac gardener, who, instead of endeavouring to clear his garden of weeds, in opposing to reason, entwines them with his flowers!" These remarks exaggerated Webster's unconventionalness. The *American Dictionary* conveyed his respect for authority and passion for social control. He defined "to weed" not only as "to free from noxious plants" but also as "to root out vice." A New York newspaper ridiculed Jeffersonian ideals as impractical by printing a poem that lampooned a gentleman farmer who thought, "To root up weeds their [*sic*] is not reason, Against the *rights of plants* 'tis treason. Each has an equal right to live." When Thomas Benton criticized the Senate's censure of President Andrew Jackson for his policies toward the Second Bank, Benton dismissed claims that Jackson had ruined the economy and turned "the streets of populous cities to grass and weeds." In emerging Ohio River Valley cities, educated citizens voiced concern about the children of wage earners who were "growing up like weeds, without benefit of cultivation." Such vituperations may have helped the three-quarters of all Americans who lived on farms understand when people were objects of contempt.[13]

As seventeen new states joined the nation and new territories were established in the first six decades of the nineteenth century, even more plants became weeds. George Perkins Marsh worried that New England's deforestation produced "barren" hillsides covered with "noxious weeds" whose seeds were washed into valley farm fields. Americans moving west encountered unfamiliar plants and spread old ones as they turned forests and grasslands into fields of corn, wheat, and cotton. The *Farmer's Cabinet* urged readers to wage "a war of extermination, against weeds of every name and nature." War against every weed, however, may not have been necessary. William Darlington, an amateur botanist who wrote about agriculture, pointed out that there were different kinds of weeds. Farmers considered weeds to be "intrusive and unwelcome individuals that will persist in growing where they are not wanted." Such harmful plants differed from plants that people did not "respect" because they were "homely" or lacking "medicinal or other useful qualities." Darlington concluded that an "old" definition of a weed—"a plant out of place"—was the best one. Yet once a plant became known as an "out-of-place" weed because it was "out of place" in a field, that same plant was likely seen as an out-of-place weed regardless of what environment it was seen in. Such entrenched impressions could deter reassessing weeds despite ongoing environmental and historical change.[14]

Northern religious leaders and abolitionists identified a variety of people as weeds when discussing the conflicts leading up to the Civil War. Theodore Parker described "the national soil" as "dreadfully cumbered with weeds of two kinds . . . Whig-weed and Democrat-weed." George Cheever warned that God would harshly judge those who cherished liberty but voted "to make their fellow-creatures . . . weeds, by fastening the chains of a perpetual slavery upon them." In a fictionalized slave narrative, a narrator described her "utterly neglected" companion as "a complete human weed." On Thanksgiving Day in 1861, Henry Ward Beecher compared the South to "an immense field of nettles . . . overrun with the pestilent heresies of State rights." During and after the war, Northerners also found weeds overtaking Southern land. A soldier from Massachusetts reported that "high, coarse weeds and rank grass" covered Manassas, Virginia's rolling plains. Republican leader Carl Schurz encountered fields "wildly overgrown by weeds" and deserted planters' houses with broken windows and "yards and gardens covered with a rank growth of grass and weeds" in his postwar travels.[15]

After the war, scientists and naturalists wrote about Old and New World weeds and expressed trepidation about the disappearance of the distinctions between these worlds. Botanist Asa Gray wondered why so many weeds were Old World "foreigners." Naturalist John Burroughs claimed that most farm and garden "outlaws" came "from over seas." Eurasian plants had "pugnacity" because farmers' battles with them meant that "their struggle for existence has been sharp and protracted." In contrast, "native weeds" were "shy" and "harmless." Burroughs was not worried by either; he wrote, "Ours is a weedy country because it is a roomy country. . . . By and by we shall clean them out." English science writer Grant Allen doubted such an inevitability, reporting in 1886 that "European weeds and wild flowers" had become widespread in the United States. He remarked, "All the dusty, noisome and malodorous pests of all the world seem here to revel in one grand congenial democratic orgy." Allen, who was disturbed by the predominance of southern and eastern European weeds, predicted that "in the cosmopolitan world of the next century the cosmopolitan weed will have things all its own way." Allen envisioned a future when "the whole earth will be one big dead-level America . . . a single uniform landscape of assorted European, Asiatic, American, African, and Australian weeds" and the plain architecture of crematoriums, Salvation Army barracks, and drainage works. Allen's view conditioned how some Americans saw their country. To counter such criticisms, journalist Henry Mann wrote that America's motley but energetic people consisted of a lower "percentage of human weeds and refuse than any other nation on the globe." Other Americans, however, agreed with Allen. California herbarium curator Alice Eastwood thought the "cosmopolitan" nature of civilization worked to reduce people and plants "to one dead level." Willis Blatchley, an Indiana scientist uncomfortable with immigration, added to Allen's "democratic orgy" assessment by contending that America's "soil seems to suit exactly those weeds which are the offscourings and refuse of civilization in all countries."[16]

The persistence of weeds sustained interest in what weeds were and what plants were weeds. Ralph Waldo Emerson challenged traditional views by redefining weeds as plants whose "virtues have not yet been discovered" and by dreaming that innovators would use "the infinite applicability" of the plants to increase "our modern wealth." However, since weeds seemed to reduce wealth—$17 million in crop losses were attributed to them in 1896—their depredations continued to be declared. Botanist Byron Halsted wrote that "pestiferous," "vile," "execrable," and "miserable" weeds asserted "their inborn rights above all others" and fought people "for the possession of the earth." According to Samuel Schmucker, weeds "had learned to take care of" themselves, and accordingly, children should be taught "to wage war upon them." Gerald McCarthy realized that so many plants were called weeds because "distinguishing of plants as weeds and not weeds is purely human and artificial." He proposed focusing scientific research on the "obnoxious" ones with "national reputation[s]." Elmer Campbell rejected defining weeds as "out of place" because "human caprice" alone could "instantly" turn "any plant in the universe" into a weed. Like McCarthy, Campbell thought weeds should be defined only as plants that were "persistently obnoxious on cultivation areas." Orin Stevens resisted this revisionism. He believed that weeds were plants "detrimental to man's interests, displeasing to the eye or of no evident value." Weeds were, indeed, "out of place." *Science* reported that botanical research indicated that "the dandelion can literally be called the king of the weeds." In 1939, botanist Edgar Anderson attempted to change the nature of this debate by emphasizing the importance of plants' "spreading" capacities. Weeds—"plants unintentionally grown by man"—were "peregrinators," along with "cultivated" plants, plants living in "man-created habitats . . . though not actually cultivated," and plants that spread quickly "even when not associated with man." However, the scientific quest to comprehend the nature of weeds continued. In 1966, for example, Oklahoma State

University botanist Jan de Wet concluded that "true weeds" were plants that had "become adapted to the permanently disturbed man-made habitat[s]."[17]

Scientists' definitions of and ideas about the plants did not preclude the employment of the word *weeds* in debates over foreign policy and wars, present and past, as well as economic corruption and crises. In promoting pesticide use to increase garden yields during World War I, American Civic Association president J. Horace McFarland condemned the "human weeds of Germanic nations" fighting "to take all the space they want." During the 1920 Palmer Raids against anarchists and communists, former Michigan governor Chase Osborn advocated internment rather than deportation "to solve our human weed problems." Looking back at the American Civil War, William Wood, a colonel in Canada's military, concluded that immigrants who had formed peace parties in the North had been "human weeds, clogging the springs of action everywhere." Theodore Bassett, the Communist Party's education director in Harlem, identified some of the most troublesome Americans as the "anti-Negro, anti-democratic terrorist gangs [that] breed like weeds in the dank soil of intolerance created by the wanton disregard of the Bill of Rights." In 1933, lawyer Arthur Wickwire indicted speculative pools as financially corrupt "weeds" that choked financial markets and inflicted "untold losses upon innocent people." In his 1934 State of the Union Address, Franklin Roosevelt listed excess industrial production, exploitation of consumers, and child labor as weeds that had to be prevented in the future by denying them "soil in which" to grow. Herbert Hoover emphasized that the nation's problems had less to do with capitalism than with individuals who abused the system, telling Stanford University's class of 1935 that "weeds of abuse will always grow among the fine blossoms of free initiative and free enterprise." Hoover argued that it was preferable to engage in "unceasing labor" to dig out such "evil" weeds, rather than to permit "the blights . . . of governmental tyrannies" to kill "blossoms." When Hoover recalled the challenges he had faced when entering office in 1929, he restated this view: "We had our share of thistles and many sprouting weeds."[18]

During the Cold War, Americans denounced communism as an invading foreign ideology and expressed disdain for their antagonists in domestic politics by abstracting both ideas and people as weeds. The University of Notre Dame's Gerald Phelan argued that "Christian verities" could protect the nation's democracy from "the weeds of atheistic communism and agnostic liberalism." Labor leader Philip Murray argued that the involvement of government and trade unions in a "healthy, balanced economy" created "an environment [in which] the weeds of Communism find scant encouragement." The belligerent *New York Daily News* often printed the Cold Warrior's creed: "The only good communist is a dead communist." This expression updated Frank Mann's 1913 declaration in "the Farmer's Creed" that "the only good weed is a dead weed," which itself was based on the

Spread it on. Or spray it on. Either way, you get a fast and easy kill of broad-leaved and grassy weeds with Hooker MBC.

MBC leaches into the soil after rain or irrigation. Attacks roots. Sterilizes the soil for a season or longer. It kills top growth almost on contact.

MBC wipes out such hard-to-kill pests as Johnson grass, bur ragweed, hoary cress, and bindweed along ditches, roadsides, and on industrial and other noncrop land.

It is very soluble in water. Lets you clean your equipment easily.

For application information on this powerful, nonselective herbicide, write for our descriptive folder. Agricultural Chemicals, Hooker Chemical Corporation, 806 Buffalo Avenue, Niagara Falls, N.Y. 14302.

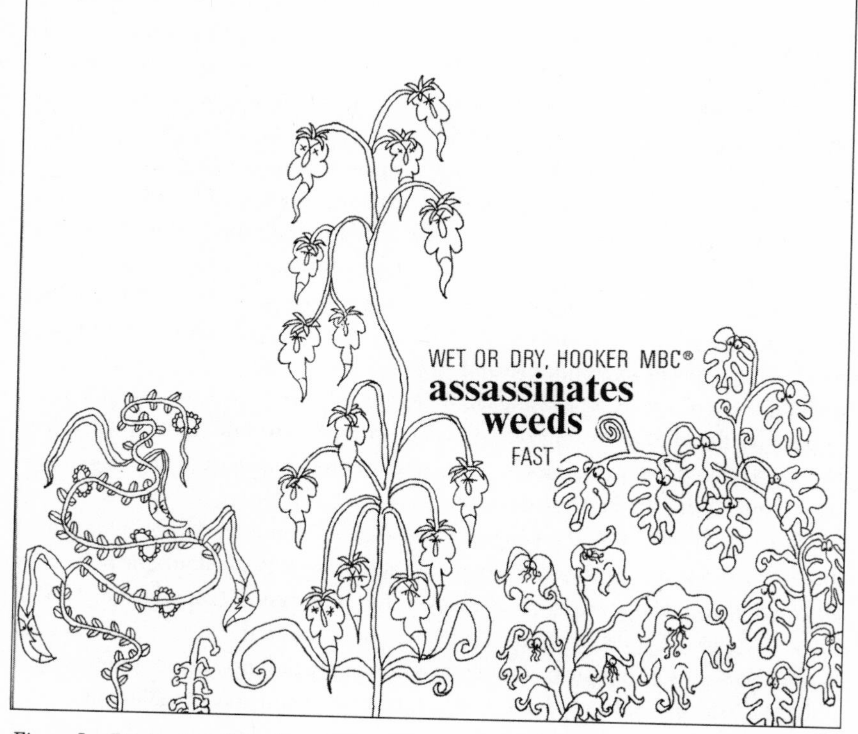

Figure I.3. By the time this advertisement ran in *Public Works* in 1967, Hooker had already buried tons of chemical by-products in land in Niagara Falls and sold the land to the city's school board, which a decade later generated the Love Canal environmental controversy. Printed with permission.

frontier expression, "The only good Indian is a dead Indian." Conservative newspaper columnist Morrie Ryskind ridiculed the inaccuracy of presidential primary polling by opining, "Noah erred when he allowed the progenitors of the pollster, a human weed, to come aboard."[19]

Weeds, when they were not irritating people working or managing land, helped some Americans grasp and express the problems of modern life, including its environmental and social challenges. As the fifth Earth Day approached, *Pasadena Star-News* editors proposed renaming this education effort "Earth People Week." Although they acknowledged protecting the earth from atomic bombs, criminals, "desecrated nature," and smog was important, the editorial writers contended that environmentalists needed to do more to clean up "the minds and hearts of some of the humans who clutter the landscape" because some of "the most obnoxious features of our environment are . . . people." The *Star-News* stance was a response to the announcement in early April 1974 that the kidnapped granddaughter of media mogul William Hearst was joining her abductors, the Symbionese Liberation Army. To the paper's editors, "humans who are like thorny weeds in our Garden of Eden" were greater dangers than smog and polluted water. Less harmful than Hearst's tormentors were the "human weeds" who created "pollution and do other things which make our earth a less healthful and pleasant place to live." The tumult of the times convinced these Californians that improving "the quality of human beings" was a necessity, even if they were uncertain of the best means. Decades of retributive justice had been a failure; the editors wrote: "We have tried eliminating [the vicious human nettles that mar our garden] in the gas chamber; we have tried terms of various lengths in prison, but as a few are removed too many more grow in their places." According to the *Star-News*, young environmentalists could do much to help out the world by shifting their energies from stewarding the earth to helping reform "unproductive" young people "likely to become criminals," since it was "possible to change what might become a weed" when a person was still young. Even as the *Star-News* editors blended the social and the natural in their use of the human weed metaphor, they conceived of social and environmental problems as distinct rather than interrelated problems.[20]

From America's colonization through the nation's modernization and beyond, environmental, cultural, economic, political, and social changes made Americans aware of all kinds of weeds. On the ground, weeds changed environments. In print, weeds informed cultural and social life. Changes in cities made the plants part of urban life as well. The fortuitous flora that grew where natural and built environments converged in and around cities were often perceived as weeds that undermined health, safety, social order, beauty, and efficiency. Weeds annoyed, inconvenienced, intimidated, poisoned, and weakened people. City dwellers believed that weeds spread infectious disease, permitted pest propagation, caused allergies, and encouraged illegal dumping. They feared weeds made cities unsafe by attracting and sheltering vagrants, drug dealers and users, and violent criminals. They saw them as blight that marred landscape aesthetics, clashed with neatly tended land, lowered property values, and degraded nature. They argued that weeds interfered with urban infrastructure. Yet some Americans were impressed and soothed by happenstance plants.

Understanding the ecology of fortuitous flora requires examining the ideas and actions of botanists, ecologists, environmentalists, gardeners, ordinary people, plant enthusiasts, public officials, and urban weed-control advocates. People who despised weeds included home owners and apartment renters; neighborhood groups, journalists, and social reformers; municipal, state, and federal government officials; and scientists, chemical manufacturers, and businesspeople. Their recurring efforts to destroy weeds are part of the historical record. City dwellers who admired, appreciated, and utilized—even defended and disseminated—these plants were less numerous and less powerful. Some urban Americans probably never noticed, thought about, or touched happenstance plants, but their unintentional inattention also contributed to the development of vegetation. Municipal, state, and federal officials' correspondence, publications, and reports, along with court cases and newspaper accounts, provide evidence regarding where and how these plants grew. The papers of scientists, ecologists, and their professional organizations, as well as academic journals and trade publications, document local landscapes, environmental change, and American environmental thought. Americans' relationships with urban herbs inspired laws, essays, work relief, scientific experiments, art, and free speech, all revealing how cities grew and changed. While cities such as Buffalo, Chicago, Lincoln, New York, St. Louis, Washington, D.C., and some of their neighboring municipalities are the focus of this study, examples from other cities, from Boston to Los Angeles, demonstrate that concern with these plants were national in nature.

Throughout the century, urban herbs were in cities as they grew outward, remained in them as their growth intensified, and continued to grow as Americans lamented their cities' declines. In the late nineteenth century and the first decades of the twentieth century, cities were unable to seize command of the ecological processes shaping urban vegetation. With the acceleration of the United States' industrial economy after the Civil War and through the strains of the Great Depression, fearful city dwellers decided that improving cities involved controlling weeds, as well as controlling the people they thought of as human weeds. Although the nation prevailed overseas in world wars, battles at home against weeds that refused to surrender informed midcentury environmental politics. As metropolitan landscapes diversified in postwar America, happenstance plants proliferated, contempt for weeds remained, and Americans sensitized to the plight of nature admired just plants in cities. People who worked to improve the livability of their cities—to make them more tolerable, humane, and beautiful—were among those who struggled to get rid of weeds, live with just plants, or win admiration for happenstance plants. Their thinking about and handling of the plants was influenced by and has contributed to the evolution and permutations of American environmentalism and the nation's nature.

CHAPTER ONE

Urban Growth and Ecological Time

One of the most perceptive guides to changing turn-of-the-century cities was Brooklyn's "urban Thoreau," Charles Montgomery Skinner. Like Thoreau, Skinner was dedicated to appreciating the pleasures provided by nature. While some New Yorkers considered Brooklyn a suburban retreat from Manhattan, Skinner did not find this urban environment ideal. Next to his home in Stuyvesant Heights were a "carpenter's yard, with its piles of lumber," a gravel dump where boys played, a board fence, and a lilac bush. The neighborhood had plenty of open "made land"—excavations or hollows filled with ashes, bedsprings, bird cages, bones, boots, bottles, cobblestones, hardware, hoopskirts, oyster cans, and spikes, all topped with pebbles, rocks, and sand. In the midst of this environmental transformation, and despite his disgust with city builders who covered "pleasant fields with unpleasant houses," Skinner claimed that a city person on a "morning stroll" to the office could experience as much nature as a lumberjack traversing eight miles of forest. The trick was to scrutinize the ordinary to discover what was fascinating. Skinner noticed how on windy days scraps of paper snagged by an "amazing patch" of cat greenbrier fluttered like "an army with banners." He gathered "plebeian" wild mustard from "mean places"—empty lots and "littered

Figure 1.1. Mulleins and child about one mile south of Chicago's Jackson Park in the early 1900s. Special Collections Research Center, University of Chicago Library.

side streets"—and transplanted it to his yard, where it "flowered copiously." He described the dandelion as an "idyllic flower . . . a creature of the sunshine; a fallen star." Skinner quipped that if Birmingham factories manufactured dandelions, "many ships would be laden with them every spring." In the *Brooklyn Daily Eagle*, Skinner or a like-minded associate opined, "How we would prize the daisy if the plants sold at a dollar!" Skinner considered happenstance plants an unbeatable bargain: "The best is the cheapest, and the very best costs nothing. . . . The beauty of nature . . . is always there for the looking and smelling and hearing."[1]

Skinner's urban nature writing amused and soothed middle-class readers who were anxious about city life. Like nature writers such as Frances Parsons and Samuel Schmucker, Skinner contextualized his observations about the environment with notes about changes in the seasons. He romanticized the virtues of the countryside for producing free, self-reliant people, and he derided the "aggregation habit" in both New York City's tenements and its opulent Fifth Avenue residences. Although he professed to prefer the country, Skinner had decided to live in the metropolis and make the most of the nature he found there. In *Flowers in the Pave*, Skinner celebrated the magic and comforts of the natural world rather than subjecting them to scientific analysis. He imagined the joy of venturing into a "plushy roadside growth" of asters and setting "a million seeds afloat." Daisies and goldenrods demonstrated to Skinner the "kindness of nature" that he was sure would "cure us of our meannesses in time," even as he wondered "what will humanity be" when cured.[2]

Migrants like Skinner and immigrants from around the world who were finding and making homes in New York City accelerated changes in its landscape. Skinner's *Nature in a City Yard* appeared one year before the 1898 consolidation of New York City—the merging of Manhattan, the Bronx, Brooklyn, Queens, and Richmond—that established a municipal entity with wide-ranging boundaries incorporating many different environments: alleys and elevators without sunlight, cobblestone streets under elevated rail lines, cabbage fields, nurseries with weeping beech and white mulberry trees to be planted throughout the city, creeks blackened with oil and sludge, and bathtubs filled with water from upstate lakes. As new buildings reached toward the sky and as new infrastructure spread underneath the ground, the space for wildflowers shrank. Elizabeth Britton, a bryologist, thought that New York City's growth demanded the protection of delicate, rare plants like the arbutus. Britton, who with her geologist husband, Nathaniel Britton, and the Torrey Botanical Club had established the New York Botanical Garden in the Bronx in 1891, founded the Wild Flower Preservation Society of America in 1902. Britton identified the "unavoidable" and "continued extension" of New York as the greatest danger to plants, although she chastised acquisitive picnickers and wildflower hunters for accelerating their destruction. Botanizers in other cities also commented on how urban growth affected plant populations. A Chicago wildflower admirer pointed out that more truck farms producing more vegetables for more city dwellers meant that the places where people searched for "wild beauties" were not "what they used to be" and that "the prettier" plants would meet "certain death." William Rich, a librarian at the Massachusetts Horticultural Society, worried that Boston's "profusion of plant life" would someday be found only in collectors' herbaria. Rich reported that yellow patches of garden loosestrife had disappeared under the new buildings constructed to advance the pursuit of knowledge at the Massachusetts Institute of Technology. In Washington, D.C., wildflower seekers noticed that efforts to fill in wetlands around the city reduced habitat for nimblewill grass and tenangle pipewort.[3]

The environmental changes that threatened some plants nevertheless created spaces that attracted plant enthusiasts. Boston botanist William Whitman Bailey wrote that "waste places and open lots about our cities" were "desirable ground for the beginner . . . [to] find any number of plants with which it is well to become familiar. They can do no harm where they are, except by circulating their seed." William Rich noticed that places near Boston's busy streets and residential avenues made the city "a desirable field of botanical observation." He found hooked bristlegrass in a front yard a few streets north of the Boston Common, and he spotted oakleaf goosefoot in a brick sidewalk near the public garden at the edge of the common. The white rays around the yellow disk flowers of a gallant soldier crawling out from under a granite wall and "growing freely" near a downtown subway entrance grabbed his attention. The "vacant lots and dumping grounds" of the South Boston flats, railway terminals, and the gravel-filled Back Bay area provided an "almost inexhaustible supply" of plants to observe. Globe thistles and Japanese knotweed, which Rich called "outcasts from the Park," were "happy and thriving" on rocky banks. In the District of Columbia, naturalists noticed alpine yellowcress "growing without cultivation" near the Washington Monument and marsh yellowcress living "in the streets and gutters." In June, ambitious botanizers could search for "fugitive specimens" of American dragonhead on ground a half mile north of DuPont Circle's traffic rotary. Theodore Dreiser wrote that tarweed and sneezeweed were among the most abundant plants in Atlanta; that squirrel-tail grass and Russian thistle stood out in Denver; and that wild licorice, spiny cocklebur, wild heliotrope, and milk thistle were among the most common plants in San Jose. In San Francisco, Alice Eastwood, curator of the California Academy of Sciences' herbarium, admired "the wilderness of plants" thriving between cable-car tracks on steep Nob Hill. She called them "oases of verdure" where nature had started "to cover over the defacements of man and once again make all beautiful," but she also lamented the vegetation's increasingly "cosmopolitan" nature.[4]

In changing cities, some plants disappeared, others flourished, and new ones arrived. Where people were moving, happenstance plants were ready to live with them. City dwellers' experiences of their urban environments depended in part on their relationships with these plants. New York's Wild Flower Preservation Society advocated saving rare plants; it did not defend all urban herbs. Britton admitted that asters, black-eyed Susans, buttercups, clovers, daisies, dandelions, goldenrods, violets, and wild carrots could be "picked in large quantities, without likelihood of extermination." Skinner accepted happenstance plants as pleasant vegetation. The plants that he listed as "indigenous" to his yard were from all over the world: aster, chamomile, chickweed, clover, daisy, mallow, oxalis, pigweed, purslane, ragweed, smartweed, sorrel, thistle, wild parsnip, and yellow-dock. They demonstrated to him "how largely vegetation prefers to consist of weeds." Skinner considered his yard's "wild corner" of buttercups, daisies, dandelions, goldenrod, and violets "the prettiest and most reliable part." He encouraged thrifty gardeners not "to buy anything. Raise wild flowers. Every vacant lot has them." Britton's attention to marvelous rare wildflowers suggested that desirable nature was shifting farther away from cities; Skinner's appreciation of fortuitous flora indicated that these plants belonged in cities and planted nature in urban environments.[5]

Happenstance plants could be vanishing or plentiful, and when they were the latter, some botanical-minded urban Americans could not tolerate them. Members of St. Louis's Engelmann Botanical Club occasionally gathered to admire "native flowers" that they had collected but also trained schoolchildren to identify and destroy "common weeds" on vacant lots. U.S. Department of Agriculture (USDA) botanist Lyster Dewey considered some urban herbs to be environmental problems because they migrated to and damaged carefully tended croplands and sullied neat rural towns. Dewey claimed thistles and burrs detracted from the "fine" lawns and shrubbery of Washington residences. He monitored a vacant lot about a half mile west of his Wallach Place home, where a wiry rush skeletonweed was pushing its taproots deeper into the ground and generating plumed seeds to ride in the wind. Dewey recommended that cities burn all coarse weeds every autumn.[6]

Dewey and Skinner may not have agreed about the virtues and dangers of the plants around them, but both were engaged with the plants growing on evolving urban land. When cities expanded their municipal borders and where residents built new homes, lines on maps, words on deeds, entries in tax ledgers, and bird's-eye views did not capture the nature of these environments. Cities were not only buildings, streets, and technological systems; they were also the ground where people's cultural, economic, and social activities mingled with ecological processes. From some time before the day when land was acquired for the creation of new city blocks to the day when people began inhabiting new houses and operating new businesses, just plants populated the land. What plants had been there, what plants were there, and what plants should or could be there changed with ecological time, as cities were built and as perceptions about plants and spaces took root. Happenstance plants complicated the development of ideal urban environments. They thrived on small and large pieces of property making up patchwork urban landscapes and the ever-shifting, rough edges shared by—yet separating—cities and their hinterlands.[7]

Adding new plants to urban landscapes, giving new meanings to old plants, and removing plants were all attempts to reshape the nature of cities. Cities planted trees in parks and along streets, and home owners established lawns and gardens. Both uprooted vegetation and tried to prevent the growth of weeds. While living within the natural world, urban Americans attempted to eradicate weeds to distinguish their cities from the past and nearby surrounding landscapes, to define urban space, and to delineate relationships between people and plants in urban society. City dwellers cared for and fought with plants to shape the ecology of urbanization. The experiences of people living in St. Louis and Washington who struggled with weeds and failed to banish unwanted plants from their cities best illustrate that idealized environments were not perfectible. In addition to being shaped by people's desires and labors, American cities were shaped by persistent past ecologies and imperfect new ecologies.[8]

Over the course of the nineteenth century, American cities grew more dense and expanded in size. Older sections of cities became crowded as propertyless individuals sought housing, building regulations and restrictions impeded construction, and cities invested modestly and haltingly in infrastructure. Cities gained territory when state and federal governments approved their annexation of neighboring settlements and acquisition of tracts of adjacent, unincorporated land. Visual representations emphasized dense construction, not the outer city where few people lived. In St. Louis, the bustling waterfront of steamboats and warehouses was at the center of views drawn by John Caspar Wild, George Hofmann, Charles Parsons, and Lyman Atwater. In the District of Columbia, the Capitol, federal institutions, monuments, and wide avenues dominated the urban landscapes drawn by George Cooke, John Bachmann, Edward Sachse, and Theodore Davis. Representing the great energy of urban life did not require illustrating the ever-changing patches, expanses, and corridors of grasses, herbs, and trees between built-up areas and that buffered cities' settled edges.[9]

The territorial growth of St. Louis made thousands of acres of vegetation part of the city. The common fields established for agriculture to support colonial governments' aspirations had been unevenly and lightly farmed. Shrubs, trees, and prairie grasses remained in and ringed the city. After the Louisiana Purchase, the design of a grid of numbered streets revealed leaders' hopes to create a speculative city of "endless growth." The roughly one-half-square-mile city of 1822 had expanded to four and a half square miles by 1841. Still, where buildings had yet to rise, there were hazel and sumac bushes and oak and hickory trees. St. Louis generated revenue to fund infrastructure development and improvement by annexing more land, selling off portions of it, and taxing its new owners. Many properties became held or traded among speculators. Since most infrastructure investments were made in established and politically connected parts of the city, newly incorporated areas were without streets, sparsely inhabited, and slowly subdivided. Where St. Louisans were not building, nature was nurturing clovers, chamomile, dog fennel, grasses, milkweed, mullein, and thistles. The extensive root systems of bitter dock, black nightshade, bouncingbet, and ironweed sustained the plants for a couple of springs. To casual observers these plants were "weeds familiar by sight but unknown by name." To a person who had lived in the city for two decades, "rank and full-grown weeds" created "wilderness, as it were."[10]

The creation of the city of Washington and the District of Columbia with fixed boundaries carved out of land in Maryland and Virginia established not only a national capital but also an urban core dotted with and surrounded by happenstance plants. George Washington's selection of the Potomac River site of the capital initiated speculation and real estate development that influenced vegetation.

Although Charles L'Enfant believed that the capital could grow quickly, settlement and construction proceeded slowly, in part because speculators purchased city lots and waited for them to appreciate in value. In 1800, Connecticut representative John Cotton Smith claimed that Pennsylvania Avenue was "covered with alder bushes." After the Civil War, residential building increased, and Congress created a board of public works that invested millions in street improvements and sewer construction. By the end of the century, building and urban development touched the city's edge at Florida Avenue and spread into the District's outer areas as miles of streetcar tracks carried workers to new houses. Although historian Tim Davis writes that during the nineteenth century the capital's "unkempt wilderness" of bushes and briars became an increasingly "tidy and mature" landscape, ubiquitous and persistent happenstance plants indicated that the transformation was partial. Fireweed, goosefoot, hogbite, and toadflax were among the hundreds of plants "growing without cultivation." Red sandspurry made itself at home in dirt streets. Although Washingtonians purchased plants cultivated in nurseries to remake nature around their homes, dense burdock covered a property three blocks from the White House, and chicory and ragweed patches abutted mansions. On vacant lots, wild onions braved winter chills, buttercups colored the spring, summer skies shone on wild carrots, and horseweed swayed in fall breezes.[11]

St. Louis and Washington were urban environments in which people's lives mingled with the life cycles of plants, especially in areas where development was accelerating. Both buying land to build on and building on the land changed the places where plants grew. Making roads and buildings destroyed plants, even as seeds deposited in the ground in seasons past waited for the chance to germinate in rich or depleted, moist or dry, acidic or calcareous soils. Seeds that traveled through the wind and in waters, that animals carried, dropped, and buried, and that journeyed through economies joined the vegetation that changed the land. The appearance and disappearance of plants of many heights, densities, and colors occurred and recurred throughout urban environments.

Where people walked, rode horses, and drove wagons, they squeezed air out of the soil and prevented many seeds from germinating and developing roots. Bright sunshine evaporated water and warmed the ground. On this bare, hard ground, bluegrass, white clover, shepherd's purse, and pigweed were among the few plants able to grow. Where people filled holes with sand or gravel, mat sandbur and knotweed attempted to keep loose ground in place. Where tansymustard and poorjoe took root and flowered, they left seeds that became new plants the next year. Where quackgrass stems and leaves protruded from the ground, the plants' rhizomes became a bit longer each year, even in dry summers. If the emerging vegetation was not trampled and mashed, the yellow petals of cinquefoils might appear in June, the white petals of flowering spurge could open in August and

September, and horsenettle's pale purple flowers could produce yellow berries by the fall. With ecological time, the diversity of plants increased, and they disorganized the once-packed ground. Mined or quarried land often lacked the nutrients to sustain more than a few plants, such as goldmoss stonecrop, rock polypody, and sheep sorrel. Tall tumblemustard and narrowleaf plantain waited long stretches of time for black locust or white birch trees to appear.[12]

On city land where farmers had once produced animals and crops and where manure and household wastes were dumped, a nutrient-rich soil allowed plant communities to grow dense as generation after generation produced seeds. Bluegrass and chickweed were among the fastest-growing plants when such ground was scraped again and again. Ground covered by crabgrass and goosegrass could be home to broomsedge and foxtail barley a few seasons later. Lambs quarters' red-streaked stems inched upward. Fleshy purslane covered bare spots. Prickly lettuce, which sometimes grew taller than people, produced tiny yellow and white flowers that opened in morning sunshine. Wild mustard might become knee-high in June and July as it worked to make hundreds of purple-brown seeds. These resilient ruderal plants were able to grow—and sometimes only able to grow—in fertile but fast-changing ground. Few survived for long if people regularly or harshly used this ground. Plants that may have predated European exploration and American settlement of a region flowered from time to time. However, red columbine or whorled pogonia had little chance of growing into large colonies and instead created ground in which other plants would flourish extensively. Plants that matured quickly and produced seeds rapidly, even if they were more ephemeral, more often endured a city's energy and expansion. Plants that botanists believed were rare often did not live long enough to reproduce in the places closest to cities' intense economic activities.[13]

The plants that grew quickly and produced many seeds, even in harsh years, did not necessarily grow into wide and thick plants. They left minerals in the ground that offered their seeds and the seeds of other plants a chance to germinate. Ragweed, smartweed, and velvetleaf began to appear and to return annually. If these plants were periodically cut, their seeds germinated with rain and sunshine. More tall annuals with many leaves resulted in fewer low plants. White flowers surrounding a single, tiny purple flower blossomed on top of wild carrots. In late summer, mulleins seemingly launched spikes of yellow flowers into the sky. Black medick's twisted black seedpods fell to the ground in October breezes. Where these plants grew close enough to shade the ground, the soil retained more moisture. Creeping buttercup and yarrow could start growing into larger patches that reduced the number of dandelions. Proliferating perennials led annuals to disappear. Pokeweed and goldenrod grew larger; if dumped on or trampled, they sent up new shoots that reduced how much ground was bare. If people avoided this

Figure 1.2. Vacant lot in southwestern Washington, D.C., in the early 1900s. Library of Congress, LC-Z71-17.

ground, chicory and milkweed could appear, and the blackberry lily's orange flowers might bloom in June and July. Tall tickseeds might top nine feet.[14]

Plants worked above and below the ground to turn minerals and water into shoots, leaves, flowers, and seeds. As plants transformed the ground, the right conditions might emerge for new species to start growing and join the rhythm of the seasons. Over a decade or two, multiflora roses, blackberries, smooth sumac, and tuliptrees could become thickets that people had to travel around. Poison ivy and Japanese honeysuckle could form colorful islands. Plants produced seeds more slowly, and opportunities for seeds to germinate became fewer. They quietly announced that trees like white poplars might become part of the urban landscape, if the land remained peripheral to people's daily lives.[15]

The acquisition of land for speculative purposes by "gentlemen" farmers and the planned transfer of intensively worked farmland to people ready to grow real estate from the property allowed urban herbs to thrive. Declines in farmers' incomes or profits when prices for their commodities fell or their production costs increased led some farmers to sell their land to real estate speculators. Real estate for urban markets was also created when farmers died and their heirs sold their property. Some farmers continued to plant crops until they could maximize the profit from selling land. After a farmer's final harvest of crops, seeds that had accumulated in the soil over the years had the chance to grow. The plants bequeathed to cities also varied with the animals farmers had kept. Where goats, sheep, cows, and geese had been plentiful, stramonium, wild chamomile, Canada thistle, and wormseed were sometimes common because these animals did not eat them and check their spread. Year after year, plants tended dozens or hundreds of acres of speculatively held land traded among speculators. Speculating in, subdividing, and developing this land—especially annexed land—permitted ecological time to pass and happenstance plants to grow.[16]

Slicing and cutting large properties into small parcels and single lots of less than a quarter acre perpetuated these plants' growth. Pumpkin patches did not become new neighborhoods in midnight moonlight. They became patches of pigweed, purslane, and pokeweed. Some speculators bought real estate from afar, site unseen. Slopes, sinkholes, or isolation delayed transforming these properties into building sites. Real estate developers borrowed money or mortgaged land to raise funds for leveling, grading, filling in ground, surveying and staking out lots and streets, and hiring contractors to lay pipes to make land into urban real estate. Landowners used advertisers and real estate agents to sell individual parcels of land to other speculators and future home owners. The number of available lots in many cities far exceeded not only the number of people who could buy land and gradually build a home but also the entire population of cities. Excessive speculation diluted demand. Depressions bankrupted builders and home owners. Property owners waited for profits to bloom among fallow fields. With each passing July that speculators waited for land values to appreciate, mulleins grew taller. Horseweed and horehound sprouted as easily as houses. The *Washington Post* described vacant lots without homes where sweet clover flourished as "weed plantations." Until landowners built something—or sold their properties to people who did—fortuitous flora owned acres of ground.[17]

New blocks with new houses appeared unevenly throughout cities. People migrating out of urban cores did not move into neighborhoods of finished houses replete with all available conveniences and amenities. The wealthy and those who could obtain large loans attempted to shape the development of whole blocks. The middle-class ideal of a detached home with land around it was often a structure with its own land adjacent to vacant properties. In parts of cities being settled by the less prosperous, workers' mutual aid societies and immigrants' ethnic societies supported families who had saved enough to build a single house. Some lots were jointly held by multiple owners who purchased land but lacked the capital to build houses and the time to travel from those areas to their work locations. In poor sections, opened streets were not always traversable streets. Deferred infrastructure construction hindered development and made home owners the neighbors of urban herbs. Clovers and dandelions gave wealthy and modest home owners at least two things in common. The fifteen years that elapsed from the time when a speculator subdivided land to when a family began construction of a house was enough ecological time for a score of ragweed plants to produce more than one hundred thousand seeds.[18]

City dwellers who had settled the greatest distances from rail transportation may have seen the most of these plants. In the late nineteenth century, Americans built commuter and streetcar rail lines to more distant areas of cities and nearby towns. Within and around cities, horsecar, steam-engine, and electric trolley companies built tracks and connecting spurs that cut up the land, advanced real estate development, and initiated more speculation. Creating and maintaining transportation corridors allowed ruderal vegetation to flourish. Depositing gravel and sand from rivers as ballast on which to construct tracks created new habitat, introduced new plants to rights-of-way, and set back ecological time. New plants appeared and sometimes spread as railroad cars carried seeds into and out of cities and as laborers working in the rail yards moved seeds attached to their trousers along the tracks and into other parts of cities. However, plants, seeds, and rhizomes transported long distances on trains did not necessarily survive more than a season or produce seeds before succumbing to dryness, spring frosts, or other environmental conditions to which they were not adapted. Maintenance divisions also periodically cleared plants from ground along the track. This work limited which species grew year after year and interfered with plants' growing, decaying, and changing environmental conditions so that other plants could become established. Bluegrass, brome, canarygrass, Johnson grass, wild mustard, and pigweed were common, although evening primrose and Virginia dayflower also lined the tracks. Sumac, scrub oak, and chokecherry often concealed these corridors. Creating streets in cities created ground where plants such as Russian thistle, puffsheath dropseed, and sixweeks fescue quickly settled. Philadelphia botanist John Harshberger observed that many plants populated both vacant lots and roadsides, although prairie plants such as black-eyed Susans and sunflowers also made use of these spaces.[19]

Plants also inhabited places where people disposed of wastes. Getting rid of stable manure, butcher's offal, dead animals, broken mattresses, shoes, and newspapers buried existing vegetation, created new land where plants could grow, and introduced new plants into cities. By the first decade of the twentieth century, a typical urban American produced "one-half to one ton" of refuse each year. The most expedient dumps were in sparsely populated areas, especially waterfronts, quarries, pits, low-lying land, and wetlands. Disposal practices varied with the makeup of neighborhoods, economic activities, household practices, and differential levels of technology and sanitary knowledge. Garbage and ashes were dumped on ground that was below grade to make new vacant lots. As sanitary officials and municipal engineers worked to develop refuse collection and disposal methods and systems, unregulated dumping persisted. Dump places covered with metal scraps, rocks, gravel, cinders, and ashes made minerals less accessible and made it harder for plants to grow. Ground covered with food scraps and manure was foul smelling but became rich soil. Where cities planted garbage, debris, yard rubbish, and sweepings as in-fill, there were soon asters and sunflowers from around the nation, as well as plants from around the world—hoary cress and wormseed wallflower from England, bastard cabbage and ball mustard from continental Europe, gallant soldier from South America, and spiderflowers from Asia. Catmint, clovers, teasel, and nightshades living in manure from the Chicago stockyards continued to grow where the material was dumped to fill excavations. Some city dwellers commented that these plants helped ameliorate deplorable messes. In northern New York City, buttercups, daisies, and thistles had transformed an "unsightly dumping" into "something well worth looking upon," which neighborhood children called "the mountain."[20]

Real estate speculation and development, the operation of transportation corridors, and waste disposal continually remade urban environments. Fortuitous flora gradually worked such lands as people gradually built their cities. Metropolitan photosynthesizers were remnants of past environments and products of modern urban economies. When urban growth occurred, happenstance plants altered places' ecologies; where they persisted, they advanced ecological time. Throughout cities and in their neighboring settlements, different areas of land were in different stages of development, with lots of vegetation in between. Many real estate developers tried to manufacture the expectation that desirable, healthy environments were plentiful beyond the centers of cities. Developer Jesse Nichols claimed that careful subdivision of land was necessary "to keep every home close to nature" and to make "every neighborhood . . . garden-like in character" with cultivated grasses and flowers. Using nature to create a "beauty spot" cost labor and time. The happenstance plants living where new neighborhoods were emerging were usually not the right kind of nature. Landscape architect Frederick Law Olmsted dismissed some of these places marketed as ideal environments as "patches of waste land and ill kept fields." Many city dwellers were disappointed with the conditions of their new environs. They did not often admire, as a *Brooklyn Daily Eagle* editor did, the "green grass, the decorative stramonium, the nodding daisy, the fern like rag weed, the twining convolvus, the royal thistle, the golden dandelion and the other kindly and beautiful coverings that nature throws over the rudeness of man's misbuilding and neglect." Just plants that thrived as cities evolved became problems that shaped Americans' relationships with the natural world.[21]

Among the reasons happenstance plants became weeds were that Americans tried to define urban space with imprecisely defined plants and that Americans tried to define an array of plants with urban space lacking much definition. Weeds seemed to make spaces dangerous or difficult to enjoy, and places that appeared to be blank spaces to fill in seemed to be inhabited by weeds. Vaguely defined plants and vaguely defined spaces reinforced uncertainties. At the turn of the twentieth century, few people shared "exactly the same definition" of what a weed was. The *American Botanist* noted that the scientist's definition ("a plant out of place"), the gardener's definition ("a plant which grows unbidden, and insists on surviving . . . no matter what"), and the popular definition ("a plant of spontaneous growth") sometimes converged and sometimes diverged. These three definitions still left other dimensions unaddressed. North Carolina botanist Gerald McCarthy wrote that a weed was "hurtful to human interests; a vegetable malefactor." Brooklyn newspaper editor Charles Skinner remarked that the "objection" to many happenstance plants—what made them weeds—was their "cheapness." Philadelphia nurseryman and scientist Thomas Meehan believed that "the simplest . . . most accurate" definition of a weed was "a plant which grows where the cultivator does not desire." Meehan's simplistic sanctioning of individual whim lacked any consideration of ecological changes and environmental contexts.[22]

Plants called weeds were in the right place ecologically, but the wrong place environmentally. They did desirable ecological work; they prevented erosion, produced oxygen, sustained pollinators, and fed wildlife. The vegetation that emerged from ecological time could only be just what it happened to be, but these plants were the wrong organisms for the environments that city dwellers wanted to create. Plants that had existed for two months, two years, or two decades in particular spaces became out of place and no longer belonged when they were in the way of cities being built. Ecological processes cultivated plants where real estate speculators wished to cultivate wealth and where home owners attempted to cultivate health, comfort, and beauty. What these city dwellers imagined to be marvelous nature was not something the unfolding of nature could reliably deliver. Many environmentally sensitive urban Americans who desired to live closer to nature became somewhat disenchanted with the natural world. Americans treated weeds as aesthetic, sanitary, safety, and moral problems.[23]

Plants influenced how picturesque Americans found the environments where they built new houses. In the nineteenth century, many New England towns and cities planted trees to create "spatial beauty." Tree-lined streets and tree-filled vistas conveyed that places were long-established environments suitable for comfortable homes. Landscape architects searched for "attractive sites of comparatively wild land" that could be arranged "tastefully" when developed. Somewhat refined places could even provide some inspiration for newly arranged spaces. Olmsted, who emphasized that "one consistent expression" was the basis of spatial beauty, held pleasant memories of his visit to a village where there were "abundant asters and goldenrods, burdocks and mulleins, and . . . blackberries in the thickets." While from a distance, some happenstance plants could lend a site charm, across the street or next door these plants were weeds that degraded beauty, taste, and social distinctions on public and private property. Undesigned land covered with such plants was seemingly absent of meaning and memory and accentuated the undone nature of cities. City Beautiful proponent Charles Robinson wrote that many Brooklynites applauded "the removal of the wilderness flagging" city hall and the addition of evenly spaced Norway maples and urns of flowers. A St. Louis Civic Improvement League vice-president stated that cutting weeds on vacant lots made the city "a better place to live in." These people believed that pleasant environments made possible and reinforced people's efforts to better themselves and their communities and that squalid environments reproduced and intensified degradation.[24]

Newspapers, civic groups, and real estate associations across the country thought that weeds marred urban environments. In Washington, the *Evening Star* asserted that weed-covered lots spoiled the "beauties of parks, parkings, and ornamental vegetation." In St. Louis, the Engelmann Botanical Club claimed that "crops of unsightly weeds" were among the city's "most unattractive features." The *Los Angeles Times* warned readers that "dusty weeds" and the "thick growth of weeds" could "shock" tourists escaping Eastern and Northern winters, restoring their health, or examining real estate. In Santa Fe, the capital of the New Mexico territory, newspaper editors discussed a City Beautiful campaign in Denver to clear "unsightly vacant lots overgrown with weeds" to inspire their readers to do the same. Although it was "far from being a 'spotless' town," they declared that Santa Fe "certainly should be a weedless city." In St. Louis, the Real Estate Exchange, an organization of large landowners, suggested that home owners could work together to clear the land and plant grass and shrubs on lots between their houses, noting that "as a rule, not even the owner's consent is necessary." Well-to-do neighborhoods collected funds for the work. The West End Residents' Protective Association, an entity that safeguarded the property values of its elite home owners, listed among its purposes, "Weeds. Cut on vacant lots." The Civic Improvement League encouraged block clubs to eliminate weeds. In the springtime, the Good Health Brigade of small Victoria, Texas, descended on vacant lots "to cut the little weeds before they get to be big ones." Although the Los Angeles realty board acknowledged that vacant lots were "disfigured by weeds" in 1903, it was sometimes left to Los Angeles children enlisted in "civic housekeeping siege[s]" to make flowers "blossom where weeds once bristled." By the time real estate developer J. C. Nichols, who established the refined Country Club District in Kansas City, encouraged his fellow businessmen to keep vacant lots neat in 1912, the inadequacies of developers' efforts to shift such responsibility to inhabitants of neighborhoods were clear.[25]

Home owners' creation of orderly, tidy, and reliable beauty in gardens in cities also accentuated weeds' aesthetic deficiencies. Garden plants, like trees, allowed some people to orient themselves in time, express personal values, or renew memories. Gorgeous plants and the arrangement of them on the grounds of American homes became increasingly important. Some plantings represented the simplicity of past times in "old-fashioned" gardens, while other gardens embraced English or Japanese designs of naturalistic landscapes by featuring plants believed to be "indigenous" to the United States. Other gardeners imitated French or Italian formalism or experimented with exotic specimens. Home owners inspired by the Arts and Crafts movement used nature to extend the home and living space by creating a comfortable outside environment for gatherings. Horticultural catalogs, gardening books, popular garden magazines, and businesses selling seeds and plants shaped Americans' ideas about appealing landscapes and guided city people's use of their properties. Plant lovers obtained and cultivated exotic, novel, and beautiful species. One writer hoped that small gardens containing Japanese barberry and Chinese wisteria could make "choked, evil-smelling" cities more tolerable. Members of the Botanical Society of Western Pennsylvania agreed that Japanese honeysuckle was one of the "best porch vines" in Pittsburgh. A writer recommended using the "very handsome" Japanese knotweed "in window and porch boxes where a mass of foliage was desired as a screen." Beyond homes, exotic plants were used to shade sidewalks, landscape parks, and conceal ugly buildings.[26]

Garden writers emphasized ordering space, and their prescriptions to conceal fortuitous flora may have reduced their readers' tolerance of them in cities. Elizabeth Strang contended that coupling the ingenuity of the American spirit with being "dutifully subservient" to the "four generals" of "design, construction, planting and maintenance" could overcome the challenges of gardening in a small city lot. Charles Skinner advocated symmetry—not the emulation of nature's randomness—in small residential spaces. Although the *Standard Cyclopedia of Horticulture* stated that the "appearance of untamed luxuriance, of careless and unstudied grace which suggests perfect freedom," in wild gardens gave "lasting pleasure," it recommended locating such gardens "against the rear of buildings." While *House and Garden* writer George Klingle enjoyed a "medley of weeds" in his yard, he admitted that these plants were "inconsistent with any respectable garden." Grace Tabor instructed *House and Garden* readers that they could create wild gardens by welcoming "every wild thing that comes in of itself" and allowing them to "multiply." Tabor warned that such gardens had to be kept unseen. A wild garden close to a "formal work of man" was "dangerously near to being unsuitably placed." These gardeners' tastes likely varied, but they shared a willingness to place boundaries defining where plants belonged.[27]

Many home owners worked to beautify their grounds with lawns, but weed-free lawns were uncommon. A fenced-in lawn had become a symbol of gentility in Northeastern states by the late eighteenth century, and nineteenth-century Americans saw lawns as a visual setting to embellish their houses. Alexander Jackson Downing, the landscape designer who popularized lawns among middle-class American home owners, wrote that "almost every man feels prouder of his home when it is a pleasant spot for the eye to rest upon, than when it is situated in a desert, or overgrown with weeds." Yet few men desired homogenous turf. They worked to get a variety or mixture of grasses and herbs to grow, many of which also grew on roadsides or in pastures. Adding manure to the ground as fertilizer deposited weed seeds in lawns. At the turn of the century, lawn enthusiasts were beginning to advocate cultivating a single type of grass. Throughout cities, lawns remained struggles to establish, grow, and maintain, especially where the ground was sandy or packed hard. Few home owners attempted to decimate crabgrass and dandelions. Their lawns were trimmed greenery, not a ground absent of all or particular weeds. Ongoing democratization of the lawn increased the acreage of mowed grasses and reduced spaces for happenstance plants in cities.[28]

Improving urban landscapes was not only labor intensive but also culturally complicated because the plants used to represent beauty and refinement could fail to impress. Some city dwellers who attempted to bring "the country . . . into the city" could not reconcile the different spatial relationships and landscape scales. Landscape designer Henry Sargent's 1859 criticism of home owners who "indiscriminately put together" various plants was echoed in Frederick Law Olmsted's 1905 statement that some people "fritter[ed] away" open spaces "with decorations." Elite tastemakers occasionally dismissed cultivated garden plants as being out of place as urban herbs. When geraniums became wildly popular in New York City, *House Beautiful* commented that the flowers had shot "up like weeds." A *Scribner's* writer defended the geranium by describing its virtues in terms that others might consider the defects of weeds: "blooming without reserve," not being "particular about its neighbors," and being "undisturbed . . . by social opprobrium which rests upon it." A *Brooklyn Daily Eagle* editorial suggested that the essential difference between a wildflower and a cultivar was not the latter's beauty or refinement, but the fact that it had "been paid for." The ridicule of the geranium typified the baseless aesthetics of economist Thorstein Veblen's leisure class. When "pecuniary beauty" superseded the beauty nature had to offer, Veblen observed, "beautiful flowers pass conventionally for offensive weeds." The ability to regard market-procured flowers used to enhance private property as untasteful and out of place was one measure of city dwellers' capacity to indiscriminately malign plants as weeds. On the ground, whether city dwellers refined more properties with plants of actual, conventional, or pecuniary beauty, they cultivated environments in which happenstance plants no longer possessed enough value to be thought desirable.[29]

Cultivating beauty around homes may have been rewarding to or appreciated by the middle class, but it was not enough to counteract the dangers of weeds abounding in cities. The sanitary and safety threats of "noxious weeds" worried public health officials, municipal leaders, botanists, home owners, and civic organizations. Weeds undermined public health by spreading disease, abetting waste dumping, fouling the air, and harboring pests. Weeds compromised safety by sheltering dangerous people. Turn-of-the-century reformers who were committed to improving urban environments also thought weeds represented the inefficiencies and imperfections of city life.

Americans who associated weeds with filth claimed weeds created unsanitary environments. Sanitarians were most concerned with overcrowded, older sections of cities where, as muckraker Jacob Riis claimed, the ground was nothing more than "a desert of brown, hard-baked soil from which every blade of grass, every stray weed, every speck of green, has been trodden out." Washington housing reformers investigated deteriorating shacks along alleys with rear yards full of "ashes, weeds, broken furniture . . . and . . . filthy rubbish." Areas with ample open space where vegetation decayed also seemed to be sanitary threats. One basis of the perception was physician Benjamin Rush's claim that "the putrefaction" of vegetation—ranging from crops like cabbage, onions, mint, hemp, and flax to "weeds cut down, and exposed to heat and moisture near a house"—contaminated summer and fall air. The scientific study of germ transmission and the advocacy of personal hygiene in the late nineteenth century did not convince all people that stenches and foul sights were not reliable indicators of the sources of diseases. Many Americans continued to rely on their senses of smell and sight to detect dangers in their environments. In 1880, St. Louis's chief sanitary officer, George Homan, identified weeds as a danger to public health. Homan wrote, "Rank, thick masses of vegetable growth, in some places amounting to jungles, were . . . without question inimical to the enjoyment of the best health by those living near to them." In 1896, Max Starkloff, St. Louis's health commissioner, defined weeds as "all rank vegetable growth which exhale unpleasant and noxious odors." That same year, the *Washington Post* announced that one "fruitful source of disease" was sweet clover that became tons of "slow rotting . . . rank vegetable matter." Two summers later, the *Post* reported that weeds "polluting the atmosphere" were increasing mortality and quinine consumption. USDA botanist Lyster Dewey wrote that the seasonal decomposition of weeds could "not be otherwise than unhealthful." However, the odor of decaying vegetation—even when amounting to what *Post* editors deemed a "six-polecat power of assault on the olfactories"—was likely unpleasant but harmless.[30]

Urban Americans suspected and frequently asserted that weeds spread infectious diseases, especially malaria, but such causation was not actually demonstrated. Weeds were thought to be dangerous because their decomposition caused disease and because they created environments where germs lived and spread diseases. Homan claimed that "noxious or mischievous vegetation" shaded soil, buried vegetation and animal matter, and created deposits that were without "doubt . . . a potent factor in the causation of malarial disease." The *St. Louis Post-Dispatch* warned that weeds infected children with typhoid, scarlet fever, and diphtheria. Even after the city's public health officials recognized that mosquitoes transmitted malaria bacteria, weeds remained, in one *Post-Dispatch* headline, "disease-breeding vegetation" that allowed pools of water to form on the ground and that contained objects like water-filled cans where mosquitoes bred. The paper warned that pedestrians and streetcar passengers risked contracting "malarial germs when passing where weeds . . . flourish[ed]." In Washington, some residents blamed the "heavy growths of weeds on vacant lots" for the increase in typhoid fever cases in the summer of 1906.[31]

There was speculation and doubt that weeds caused hay fever. Hay fever was, as the *Washington Times* observed, a "disease with many vagaries." Botanists lacking medical training, not physicians, seemed most sure that exposure to pollen caused hay fever. Lyster Dewey claimed that weed pollen triggered hay fever and asthma. Charles Pollard of the U.S. National Museum thought ragweed was a "disturbing factor" causing hay fever and therefore a "vegetable pariah, to be combated and uprooted." In contrast, one physician hypothesized that pollen was benign but carried microbes into the nose. District of Columbia health officials believed that the link between weeds and hay fever had "not been demonstrated" and that the symptoms of hay fever were "due to idiosyncrasy on the part of the rare individual rather than to any relation between weeds and the human species generally." A St. Louis health commissioner considered ragweed just one of the city's "poisonous weeds" and suggested it aggravated people's existing "nasal diseases" rather than caused a new one. Two researchers informed St. Louis physicians who read the *Courier of Medicine* that there was little pollen in the air when hay fever symptoms were experienced and that hay fever suffers had fewer pollen particles in their nostrils than other people. *Post-Dispatch* readers did not regularly complain about ragweed hay fever until the 1910s. Increasing awareness and prevalence of hay fever may have been most irritating to boosters of Southwestern cities. Santa Fe newspaper editors warned that "the prevalence of hay fever [was] destructive of the reputation of health resorts" and that the "disgraceful" conditions of the city's vacant lots could repel "tourists, sight-seers and healthseekers." They called for "a war on weeds . . . [to] be waged unceasingly." Perhaps the most serious danger posed by plants—cultivated and uncultivated alike—came from poisonous berries, which, if consumed, could sicken or kill children. Poison ivy was another health threat. To avoid contact with "poison weeds," some St. Louisans abandoned sidewalks to walk in the streets. Botanist Charles Saunders deemed poison ivy to be a "public nuisance," like a "vicious dog . . . at large," and declared that like stray dogs, it could be exterminated only through "imposition of fines and penalties upon property holders."[32]

While weeds were not commonly presumed guilty of polluting the air with pollen, they were accomplices, in the eyes of many city dwellers, in polluting the land. St. Louis police court judge Theodore Zimmerman declared that weeds were "the witches' cauldron into which much of the loathsome filth and offal of a great city is cast to stew and simmer." John Wight, president of the District's commissioners, claimed that weeds' "inherent properties" facilitated "clandestine deposit of putrescible . . . matters" that imperiled public health. Concern that vegetation had caused, allowed, or invited people to act improperly and invited future transgressions reflected city dwellers' moral environmentalism. Although happenstance plants did not coax or pay people to pollute, they seemed to conspire to cover up the problems of urban growth by creating environments that absorbed wastes generated by growth. "Vigorous growers" indicated where dumping had occurred and where it might later recur. Weeds also seemed complicit in creating discomforts that irritated outer city residents. St. Louis home owners felt that weeds made the air "so heavy . . . as to be stifling," and they claimed that mosquitoes living among weeds made it "impossible . . . to remain outside after sunset."[33]

Just as weeds seemed to make pollution possible, they seemed to sustain social ills. In contrast to built-up and rundown older parts of cities, places full of "sorrow and distress" that New York poet Charles Hanson Towne likened to "foul weeds," new outer-city environments were supposed to be environments of safety and order. However, crime and suffering existed there too. Criminals used tall vegetation to their advantage. One August evening in 1900, a man shot and killed an adversary in a St. Louis alley; he then ran through the streets into "a vacant lot . . . where there was a lot of weeds" to escape a pursuing policeman. Some people were reluctant to travel through the "remotest" parts of St. Louis's grand 1,371-acre Forest Park because this "jungle of wilderness" created "fear of molestation." In Washington, Commissioner Wight declared that weeds were "conducive to immorality by affording resorts for idlers and others . . . [and] thus facilitating depredations on contiguous property and other reprehensible acts." District police reported that "unimproved" lots covered with weeds and high grass were "hiding places of criminals and vagrants" and "characters of evil repute." The *Post* seconded the statement of the *Omaha World-Herald* that "day and night" weeds were "a menace to men and women, especially the latter." On Chicago's South Side, police hid in "high weeds" to ambush a highwayman who "darted" out of "thickly overgrown" prairies to rob passersby. However, in some instances, these plants could be useful to escape violence. When one black Chicagoan attempted to transfer street cars during a 1920 race riot, an Irish American gang tried to attack him. He outran them and wound up near the stockyards. He "lay down among some weeds" behind a fence to rest for a couple of hours before working his way through Canaryville to safety around State and 37th streets.[34]

Urban Americans who increased the distance between their homes and workplaces by building residences outside bustling city centers were unable to distance themselves completely from the poor. The population growth and congestion in the older built-up areas that the middle class worked to escape also forced some poor Americans and impoverished immigrants who could not find housing to squat on vacant land. In addition, migrant laborers who lacked money to room in boardinghouses or who were moving between cities in search of work camped at the edges of cities on spring, summer, and fall evenings. Happenstance plants complicated the unexpected encounters and relationships between the humble and the wealthy. San Antonio officials banned brush and weeds within three miles of the city on vacant lots that could become "a rendezvous for tramps." The *Post-Dispatch* described one of St. Louis's "weediest and most neglected open spaces" as a "refuge for squatters, who live there in huts and hovels." The paper opined that "vile smelling and generally obnoxious" tramps and weeds both "delight[ed] to infest the vacant lots of cities." In 1914, St. Louis police arrested a homeless family for sleeping on a vacant lot and turned the children over to the courts. Vegetation seemed to be dangerous nature when poorer people who lived near urban herbs were feared by more comfortable classes.[35]

Weeds seemed to create urban environments where desperate mothers abandoned newborns and infants. Lee Meriwether, a U.S. Department of Labor agent, "found a wee specimen of humanity" when he investigated a "smothered, crying sound" emanating from "weeds near the sidewalk" in a poor part of Providence. When Brooklynite Oliver Dahlstrom, on a June evening, searched for the source of a "whining sound," believing it to be a cat, he instead found a seven-week-old boy "hidden among the tall weeds." Two St. Louis men rescued an infant from a desolate, weedy lot on "a new street made through a prairie by a real estate company." After an afternoon August thunderstorm, a huckster found a crying infant, whom the *Post-Dispatch* called "Baby Alice of the Weeds." Parents who abandoned babies may have hoped that found infants would be adopted and receive better care than they would in the poorhouses and foundling homes, where children often perished on diets of sugar and watery milk. Samaritans, however, probably infrequently came to the rescue. Philadelphia authorities recovered dozens of infants' bodies from vacant lots each year. Tiny bodies were not the only ones recovered from vacant lots. Newspaper stories about missing spouses and anonymous city dwellers found in vegetation revealed that some people suffered lonely deaths. Such news may have alerted readers to the dangers of weeds.[36]

While land covered with weeds created myriad problems, it also inspired reformers. Optimistic and ambitious Americans imbued with turn-of-the-century conservation and progressive social reform ideals saw these lots as wasted land to reclaim and weeds as symbols of the challenges of improving urban life. In many cities, reformers converted vacant lots into "potato patches" to make work for and feed the indigent. One commentator believed these gardens could occupy and rehabilitate "discontented and potentially troublesome people" by abating their hunger, giving them pleasure, and providing mental and physical invigoration. Lyster Dewey thought that conservationists' commitment to obtaining natural resources' "highest possible value in use to the people" should include transforming weed-covered land into playgrounds. Boston social worker Frances Smith compared the perennially recurring adversities of the families whom she cared for to the weeds in a garden. Smith challenged her colleagues to strive "season after season" to pull up "the weeds of degradation and destitution" and to cultivate "the thrift, self-dependence, industry, virtue, health, as well as the intellectual and social natures of our poor friends." Jane Cunningham Croly, who proposed creating a State Industrial School for Girls in New York State, lamented that the "elements of . . . divine womanhood and motherhood" within "poor neglected, untrained girls" were "hidden by the rankest weeds and growth." Arthur von Briesen, president of the Legal Aid Society, worried that when individual self-interest triumphed over "high moral principles," as "weeds keep down the most precious flowers," the result was an intolerable disrespect of the law and disregard of "the rights of . . . neighbors." Weed-covered vacant lots and weed-studded declarations inspired reformers to invest energy and time in helping the poor to overcome their impoverished environments.[37]

Newspapers characterized weeds as evidence that the middle class had to take responsibility for their properties, their neighborhoods, and the environmental quality of their cities. The *Post-Dispatch* encouraged city officials to prosecute property owners who failed to cut "plain . . . American city-bred weeds" to emphasize the "virtue in the ordinance." *Post-Dispatch* writers claimed it was "the legal privilege and the moral duty of all citizens to report the existence of weeds on any vacant property." In a *Post-Dispatch* cartoon, tall weeds with leaves reading "smoke," "dirty streets," and "dirty alleys" represented some of the city's other environmental and social problems. In 1913, *Post-Dispatch* editors wrote that weeds persisted in part because there was "too great a dependence on government and too little individual responsibility." *Washington Post* editors admired the attitude of Kansas City citizens who believed weeds demonstrated a "negligence and carelessness which are wholly out of place in a city with anything like metropolitan pretensions" and hinted that the nation's capital should not be outdone by a Midwestern city. They expressed embarrassment that the District's "lustiest stalks" of wild sweet clover were at least two feet taller than the clover in Indianapolis. In 1906, the city's struggle with weeds led the *Star* to encourage readers to remove weeds themselves "in a spirit of self-help and public service," work that was "certain to redound to their benefit." A leading Chicago African American newspaper, the *Defender*, criticized South Side city dwellers with "weeds in front of houses so high that the children play hide and seek in them" to emphasize the relationship of individual property to the community as a whole. Such people may have dreamed "of going over east to live," but editors declared that "where they belong[ed]" was an area dominated by railroad tracks and factories south of the Loop. The *Defender*'s criticism revealed emerging class differences among African Americans as community leaders and the "aspiring class"—self-identified "refined people"—emulated middle-class white propriety and condemned lower-class behaviors that they believed inhibited their drive for racial progress. All of these newspapers' propositions that citizens must share the responsibility to improve urban environments with their municipal governments had been articulated by nationally recognized sanitary engineer George Waring in his efforts to clean up the streets of New York City during the late 1890s. Environmental quality depended on both the energy of individuals and community groups and the resources and expertise of government.[38]

To the extent that weeds were like horse stables, smoke, and snow in creating problems for city people, controlling weeds was one of many municipal housekeeping efforts that Progressive environmental reformers believed would improve urban life. Urban Americans who advocated eliminating weeds from cities believed that weed-free cities would be healthy, safe, orderly, and comfortable. Protecting new residential environments required vanquishing disorderly weeds that seemed to exacerbate the pollution and disarray of cities. Eradicating weeds required regulating individuals' behaviors and urban economies, as well as severing these plants from the natural world. Municipal governments used weed laws to reorder the relationships between people and plants. They worked to remake urban nature by liberating their cities from ecological time.[39]

Municipal governments, state legislatures, and courts used nuisance laws to regulate urban environments to encourage economic growth, protect health and safety, and improve environmental quality. Cities' weed ordinances made speculators, real estate businesses, and home owners responsible for the plants on their properties and established expectations that municipal officials would manage them when property owners failed to do so. While these laws were necessary to discourage property abuses that became environmental hazards, the laws alienated city dwellers from happenstance plants and rejected what they contributed to urban life.

Weed laws written to protect city people's health were distinct from state weed laws to protect farmers' crops. States and agricultural towns passed weed laws to discourage property owners who lacked the "private interest or public spirit" to keep plants that interfered with crop production out of their fields. For example, in 1861, Battle Creek, Michigan's Common Council approved an ordinance that required cutting "thistles, burdock, or other noxious weeds" on land adjacent to public streets, lanes, or wagon tracks each month from May to September. Preventing weeds from reaching roadsides along which they could more quickly spread to fields protected the local agricultural economy from a public nuisance. Compliance with these laws also potentially reduced private nuisance lawsuits in county or state courts and reduced the costs of eliminating weeds from public roads and liability for damages to private land caused by roadside weeds. Battle Creek's law did not mention people's health, safety, and comfort. State laws typically identified particular plants that reduced crop yields by their common names, such as Russian thistle and milkweed. A USDA survey of weed laws in the 1890s indicated that Midwestern states such as Kansas, Minnesota, and Wisconsin banned the most plants. The only weed that Missouri and Maryland prohibited was Canada thistle. Although the USDA advocated that only the five "most injurious" plants to a farming community be regulated, horticultural experts like Liberty Hyde Bailey and Thomas Meehan thought the laws were unnecessary because responsible farmers destroyed weeds when they tilled soil. Meehan, an advisor to Pennsylvania's state agriculture department, commented, "Weed laws abound, but the weeds laugh and grow fat on them." People like Meehan seemed to worry that public officials paid to control weeds failed to destroy all weeds in order to secure ongoing appropriations for their work.[40]

State weed laws and capital cities' weed laws were distinctive in presenting plants as different problems in different environments. The tolerance for and management of nature diverged in state laws and laws written for their capitals. In 1884 and 1893 Ohio's General Assembly banned allowing burdock, Canada thistle, common thistles, cocklebur, daisies, sweet clover, teasel, wild carrots, and wild parsnip to go to seed near highways, property lines, or partitions on improved land. Lawmakers also required townships to keep weeds off roads. In 1888, Columbus prohibited property owners from allowing "any thistles, burdock, jimson weeds, rag weeds, milk weeds, mullens [*sic*] or any other weeds of rank growth" to produce seeds. The Board of Health was empowered to declare them "prejudicial to the health of the neighborhood" and to remove them like other "causes of disease." In 1891 Nebraska legislators required landowners to mow Canada thistles on their property near highways; in 1895 they declared Russian thistles to be a public nuisance. In 1889 Lincoln officials decided no landowner could "allow or permit weeds to grow, or remain when grown." In 1885 the Indiana General Assembly made it illegal for any person, including supervisors of roads and railroads, to "knowingly allow Canada thistles to grow and mature." In 1892 Indianapolis declared that "the growth of weeds or noxious plants within the City . . . [was] injurious to public health." Persistence of these weeds in Indianapolis led the Indiana Academy of Science's Willis Blatchley to describe the plants as "a big crop of the vilest weeds . . . dense thickets through which a person can scarcely force his way." Dangers to public health, safety, and comfort distinguished city weeds from their biological brethren and country cousins. States targeted weeds that interfered with the business of farming; cities barred all weeds because they jeopardized public health. In rural areas, weeds were banned along public highways, but likely thrived unnoticed on large properties; in cities, weeds were not allowed anywhere and were obvious on vacant lots.[41]

Cities were environments that produced new nuisances, and cities were changing environments where established practices could emerge as new nuisances. Cities both generated weeds and changed in ways that transformed weeds into nuisances. Cities first had to determine whether something was a nuisance. Where and how something occurred and its effects were the contingencies that differentiated a nuisance that caused or risked physical injury from a "trifling annoyance." Public nuisances—whether slaughterhouses, factories, or weed-covered ground—harmed the welfare of the many and became more widespread in increasingly populous cities. Per se nuisances were unacceptable property uses "irrespective" of location. Legal scholars conceded that even phenomena that were nuisances "in their nature" might not be judged so by all "impartial minds." U.S. assistant attorney general James Boyd's complaint to the District's commissioners about weeds in his P Street neighborhood east of Du Pont Circle apparently helped to resolve District of Columbia city attorney Sidney Thomas's doubts about whether weeds were nuisances and weed ordinances were legal. The "flood" of annual summer complaints about these plants compelled St. Louis health commissioner Max Starkloff to craft weed nuisance legislation.[42]

Declaring urban herbs to be a public nuisance assigned the plants a legal status that denied they belonged in the natural world. According to Horace Wood, a New York nuisance law expert, people had to create nuisances. Deciding that weeds were nuisances meant that a property owner's ability and responsibility to control land determined the nature of that land, as well as rejected ecological processes as irrelevant. Property owners were responsible for making weeds and creating nuisances, whether the land was without weeds when purchased or the weeds on the land at the time of purchase were not removed and prevented from returning. Classifying weeds as nuisances, like the dense smoke that rose from powerful locomotives, the foul fumes released by boiling animal organs to produce fertilizer, and the cacophonous crashes of sledgehammers smashing against anvils, severed the plants' roots from the earth and transplanted them into a legal framework in which only human responsibility for environmental conditions mattered. The plants were not creatures that changed with ecological time, but green garbage strewn about by irresponsible property owners. As nuisances, weeds were both unlawful and unnatural, a pollution problem of unhealthy, disorderly environments. While it was erroneous to consider plants that predated the existence of primates "unnatural," in another way, weeds were synthetic objects manufactured by urban Americans in real estate markets, health departments, courtrooms, and newspapers.[43]

Weed nuisance laws indiscriminately denigrated many plants as weeds. Camden, Cincinnati, Columbus, Denver, Indianapolis, Lincoln, St. Louis, San Antonio, Washington, and Wilmington enacted weed ordinances between 1888 and 1903. In 1893, California's legislature authorized municipalities to "condemn, as public nuisances, any or all weeds whose seeds are of a winged or downy nature, and are spread by the winds." Seven of the nation's ten most populous cities—New York, Chicago, Philadelphia, Detroit, Cleveland, St. Louis, and Pittsburgh—had established weed laws by 1920, as had Birmingham, Buffalo, Cheyenne, Kansas City, Louisville, Salt Lake City, Seattle, Toledo, and Wichita, and future Sunbelt metropolises like Atlanta, Houston, Los Angeles and Tampa. Less officially but more moralistically, in the mining town of Pueblo, Colorado, the Commerce Club and the Colorado Fuel and Iron Company circulated their "ten commandments for a clean city," the ninth of which was, "Thou shall wage continual warfare on weeds, dandelions and untrimmed trees." Legislation varied from city to city, but weeds were described as injurious, noxious, offensive, rank, and unwholesome. While ordinances occasionally mentioned some plants like burdock by common name, they more often banned vegetation that exceeded a certain height—"4 inches" in Washington, D.C.—or vegetation that was vigorous—"rank growth" in Columbus. A few cities indicated in or by what month or season land was to be cleared. The malleability of nuisance laws and the expansiveness of municipal police power were useful for attempting to regulate happenstance plants, but exercising this power provoked legal challenges. When property owners in

St. Louis and Washington contested weed laws, courts upheld the laws that made Americans responsible for destroying weeds in changing urban landscapes.[44]

On 10 July 1900, St. Louis sanitary officer Peter Best investigated a lot of "four to five feet tall" happenstance plants near the intersection of Vandeventer Avenue and Olive Street. Best issued a "weed statement" to attorney Smith Patterson Galt. Galt allowed the weeds to remain and received a second citation for violating the city's weed ordinance. On 30 August, the *Post-Dispatch* published the names of prominent St. Louisans who had violated the law, including Galt; past St. Louis mayor and past Missouri governor David Francis; Henry Hitchcock, a corporate lawyer and a founder and past president of the American Bar Association; local banker and financier August Gehner; and a restaurateur who ran the city's "most famous oyster house." Galt, an attorney for railway companies, corporations, and trusts, was a member of the Civic Improvement League. He acknowledged that cutting the weeds would have been simple. However, rather than adhering to the law and demonstrating its necessity, Galt fought the charge and worked his case up to the state supreme court. He stated that the city had created the law to make "positions as 'inspectors' for more bum politicians at the public crib." Galt protested that the ordinance failed to describe the vegetation on the parcel, which was located a few blocks southwest of his Vandeventer Place mansion. City attorneys Charles Bates and William Woerner countered, "We are not concerned with general ideas or definitions of what is meant by 'weeds,' but only with the term as used in the ordinance."[45]

St. Louis v. Galt established the validity of urban weed-control ordinances. Neither party could rely on precedents about weed laws. Nor did the *Century Dictionary* or *Webster's International Dictionary*, both quoted in the proceedings, define a weed as a nuisance. City attorneys argued that whether the plants were weeds and whether weeds were nuisances depended on the probability that they were dangerous. They stated that only the knowledge that rank vegetation could be "indisputably and universally . . . unobnoxious and harmless" was a basis to rule against the city. Judge William Marshall upheld St. Louis's ordinance. As nuisances, weeds were not natural, but they were naturally dangerous in the city. Surprised that Marshall decided that the city had declared something to be "a nuisance, which is not a nuisance," Galt defended the "uncultivated vegetation" on his land as beneficial plants in the "economy of nature" essential to "the preservation of man upon the earth." Galt argued that "weeds" did not endanger health—they sustained it. Galt may have developed this argument from conversations with scientists. As a trustee of the Missouri Botanical Garden, Galt attended the garden's annual banquet at least twice before his state supreme court appearance. At the thirteenth banquet, held in May 1902, Galt had the opportunity to meet University of Illinois ecologist Stephen Forbes, University of Wisconsin limnologist Edgar Birge, and University of North Carolina botanist and state geologist Joseph Holmes. Galt agreed to speak at the January 1904 banquet on condition that he could address the audience before inebriated men had fallen asleep or received rides home. There may be no record, or memory, of what Galt said, if anything, about weeds that evening, while the men dined on clear green turtle soup, beef tenderloin, and roast quail.[46]

Abstracting weeds as nuisances meant disregarding the nature of particular plants. What distinguished benign or beautiful nature from unacceptable nuisance was not the biology or chemistry of the plants, but the plants' dangers and virtues as described by the mouths of public officials and the right to flourish granted by St. Louisans' hands. Galt claimed that the sunflowers, which he called the "queen of our mother's garden," occupying his land were not "rank vegetable growth." Prosecutors claimed the city had no ill will toward his sunflowers. However, sunflowers grew on vacant lots, near dwellings, and along roadsides. They grew in sunny spaces covered with cinders and rubble where nothing else would grow. Sunflowers seemed to be weeds. Two summers later, *Post-Dispatch* editors demanded the destruction of sunflowers that were a "forest of weeds." The paper continued to advocate "making war" on "the outcasts of the vegetable kingdom." Editors claimed that goldenrod, the state flower, should not be annihilated because it was an "innocent and beautiful" plant, although some botanists considered it a common weed. At the same time, some plants possibly being destroyed were also being welcomed to one of St. Louis's most exclusive private residential streets—Vandeventer Place—where some of the city's most successful lawyers, industry moguls, and old-money families resided. Landscape architect George Kessler's predilection for yarrow disregarded botanists' assessment of the plant as an ordinary weed, and his fondness for the goldenglow coneflower was not dampened by its regular growth on marshy ground feared to spread malaria. Kessler also planted the shrub *Berberis thunbergii*, Japanese barberry, around homes and gardens. These elite St. Louisans saw the plant's green leaves turn orange, red, and purple in the fall; they might have marveled at how the plant's spring yellow flowers provided red berries that could brighten the winter. As some St. Louisans fought weeds rooted in the past, others sowed the weeds of the future.[47]

Distinguishing plants as weeds, however arbitrarily, helped shape place. In 1894, *Meehan's Monthly* asserted that there was "a vast difference between a weed and a wild flower." The vast difference between them in St. Louis emerged from the real estate market, perceptions of refinement, and rows of type in city ordinance books that scrambled ecological time. In spatial terms, *Meehan's* located flowers in gardens, weeds on vacant lots, and wildflowers well beyond cities. Eradicating weeds and planting ornamentals blurred the past and modernity. A weed-free urban environment, according to the *Post-Dispatch*, helped make St.

Louis "a trim and up-to-date city." Yet modernizing and transforming St. Louis's urban landscape was not simple because some plants used to improve the city's present and future referenced the past. Vandeventer Place featured a beauty resulting from the "mellowing touch of time" and from not rigidly following "a rational planting of trees and shrubbery." Engelmann Botanical Club members thought that overcoming the city's degraded environment required bringing "as much as can be brought of the fields and woods into the city"—plants that once but no longer existed where St. Louis had grown. Majestic and rare plants conferred a respect on land that the happenstance plants—the beginnings of new fields and forests that seemed too disorderly to tolerate—could not. Perennially present fortuitous flora were what Edward Relph calls a "time edge," a boundary between two eras and two landscapes. In St. Louis, removing the ragged past's persistent and ubiquitous transient plants cultivated a permanent, orderly future.[48]

Eliminating weeds from the District of Columbia supported remaking the rough landscape of the nation's capital. The poorly maintained National Mall, obstacles created by railroad tracks, congested streets, and an inaccessible and polluted Rock Creek Park marred what was supposed to be a grand city. The 1901–2 Senate Park Commission Plan was the most comprehensive strategy to transform the landscape with City Beautiful ideals, but it was preceded by many bills passed by Congress in 1899 to increase the city's health and beauty, including laws regulating smoke emissions and banning weeds. District of Columbia attorneys cited *St. Louis v. Galt* as a precedent to defend the District's legislation in 1906 from Washington real estate developer Galen Green. In legal proceedings, Green admitted that his property was weed covered but denied that it was filthy. Unlike Galt, who challenged St. Louis's authority to regulate weeds as nuisances, Green and his counsel, John Ridout, a fellow real estate speculator, questioned the District's provision to obtain Green's compliance through criminal penalties but a nonresident landowner's compliance through taxes. A District of Columbia appeals court judge agreed the penalties were flawed but upheld the District's right to control weeds.[49]

In Chicago, an attempt to assess owners for the cost of cutting weeds on their vacant parcels failed without the legitimacy of nuisance law. In 1901, N. C. Van Slooten, Cook County's commissioner of Canada thistles, vowed to eliminate thistles that were "as thick as the hair on a dog" by fining and arresting landowners, regardless "of who" they were. Although another county official did not think the county could regulate land in incorporated cities, Van Slooten sought $1 from owners of vacant parcels to pay for the work, including money from twenty-two hundred Chicago landowners whose land was "blue with thistle blossoms." In 1903, Van Slooten declared that the state's Canada thistle law supported "ridding Chicago" of "poison ivy, burdock, and cockleburrs" and other "weeds declared to be noxious." Much of this land was in "outlying wards," which the *New York Times* joked consisted of remote "untraversed prairies" seen only by "naturalists and hermits." The $31,000 that he charged twenty thousand property owners in 1903 compelled the Chicago Citizens' Association, a realtors' association, and Cook County's finance board to examine his office. Among the allegations were that owners had not been notified about the condition of their properties, which in some instances were reportedly not cleared of the weeds. Investigators informed *Chicago Tribune* readers that the commissioner's actions appeared to be a scheme to amass money by extorting small sums "under the cover of a useless law." Cook County commissioners declared Van Slooten's office vacant in October 1903, after they had decided that the state's 1872 and 1885 amended thistle law should not have been extended to other plants. Ultimately, the Illinois Supreme Court determined that weed cutting was not a "local improvement" that could be made by the government through special assessment or taxation.[50]

In the Van Slooten controversy, as in *Green,* the means of fighting weeds—not the government's police power to make weeds illegal—was challenged. Van Slooten's antagonists were interested in whether public or private entities should control metropolitan photosynthesizers, not in defending the desirability of such vegetation. Refined Chicagoans had little fondness for weeds. The Ravenswood Improvement Association pledged that "battle will be waged on weeds." North Edgewater's Improvement Association hired laborers to remove poison ivy and other plants. The South Park Improvement Association, which received funding from Marshall Field and the University of Chicago, hired workers to cut weeds. Van Slooten's escapade revealed the problems of applying a rural or state weed law to an urbanizing area. The dispute emphasized that making weeds a danger to public health was crucial to a viable weed ordinance intended to remake the urban landscape. Ten years after Van Slooten was dismissed, the city of Chicago declared such plants a nuisance.[51]

Galt and *Green* secured the public nuisance principles of municipal weed-control laws. *Galt* not only bore an enduring definition of weeds in cities as legal nuisances but also proved compelling to legal minds grappling with property use and environmental regulation. The city of St. Louis's attorneys and other Missouri Supreme Court justices cited Marshall's opinion in *Galt* to uphold other nuisance regulations for dairy products and billboards. Cities across Missouri, from Sikeston in the southeast to Kansas City on the western border, cited *Galt* in dealing with nuisances including the running of cattle, the location and operation of ice-manufacturing businesses, and the storage of automobiles in need of repair. Beyond Missouri, the Maine Supreme Court considered *Galt* in an examination of conservation legislation that restricted timber cutting to protect state waters. *Galt* was cited to defend the use of police power in protecting Montpelier, Vermont's water supply and regulating building heights in Milwaukee. While many people dismissed weeds as "useless," the plants in *Galt* and *Green* sustained an ongoing dialogue about individual responsibility for the collective good and the use of collective power to compel individual responsibility. In 1911, Missouri Supreme Court judge Archelaus Woodson remarked that *Galt,* which had become "generally known as the 'weeds' case" in the legal community, was important for Marshall's reassertion that "the rights preserved to the individual by this section are held in subordination to the rights of society." In turn-of-the-twentieth-century American cities, the individual's rights were many, but not infinite. Nature in these cities was ubiquitous, but it was also increasingly confined and narrowly construed.[52]

Although the courts protected the power of cities to fight weeds, they could not endow cities with the power to eliminate them. The authority to regulate people's relationship with the natural world was not enough to transform urban nature. In St. Louis and Washington, municipal officials could not wield legal mechanisms, deploy staffs, and employ available technologies to cultivate weedless urban environments. In some regards, they may have exacerbated the problems weeds created. The ecology of changing cities that allowed happenstance plants to thrive persisted.

Since "vigorous growers" could occupy land almost anywhere in St. Louis and Washington, officials' greatest power over them was delineating where the plants were illegal and identifying which plants were to be destroyed. In St. Louis, whether a patch of vegetation was considered a field of illegal weeds depended on whether the land was part of a platted, city-approved subdivided block. In Washington, the weed law applied to "any land in the city of Washington . . . or in the more densely populated suburbs of said city." In both cities, geographic distinctions exonerated vegetation that otherwise appeared dangerous to citizens and produced seeds to distribute throughout the city. In St. Louis, the subdivided block requirement resulted in a controversy over nearly two hundred acres of land that philanthropist Henry Shaw bequeathed to the city as a garden and park. When St. Louisans complained about the vegetation in 1905, Shaw's trustees initially ignored their pleas because the law excluded unplatted land. However, they eventually acquiesced and destroyed the plants to "prevent any sickness that may spread or originate from the weeds." Arbitrarily decided heights determined plants' legal status. These cities' laws defined weeds as unrestrained vegetation growth—"over one foot" in St. Louis and "more than 4 inches" in Washington. In 1902, District of Columbia health officer William Woodward requested that Congress amend the height from four to eighteen inches, reasoning, "A weed 6, 12, or even 18, inches high is not likely to afford a hiding place for criminals. Neither is a weed below 18 inches likely to afford a place of deposit for offensive matters." By

1910, Woodward argued for "raising the permissible height of weeds to 2 feet" and estimating the area of weed growth, permitting no more than "10 per cent of any lot or parcel of land" or "unbroken area . . . [of] 100 square feet of land" to be covered with weeds. Woodward's pleas were ignored. These geographical-biological criteria to classify plants as nuisances were abstract, unecological, and unrefined and created more nuisances than either city could control.[53]

The number of eyes and hands available to city officials influenced how many of each summer's weed nuisances were found and eliminated or remained to aggravate residents. Paperwork preceded purging plants. In both cities, officials usually began issuing citations in June. July and August were the busiest months for inspectors; St. Louis inspectors issued 5,675 citations in July 1900. They continued the work in September and sometimes October. Seeing weed-covered land was easier than seeing the weeds destroyed. Inspection, identification and notification of property owners, reinspection, and court judgment preceded clearing a property. The District's nine inspectors found it "impossible . . . to enforce the weed law as it should be." Neither city was able to establish a separate weed-cutting unit, so the surge of summer weeds increased the workload of inspectors, who also examined privies, yards, wells, tenements, and plumbing. The number of lots cleared or the height that weeds reached each year varied with strained health department budgets. Costs of abating weed nuisances included placing notices in newspapers, summoning nonresidents to court, and clearing land. St. Louis officials noted that complying with the law cost more than cutting the weeds, while the District's commissioners were unable to persuade congressional appropriation committees to fill their annual budget requests. Health departments and the contractors they hired lacked technologies to overpower weeds. Poorly maintained machine mowers were ineffective and inefficient, and herbicides were too expensive for wide-scale use. Detecting violations, prosecuting landowners, and clearing land cost much, while ecological time efficiently produced urban herbs without cost.[54]

As more blocks were platted and eligible for city services, more parcels became potential sites of weed nuisances. Although more buildings appeared in these cities each year, this development did not guarantee that the number of weed nuisances would decline each year. The number of nuisance violations issued fluctuated. After their victories in the courts, St. Louis and Washington officials initially issued more citations. In post-*Galt* St. Louis, citations climbed from 1,712 in 1903 to 5,007 in 1906; were fewer than 3,000 from 1907 until 1913; and rose to 4,927 and 3,829 in 1914 and 1915. In Washington, there were only 12 violations in 1904, but 476 when *Green* was decided in 1906, more than 400 in 1907 and in 1908, and 1,275 in 1910. The number of nuisance violations that these cities issued varied year to year for many reasons, including the initiation of home construc-

Figure 1.3. South side of St. Louis's Morgan Street, about one-half mile north of Forest Park, in August 1905. Reprinted with permission of the *St. Louis Post-Dispatch.*

tion, the addition of newly subdivided blocks, and the sensitivity of people moving into these areas.

Regulating weeds as nuisances enhanced city dwellers' awareness of weeds and expectations the plants would be eliminated. The inability to address all weed problems and the return of weeds each year annoyed home owners. In 1900, haphazard enforcement and control led one St. Louisan to inquire whether the law actually existed. In 1905, the *Post-Dispatch* announced that a "Weed Patch Is Wonderful" on Morgan Street and that "Etzel [Avenue] Needs Scythe." In Washington, a citizen who rode streetcars around the city in 1906 observed "acres of rank vegetation, weeds and grass and shrubs." The *Star* reported in 1910 that there were weed-filled lots that "would hide a regiment of National Guardsmen in broad daylight." After seeing a stretch of U Street northeast of Dupont Circle, a woman wrote, "It is an ugly sight to see parts of a great city grown shoulder high in weeds." District health officials selectively enforced the law, most frequently in the best neighborhoods. During the 1910s, they attempted to respond only to specific complaints about specific properties made by persistent individuals. Taking action against violators generated more complaints and more work when those violators protested the conditions of properties near their land and lodged complaints about them. Protracted struggles led the *Star* to refer to the work in 1924 as "the annual battle of police vs. weeds." Nearly thirty years after St. Louis passed its 1896 weed ordinance, the health department issued 2,483 notices for violations of it.[55]

While weeds were nuisances to residents, weed laws proved to be nuisances to health departments. Although as early as 1885 St. Louis officials commented that nuisance perpetrators had "too many opportunities to evade" compliance, such tactics hindered weed control in ways that in 1906 frustrated health official William Winn, who called "the weed question . . . the bugbear . . . of this department" and "the weeds and the laws governing them . . . unmitigated nuisances." The inability to control the weed nuisance led District of Columbia officials to worry about the "needless friction between the health department, complainants, and land owners," as well as the "unjust criticisms of the District government," both of which likely made summer in the humid city even less comfortable. In St. Louis, some property owners constructed wooden nuisances that exacerbated the weed nuisance problem. City councilors and the commissioner of public buildings claimed that billboards "sheltered and promoted the maintenance and growth of underbrush and weeds." In cultivating legal definitions of weeds, St. Louis and Washington officials (and the workers whom they hired to clear land) had become the land's cocultivators, although they did not own the land and removed plants from land often owned by people who did not live in the immediate vicinity or the region.[56]

Continually buying, digging into, dumping on, landscaping, moving across, and selling land transported plants and gave them new places to grow. These disturbance dynamics sustained plants and could not be overcome by occasionally brandishing hoes and scythes. While newer areas were being built, people were producing the seeds of future development and more happenstance plants elsewhere in cities and beyond cities. Successful speculators held land, sold it off, and invested their profits in larger acreages of city land or agricultural land to which they would lure people who otherwise might have purchased vacant city lots. Failed speculators held land for too long, missed opportunities to profit, and waited for buyers to cover or minimize their losses. "Neglected" vacant land on which vegetation grew profusely could result from these dealings. Municipal governments had limited influence over these economies. Making weeds illegal was simple; finding effective penalties to compel property owners to maintain land in a condition tolerable to residents was harder. Placing special taxes on these properties to pay for weed cutting in some instances may have increased the prices of lots and made them less affordable to prospective builders and buyers. Yet failing to control weeds on vacant lots could produce seemingly nasty environments that repelled investors from a block and even pushed people beyond cities' limits. Undeveloped land may not have been a problem if it was truly empty space and lifeless ground. Instead, metropolitan photosynthesizers emerged.[57]

Both cities also perpetuated weeds on properties where they succeeded in compelling vegetation destruction and on properties that city officials eventually cleared. By continually destroying plants that worked the soil and allowed vegetation change, impatient urban Americans cultivated ruderal annuals and perennials. Intense, regular disturbance reproduced problems with weeds year after year. City workers destroyed weeds without attempting to discern how different patches of vegetation were changing. The ecological processes that might have produced more pleasant landscapes were replaced with an urban variation of farm-field clearance. Regularly cut vacant lots, mowed lawns, and well-tended gardens of ornamental plants may have pleased city dwellers who wanted all ground rigorously controlled, but they inhibited the interplay of species that changed the land. Both destroying weeds frequently and being unable to prevent happenstance plants from growing made these plants perpetual city inhabitants.[58]

When New York City enacted weed-control legislation in the summer of 1915, the *New York Times* reprinted a column from a Columbus, Ohio, newspaper entitled, "What Is a Weed?" In 1902, novelist Theodore Dreiser had answered the perennial question with Emersonian enthusiasm—"the next big thing." Although Dreiser was aware that many Americans denounced weeds as "impudent," "outlaws," and "bandits," he predicted the plants would eventually become "indispensable." Dreiser imagined a "not far distant" future "when the poisonous weed would have been mastered and applied, and the most useless weed put in its place and made to do serviceable work." Long "misunderstood" and "not understood," weeds—even ragweed, jimson weed, burdock, and sneezeweed—were subjects of "investigation of a most profitable order." The plants already provided inspiring lessons in "how wonderfully life prevails even in the face of great hardship" and how to "make thorough use of the poorest opportunity." Although Dreiser claimed researchers would discover "their place in nature," Americans decided that cities did not have space for weeds. Urban Americans did not recognize happenstance plants, as Charles Skinner may have wished they did, as "a little . . . wild nature, in our towns." If they "let the weeds alone," as Skinner advised, it was likely not for the "food for sight" that Skinner promised, but because of unawareness of, indifference to, or resistance to city laws. By 1915, many New Yorkers seemed unable to sense a vacant lot of "dandelions, buttercups, butter-and-eggs, asters and daisies" as "an open space in itself . . . that let a breath of air into the view." As city dwellers settled into the twentieth century, many did not want to make room for what botanist William Bailey had once called "weed gardens"—land home to happenstance plants that made it possible for "the poor man . . . [to] have his botanic garden" and that benefited "those who cannot take long walks, and who yet are interested in nature." City dwellers had not proved that weeds were useless plants so much as shown that they thought they did not need them and felt they would be better off without them.[59]

The ceaseless changes of urban life resulting from real estate development, cities' production and distribution of waste, the perceptions and comprehensions of disease, evolving aesthetic norms, and municipal laws and operations allowed fortuitous flora to flourish and made them into and sustained them as urban environmental problems. Weed-control advocates believed that distinguishing between city vegetation and cultivated plants put distance between filthy, dangerous places and healthy, safe spaces, as well as between an unkempt past and an ordered future. They did not appreciate the labor of plants working to occupy city land. Instead they used laws, bureaucracies, and culture to formalize and institutionalize animosity toward weeds. Outlawing and destroying what they deemed transient nuisance vegetation reflected their desire for the production of stable landscapes amid the dynamic city-building process, in which property owners often sought the "highest and best"—which they often considered the "most profitable"—use for their land. By not accepting weeds as natural—by categorizing weeds as a manmade nuisance—weed-control advocates discounted and obscured the ecological forces enveloping their cities' fragmented landscapes, which regenerated weed nuisances season after season until concrete, buildings, and hardy lawns blanketed city blocks. Such development also would produce its own weeds. Wealthy Bostonians derided three-decker apartment buildings, often inhabited by working people, as "Boston's weed." Some city dwellers erected inexpensive cottages from Sears, such as the three-room "Goldenrod." Historian Kenneth Jackson later described the interwar period as one in which "steel skeleton skyscrapers grew like weeds on the urban landscape."[60]

CHAPTER TWO

Human Weeds in Industrializing America

In the first decades of the twentieth century, happenstance plants colored Chicago and neighboring cities, marked their edges, and inhabited the spaces between them. Botanists estimated that "weeds and wild grasses" such as cocklebur, dog fennel, green foxtail, and sunflowers covered nearly 40 percent of Chicago in 1925. Thousands of Chicagoans had seen galinsoga, which seemed to be "the one plant that . . . prosper[ed]" in apartment buildings' shady courtyards. Sow thistle lived in the alleys. Goldenrod and horseweed periodically grew near drainage canals visited by adventurous boys and wandering dogs. Prickly lettuce, which was very common in waste places, was gathered "by the cart load" by Chicago's "foreign population" as a salad green. In a west-side neighborhood, fortuitous flora reached "ponderous proportions." Even some ironweed, rosinweed, sky-blue asters, golden ragwort, and blazing star remained from the city's prairie days. Russian thistle tumbled around the city and beyond it, occupying city lots, growing "annually by the thousands" in Gary, and inhabiting large areas of Lake Michigan's sand dunes.[1]

These happenstance plants provided University of Chicago faculty and their students with opportunities to see how metropolitan environments were chang-

Figure 2.1. *(above and opposite)* A pair of photographs of Main Street in Gary, Indiana, circa 1909. Special Collections Research Center, University of Chicago Library.

ing and developing. Although these areas were emerging across the nation, their ecological processes may have been studied more intensely in Chicago than anywhere else because one-time University of Chicago philosopher John Dewey had identified the "evolution of environments" as a research frontier. The university's botanists, ecologists, zoologists, and sociologists, their students, and their associates produced knowledge about the region's ecology in part by wondering about the similarities between natural places and urban environments, as well as the similarities between plants and people. Ecologist Henry Cowles called Lake Michigan's sand dunes "a marvelous cosmopolitan preserve, a veritable floral melting pot," because species from "diverse natural regions" were "piled . . . together." After working with Henry Cowles at Woods Hole in the summer of 1902, Canadian teacher George Fuller—who eventually earned his doctorate at Chicago and joined the faculty—compared "hundreds of plants jostling and crowding upon one another" to "all the strenuous efforts of the inhabitants of a great city." Sociologist Robert Park thought that the city was "like a plant" because over time both grew "in size" and became "more highly organized." One way that Chicago-trained

sociologist Roderick McKenzie taught ecology was by assigning a botanist's description of a forest as a plant community that was "just as simple and understandable but with its multitude of activities just as complex, just as inevitable in its structural make-up but with its succession of life problems just as intensely interesting as any city or other community dominated by the genus of bipeds to which we belong."[2]

Chicagoans contemplated how the process of succession made the changes in plant and human communities analogous. Cowles defined succession as one plant "society" replacing another "as the years pass by." He explained that as water and topography made environments less extreme over time, the "changes and counterchanges" of plant societies demonstrated a "mesophytic tendency." Park proposed that succession could be used to describe series of changes generally and changes that were "orderly" in particular. He thought that succession made changes in human communities seem "identical in form" to changes in plant communities. These men recognized that there were important differences between these communities. Ecologists saw little evidence that plants changed their environments,

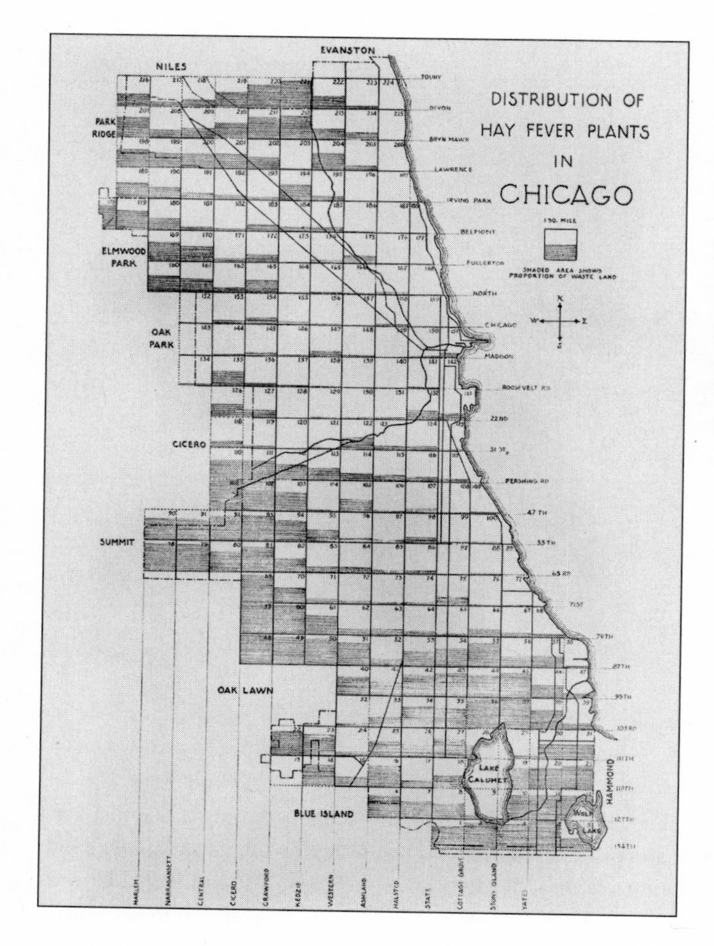

Figure 2.2. This 1933 map indicated the amount of land covered with weeds in square-mile sections of Chicago. Reprinted with permission of the Illinois Medical Society.

while sociologists emphasized that people's interactions with each other allowed them to shape their environments. However, this differential power to influence the environment was less important than the analogical outcome of succession—increased order.[3]

Discerning the resemblances between vegetation and urban communities was accompanied by proposing resemblances between ecological and sociological inquiry. Ecologists sometimes referred to their work as plant sociology, and sociologists called their field human urban ecology. Cowles evaluated the influence of environmental change on "plant societies." Helen MacGill, a sociology student, noted that plant ecologists and sociologists perceived similar patterns. She wrote that ecologists had established the fact that some varieties of plants "occur relatively late in the development of the community" and that sociologists were discovering that certain institutions "require a high degree of intensive human settlement before they can live in the community." Park recognized such analogues because he thought that people and plants experienced a "vital inter-dependence" in their respective communities.[4]

Since Chicago's metropolitan environment contained much vacant space where plant communities and human communities were changing, ecologists and sociologists shared some similar research strategies. Cowles sometimes made "nearby vacant lands his ecological laboratory." When such spaces were "border lines or zones of tension . . . rather than at the center of the society," Cowles thought ecologists could "best interpret the changes that are taking place." When sociologist Ernest Burgess diagrammed expanding cities, he conceived of a "zone in transition" with slum housing and factories as an "area of deterioration" ringing a downtown where immigrants, migrants, and criminal elements mixed and where people's struggles to adapt stimulated the city's outward growth. Roderick McKenzie's ecological approach included investigating transitional areas of cities where marginalized people could be found. One "Area of Transition" mapped by Burgess students was a stretch of land between a railroad line and State Street, running from 18th Street to 39th Street, which had several vacant lots and was becoming home to Southern African Americans during the Great Migration. Another Burgess student described the tensions of succession arising from the influx of people into the South Side, which was like a "'Gothic' invasion." The student reported that the relocation of longtime inhabitants' cultural institutions had left "the community . . . denuded. It stands like a battered fortress in an open prairie." Chicago's metropolitan expansion created changing environments that altered the plant communities observed by the natural scientists and the conditions of human communities monitored by the social scientists.[5]

Despite their common ground, Chicago's scientists and sociologists may have never explored or written about Chicago together. They did not synthesize a collaborative methodology from their analogical musings. One reason was that some of the plant scientists' most important studies about succession and the region's plants were completed before sociologists Robert Park and Ernest Burgess began teaching in and studying Chicago. In a somewhat successional process, the scientists' work pioneered this intellectual terrain, and the sociologists later interacted with and adapted their ecological ideas to analyze urban life. Moreover, by the time the Chicago sociologists began executing their research program, Cowles and his students had initiated new lines of inquiry and invented new analogies that the sociologists did not adopt. Instead, the sociologists advanced their understanding of succession with the writings of University of Nebraska ecologist Frederic Clements. However, while the plant ecologists' and sociologists' theoretical interests diverged, the ecological and sociological analogues they had developed continued to inform their interpretations of metropolitan nature. For example, George Fuller used the sociologically grounded conceptions of plant communities to determine what plants belonged in the Chicago region. Fuller contended that species composing natural plant communities had "clearly defined associations," and he denied that "weeds or ruderal plants" were a part of Chicago's vegetation, even if they inhabited the metropolitan environment. Herman Pepoon, a local botanist who associated with the university's scientists, commented that *Commelina communis* was "one of the most pernicious of city weeds, growing, by preference in shaded places." In offering common names for the plant—Asiatic dayflower and wild wandering Jew—that referenced foreign places and mythological transients and in describing its habitat as an urban environment common in Burgess's "zone in transition," Pepoon seemed to assert the social boundaries of urban life in ecological terms.[6]

The analogies that Chicago scholars proposed to discern their common interests were not strong enough to overcome their uncertainties about human relationships with the environments developing in the region. John Dewey's statement about the significance of the "evolution of environments" did not detail how people should manage this evolution. Dewey contended that the environments people were creating sustained their evolution. However, Dewey's ideas were not inspired by his investigations of the Chicago region; they were a response to the ideas of Thomas Huxley, a British scientist whose writings had profoundly shaped Dewey as a young man. Contrary to the progress Dewey speculated was resulting from the coevolution of people and their reworkings of the natural world, Huxley saw all creatures—including people and "the humblest weed"—as unique products of "the cosmic process" competing to shape the planet. To Huxley, human progress arose from the labor people expended in creating and maintaining environments—Huxley used gardens as a quintessential example—that other organisms of the cosmic process continually worked to undermine. These conflicting views had conflicting implications for sociologists and ecologists. The tension in these views created fewer problems for sociologists, who were primarily interested in analyzing and addressing problems caused by social relationships. However, to the ecologists, it could seem that as people in their region reworked nature, they produced environments that confirmed Huxley's argument that expanding space for humans required eliminating competing "works of nature."[7]

Dewey's pragmatic view that changing environments were possible determinants of evolution and expressions of humans' ongoing evolution may have troubled Chicago scientists when they investigated the ecologies of the region's industrial centers. The vegetation from different parts of the continent that became established in the region during the geological tumult of ancient North America was being obliterated, especially as the U.S. Steel Corporation transformed the

Figure 2.3. These "tramped trees" on an edge of Gary were evidence of city growth to Victor Shelford, circa 1909. Special Collections Research Center, University of Chicago Library.

area. Ecologists documented how factories and people both destroyed trees. In Buffington, just west of Gary, pulverized slag particles released by a cement factory injured pine trees. On the edges of Gary, migrant workers stripped trees of their branches and limbs for fuel. To Victor Shelford, a young ecologist who read widely in sociology, studied zoology with Charles Davenport, and was mentored by Henry Cowles, metropolitan growth had produced flat, treeless Gary. Shelford saw "succession with city growth" as ecological destruction. Along Main Street's barren expanse of concrete, utility poles outnumbered the few remaining trees. Beyond the central commercial hub "succession with city growth" resulted in poles suspending power and telegraph lines that towered over a few remaining trees near railroad tracks. Shelford saw that urbanization was not contained within city lines but extended well beyond them, with infrastructure development and resource exploitation by people on the margins of urban economies. Although ecologists did not consider Russian thistle part of the region's vegetation, they may have seen Russian thistle next to railroad tracks and smoking factories as the natural vegetation of Gary. Similar environments were evolving between the university and Gary. When a Burgess student investigated urban industrial growth in

Figure 2.4. Russian thistle in Gary, 1910. Special Collections Research Center, University of Chicago Library.

southeastern Chicago along the Calumet River, he reported finding patches "overgrown with grass and weeds" arising from unpaved streets, as well as surrounding wood-frame homes erected for mill laborers.[8]

The University of Chicago's faculty and students were not the only Americans who attempted to understand urbanization with ecological analogies. For example, Charles Saunders used his observations of vegetation change on Philadelphia's vacant lots to comment on demographic patterns. Saunders saw an "urban flora . . . largely of foreign origin," remarking, "The foreign plant immigrant . . . is apt to be less fastidious, and will often live and spread even amid ashes and rubbish." In contrast, Saunders claimed the "average native born American plant, like the redman, finds city life too circumscribed for it, and leaves when bricks and mortar come." Saunders's equation of people and plants—the "redman" and the "native born American plant"—turned his description of plant distributions and dynamics into a statement about the boundaries of community and belonging. It also foreshadowed Saunders's future, as he would soon relocate to Pasadena, California. In the first decades of the twentieth century, Americans analogized people and plants—frequently with the term *human weeds*—to express their concerns about disorder and disarray. These so-called human weeds—and the plants associated with them—threatened to change American life. Vegetable weeds and so-called human weeds were used to characterize cities as environmental problems that were undermining the nation. Controlling and eliminating these weeds were necessary to guide the evolution of cities.[9]

The contemptuous expression *human weeds* compared people with plants to assert that they did not belong. The analogy assumed that some people and plants were excessively mobile and common, without purpose, or harmful and destructive. People were also tacitly regarded as weeds when they were conceptualized as objects to be weeded out. Comparisons of transient people with weeds—which were typically based on spaces, mobility, and behaviors—were particularly common. Although descriptions of vagrants were "necessarily vague and uncertain," to the extent that vagueness made analogizing plants and people easier, Americans who identified people as human weeds expressed derision without precision. Such analogies often explained little about the people objectified as human weeds. However, the analogies did reveal the thoughts and feelings of Americans who feared urban environments advanced social and cultural degeneration.[10]

In the nineteenth century, Americans discussed an array of human weeds who were damaging, unwanted, or useless. In Connecticut, the fraudulent transfer of a pauper from East Haddam into Marlborough was compared to one farmer sowing the seeds of a "pernicious weed" in another farmer's field. A poet jested that unmarried, pleasure-seeking men who had no children failed to advance the "world's good"; they were "human weeds" less useful than cabbage and sheep. Demographic changes led naturalist John Burroughs to describe "human weeds" as "seedy" because "they increase and multiply over the more valuable and highly cultivated plants." Journalist Henry Grady remarked that the "red man was cut down as a weed, because he hindered the way of the American citizen" during the nation's rapid geographic expansion. In 1893, Sacromentans may have laughed about the "rank growth of cads" who thrived "best" in Gilded Age New York. These native-born sons of privilege were "human weeds [who] generally worship wealth and have no higher aim than to be considered above earning a living." Lawyer William Hopkins lectured Fitchburg High School students that becoming overspecialized and engaging the world only in terms of one's narrow interests risked becoming a snob—"a cultivated human weed . . . admired only in his own narrow circle" that "the rest of mankind would willingly throw . . . away."[11]

The botanomorphization of Americans who were continually searching for work stretched throughout the nineteenth century into the twentieth century. A colonial governor in *Knickerbocker's History of New York* vilified vagrants and beggars as "dandle-lions" [*sic*]." Burroughs wrote that "weeds" were "the tramps of the vegetable world" because both weeds and tramps seemed abundant along public roads and railroads. "Industrial and moral" Duluth, Minnesota, reportedly had "no place for idlers" because its "wide-awake community naturally weed[ed] itself of them." Brooklynite Charles Skinner detested tramps and "drunken, corner-loafing, wife-beating, non-washing, swearing, insolent creatures" enough to opine that "the weeding of the human race cannot begin too soon." People on the road may have considered "the weeds" as a satisfactory place to rest without being disturbed when they had no alternatives. *Industrial Worker,* the International Workers of the World newspaper, published a photograph of migrating workers gathered in one of their "jungles"—a patch of plants on rocky ground beneath an elevated railway. Ray Stannard Baker's fictional account of tramp life demonstrated the persistent vague comprehension of transient Americans as weeds. Baker's narrator applied Emerson's definition of a weed to a tramp—"a misplaced man, whose virtues have not been discovered." A tramp's robbery of the narrator implied why these men were misplaced and why their virtues were not discovered. However, as industrialization and depressions continued to displace Americans, some reformers tried to distinguish among tramps, hobos, and bums, which to one reformer were "three species of the genus vagrant." In a Colorado railroad town, leaders attuned to the plight of the unemployed tried to identify which men were willing to work for meals or lodging when they were "weeding out the tramps."[12]

Twentieth-century Americans anthropomorphized weeds as tramps and vagrants to emphasize that both were out of place and undesirable. *St. Louis Post-Dispatch* editors were disgusted by "vile smelling and generally obnoxious" jimson weed, which "like a tramp . . . delights to infest the vacant lots of cities." Willis Blatchley, an Indianapolis entomologist, geologist, and natural history author, described weed seeds as "constantly traveling . . . hoboes in hay, bedding, packing, [and] shipments of fruit." University of Michigan ethnologist Melvin Gilmore, who studied with Charles Bessey to earn his Ph.D. in botany at the University of Nebraska, characterized "exotic species" from other parts of the world as "vagabonds and tramps" that "introduced themselves fortuitously, coming as furtive and undesired immigrants, stowaways." In his "indictment against our city weed population" for injuring people's health, Purdue University's Albert Hansen declared ragweed the "blackest villain" and poison ivy the worst of the "plant waifs." Frank Thone, who studied with Henry Cowles and George Fuller to earn his doctorate at the University of Chicago, claimed that pokeberry had become "a vagabond and a weed" after being "robbed of all its possible occupations" as a dye or food. Botanist Oren Durham thought that ragweed plants were "just like undesirable people" because they defied "race suicide" with their "amazing" adaptability and their "discouraging" fertility. A *Chicago Tribune* columnist claimed the black-eyed Susan was a "vagrant that has hoboed its way into Chicagoland from the west." These analogies reinforced people's association of tramps with weeds.[13]

Identifying plants and people as weeds labeled what and who was unwanted or no longer belonged in cities and modern spaces. Although some social reformers wanted to assist and uplift human weeds, others wanted to control or eliminate them. Intervening in the lives of human weeds—people who had little and who suffered much, whether from disease, injury, misfortune, or poverty—potentially reduced the deprivation that might be suffered by future poor people as well as potentially enhanced the quality of life of the middle class and elites. Some Americans believed that "weeding out" human weeds could rework the ecologies of their changing cities. However, human weeds persisted in urban life partly because their detractors were eager to "weed out" unwanted people but did not fully comprehend "weeding out" as an ecological phenomenon.

Polluted environments inspired Logansport, Indiana, physician Robert Hessler's concern for "human weeds." In 1903, the enthusiastic naturalist tramped around Lake Michigan's sand dunes; gazed at "wild plants in their native homes"; and ate "little else" but sand cherries, wild grapes, and huckleberries. Closer to his home in Logansport, he walked about and monitored "new weeds." On one walk, Hessler followed the Wabash Railroad tracks away from Logansport. When he climbed a steep embankment that reminded him of a space like a volcano's "desert . . . of black cinders," he found only sheep sorrel and pondered the "precarious" existence of people who lived in marginal areas. The experience was one that Hessler analyzed to comprehend the ecological and social dimensions of urban growth, and he used it to explain the plight of "human weeds" to fellow members of the Indiana Academy of Science and reform-minded *Survey* readers.[14]

Hessler stressed how the production of environmentally degraded spaces shaped the development of plants and people. When city building replaced forests and meadows with concrete, stone, and gravel surfaces, when nutrients in the ground became scarce, and when air pollution intensified, the happenstance plants capable of tolerating this stress often grew too slowly and too sparsely to green bare earth. In Logansport, such environments emerged from an economy invigorated by the Wabash Railroad and Pennsylvania Railroad lines that connected Logansport with Chicago and St. Louis to the west and with Toledo, Dayton, Cincinnati, Detroit, Buffalo, and Pittsburgh to the east. Hessler's descriptions of squalid areas suggested that the city's environmental stewardship had become less effective since the state Board of Health praised Logansport's efforts to banish diphtheria in 1886. Hessler compared the struggles of these plants with those of the people who survived in these habitats. Logansport's degraded areas led Hessler to wonder "how long human life" could endure near railroad yards, where "smoky air . . . killed the trees and . . . only a few weeds were able to grow." Each urban space had "its weeds" and "its corresponding class of people." In the "Shanty

Town" area, Hessler remarked, "weeds flourished among the human habitations. The people themselves, like the weeds, were of the neglected kind." Hessler commented that riverfronts were "infested by a class of people known as 'river rats,' a highly undesirable class; human weeds, so to speak." To Hessler, the most adaptable people were like ordinary plants. He wrote, "Common weeds find conditions favorable almost anywhere and flourish, especially in neglected places. Shall we say that human weeds also thrive almost anywhere . . . ?"[15]

Hessler used analogies to describe the ecological dynamics of city dwellers. He claimed that "people, like herbaceous plants, but unlike trees, are more or less constantly moving about." People living in Logansport's exclusive residential neighborhoods, places with dusty streets, or shanties all attempted to move to improve or protect their health. The wealthy cultivated places to which they retreated. However, these introverted spaces facilitated the development of diseases in poorer places, which eventually spread. Hessler worried that urban industrial environments increased pollution and disease rates, homogenized urban space, and limited the kinds of people who could survive. Immigrants, migrants, and native residents all competed for increasingly "unsanitary" niches. Ongoing growth undermined the "balance of nature," what Hessler considered an ideal population size free from ravaging pests and diseases.[16]

Hessler also thought diseases were like weeds, and he used this analogy to educate his patients and Logansport newspaper readers. Just as weeds eventually spread into cultivated fields, diseases eventually spread among human populations. When rabies appeared in Logansport, Hessler advised great caution around unmuzzled dogs to avoid being bitten and infected. He applauded a local law to combat the disease because "new diseases like new weeds if left unchecked produce mischief." Hessler was not alone in likening diseases to weeds. From 1903 through 1905, New York City's Scott & Bowne ran advertisements nationwide for its cod-oil-based—and allegedly morphine-containing—Scott's Emulsion. The company marketed the concoction as a remedy for consumption, a disease described in their advertisements as "a human weed flourishing best in weak lungs." However, Hessler stated that such dangerous patent medicine manufacturers did not "thrive in clean communities."[17]

Weeds in cities—whether plants, people, or microorganisms—inspired Hessler's euthenic reform vision. Hessler, who had served as the president of the Indiana Academy of Science in 1906 and worked at the Indiana State Hospital for the Insane, asked his fellow scientists, "Why is it that 'human weeds' are given such an undue amount of attention . . . [in] asylums . . . [?] Why must a man wait until he becomes insane or a pauper or a criminal before being housed under sanitary surroundings?" He advocated "attention to cleanliness" to reduce "weeds and diseases and ill health." Hessler also recommended providing city dwellers with "pure water, good food, good air and clean homes" and educating immigrants. In Hessler's interpretation of the Progressive conservation creed of providing the greatest good to the greatest number, the greatest good could be accomplished by improving urban environments to counter the degradation of human life, which in the long run could reduce how numerous the greatest number was. Hessler's belief that environmental management could prevent people crowding into cities from becoming human weeds diverged from the eugenic policies emerging in Indiana, such as a 1905 law prohibiting alcoholics and people deemed mentally weak from marrying and a 1907 law authorizing sterilization of incarcerated persons.[18]

Robert Hessler's hopes perpetuated a nineteenth-century reform ideal that environmental transformation could transform society. Similar ideals were echoed in editor, poet, and conservationist Robert Underwood Johnson's praise of the delivery of water from the Catskills to New Yorkers as a "heavenly ministry" to the "city's human weeds and flowers." The National Tuberculosis Association informed parents that they could prevent a child from becoming a "human weed" by teaching the "health commandments" of breathing fresh air, eating well, washing hands, and brushing teeth. Hessler's ecological analysis of human weeds was unique, but concerns about human weeds were widespread. Discussions about human weeds were nourished by horticulturalist and "humaniculturist" Luther Burbank's ideas about the malleability of the "human plant." Burbank believed that because "within both plant and man are the same possibilities for good or evil . . . the task is to eradicate the evil by substituting the good." Decent environments for and the proper care of children were essential to cultivating this good. Burbank wrote that "a dusty factory or an unwholesome school-room or a crowded tenement" harmed children, and he warned that cramming "little brains with so-called knowledge" without instilling "honesty, fairness, purity, lovableness, industry, thrift, what not," risked letting "weeds grow up in a child's life" that could destroy their futures. Burbank emphasized that "human plants and human weeds" were "always subject to improvement or deterioration according to environment." Although Burbank, like Edison and Ford, was a popular face of scientific progress, some of this popularity seemed to result from perpetuating older reform and spiritual rhetoric. In Bristol, England, John Cary described begging children as "young Plants," which when tended properly improved the "Face of Our City." A poem by Laman Blanchard celebrated the life of a pastor who delighted in "Finding immortal flowers in human weeds."[19]

Burbank repurposed reform language in service of the future. Although Burbank thought the nation was "producing too many human weeds," he also regarded masses of human weeds as a somewhat natural condition. When Burbank pointed out that "we are little more than a field of wild human weeds in which, here and there, is a superior type," he suggested that human weeds were normal, ordinary people. After Burbank supposedly remarked that if Americans managed plants with the carelessness with which they raised children they would "be living in a jungle of weeds," Elbert Hubbard quipped, "The actual fact is that socially . . . we live amid the pigweed and the purslane." Burbank thought that applying breeding techniques to human reproduction could improve future humans. He predicted that "the human race will in time find a way" to manage population growth. And he hoped that "the weeding-out process will . . . by selection and environmental influences . . . leave the finest human product ever known . . . the American of the future." Burbank perceived the highly hybrid and "crossed" nature of America—"a nation with the blood of half the peoples of the world in our veins"—as a source of vitality.[20]

Burbank's method of improvement involved "heredity, environment, selection and crossing." He believed that the best environments allowed individuals to show their greatest potential, but his work also indicated that inherited characteristics restricted their potential. Burbank recognized the significance of the interaction of genetics and environment. For example, he remarked that "just as an evil environment will undo a good heredity, so a good environment may be all powerful in overcoming a bad heredity, and helping to develop latent good and desirable traits." Even where genes seemed to be the dominant factor, Burbank saw genes as the outcome of evolutionary time. Burbank pointed out that "environment is more powerful than heredity, for environment creates heredity and heredity is the sum of all past environments." Burbank's knowledge that all cultivars had been produced from wild plants led him to conclude that "if we regard human weeds from this point of view, the outlook is not hopeless." In Burbank's analogies, all humans were the same species of plant; he conceived of human weeds as weak individuals but did not dehumanize them as nonhuman vegetables. Some scientists and doctors saw where the analogy ended and the problematic reality of grand designs began. Burbank conceded that establishing a eugenic society would be difficult because while plant breeders had the liberty to destroy individuals "not up to standard and save only the best," the same could not be done with ordinary Americans. Hessler similarly stated that plant breeders' constant elimination of the unfit could not be employed by men since they did not "seek the destruction of those not adapted." The more aggressive William Howell, a Johns Hopkins physiologist, remarked that "we should be glad to have [these 'weeds in the garden of civilisation'] eradicated by any means that does not offend our sense of humanity or endanger those bonds of sympathy which hold society together."[21]

Burbank's accomplishments both frustrated and inspired proponents of eugenics. His ability to turn "insignificant desert weeds" into "floral marvels" allowed physician John Kellogg to imagine a "New Human Race." Stanford's David Starr

Jordan stressed that Burbank had once acknowledged that selecting the best individuals over generations was "ten thousand times more important" than environmental conditions in improving a cultivar. Reformers who believed inherited characteristics to be the most important influence on future generations thought attending to "stock" and "germplasm" could produce superior individuals and lead to permanent social reform. Race-focused eugenicists likely did not accept one physician's claim that because "the human family contains but a single species," all people could become decent citizens under the right environmental conditions. Biologist Charles Davenport argued that "man is not a single homogenous 'species' but is composed of numerous elementary species . . . races." In his view, human nature was dictated by race, and this race dictated in which environment people belonged. Davenport worried about the purity of "elementary species" because of "the extensive, almost universal, hybridisation that is going on in mankind." The belief that eugenics protected and enhanced a race had a corollary; there were inferior races of human weeds. Yet this ideal future race was essentially the reclamation and protection of a past one. Burbank's ideas about the eclectic nature at work in producing fascinating and valuable cultivars conflicted with Davenport's eugenic notions of Anglo-Saxon purity. For Americans who were enthralled with the purity of their birth, who believed that the windfall of natural capital that had energized the nation's evolution proved their superiority, and who feared that declining native-born white birth rates, increasing immigration, and spreading poverty threatened "race suicide," human weeds were agents of deterioration. Eugenicists had decided that the "spring in nature and in human life" once preached by Unitarian minister Minot Savage had passed. Savage remarked that from a "civilized and cultivated" yet "supercilious" point of view, "a great majority of the growths of humans are weeds." However, Savage insisted that there was not "an essential human virtue that" one could not find "down among these human weeds." Eugenicists no longer possessed the time to cultivate whatever virtue biology had left such people.[22]

Perhaps the most vehement enemies of human weeds were Yale University geographer Ellsworth Huntington and American Eugenics Association secretary Leon Whitney—two descendants of English Protestants who had courageously sailed across the Atlantic, brilliantly established a republic that was "a golden mean between autocracy and democracy," and audaciously worked their way across the continent to the Pacific. However, despisers of human weeds were many and diverse, since eugenic thought and advocacy were not bounded by gender, race, creed, occupation, nationality, ethnicity, or political orientation. Many proponents of eugenics believed they were superior to the human weeds around them, but how this conviction tinged their everyday interactions with human weeds is probably an irrecoverable dimension of the past. Although the human weed slur is a possible indicator of their demeanors, its significance is as a conception of ecological dynamics. Eugenic advocates referred to increasing numbers of unfit people, typically in cities, as human weeds. The number of human weeds was either intolerable or bound to become so without action. Eugenic reformers believed these human weeds overcrowded environments, exhausted philanthropic resources, caused crime, and contaminated Anglo-Saxon stock. "Weeding out" referred to a process that was injuring native-born Americans but that, if seized and commanded, could benefit the nation by preventing the uncontrolled population growth of undesirable people. Weeding out was both a demographic phenomenon of cities injuring society and a strategy to save the nation.[23]

Antagonists of human weeds found these people out of place for reasons ranging from their ancestors' origins, to their reproductive abandon, to their dispositions, to their transgressions, to their seemingly empty existences. Huntington and Whitney used the ancient geographic origins of phenotypes to determine who belonged where in the present. They speculated that if England were "a tropical jungle, the modern type of Englishman might be a human weed—a being out of place." This logic was one basis of their view that they would "be better off without . . . Negroes and their problems." The concern that some people produced more children than they could support—and that their children would go on to do the same—was expressed in Burbank's statement that "what we may call human weeds have the same tendency to overrun the earth, crowding out better specimens, that plant weeds have for taking possession of the land." Huntington and Whitney argued that behavior exposed "human weeds." Although they acknowledged that truly "conscientious people" were preciously few, Huntington and Whitney believed that people who disregarded "the general weal" were "reversions to a primitive type so far as conscience is concerned." Such "human weeds" were "out of touch with those laws of conduct . . . which mankind has for ages been developing." Although Huntington and Whitney considered the "obviously insane . . . and the obviously feeble-minded . . . rather harmless weeds," William Howell stated that people with extreme forms of "lunacy, feeble-mindedness, habitual criminality and pauperism" were weeds detrimental to society. Burbank's rather pedestrian analogy "our criminals are our weeds" was surpassed by the words of a Chicago writer who described the criminal underworld a "human weed patch" of "haughty jimson weeds" and a "dense undergrowth" of dock, dog fennel, mullein, pennyroyal, and smartweed. Referencing H. G. Wells's novel *Secrets of the Heart*, Margaret Sanger stated, "Mr. Wells speaks of the meaningless, aimless lives which cram this world of ours, hordes of people . . . who have done absolutely nothing to advance the race one iota. Their lives are hopeless repetitions. All that they have said has been said before; all that they have done has been done better before. Such human weeds . . . drain up the energies and resources of this little earth." At its broadest, the human weed analogy referred to no one in particular, or just about everybody. A proprietor of the Upstairs Clothes Shop in Deseret, Utah, counted among the state's human weeds "calamity howlers, knockers, pessimists and . . . boosters for every place on earth except the one in which they live."[24]

Just as there were many reasons people might be denigrated as human weeds, there were many forces generating human weeds. Analysts regarded the dynamic American economy as a powerful force that elevated some people and displaced others. Competition affected all Americans; the production of fine flowers also resulted in the weeding out of others, who in their failures were human weeds. William Sumner wrote that the millionaire was the "finest flower of a competitive society." Such flowering could be ephemeral, since, as Thorstein Veblen observed, the elites who reverted to their "non-predatory human nature" or lacked a vigorous "pecuniary temperament" were often "weeded out" and lost "caste." W. E. B. Du Bois described the liberation of Southern slaves as an "economic and social revolution" that created "social grades," in part by "a weeding out among the Negroes of the incompetents and vicious." Columbia University's Samuel Lindsay observed that "industrial processes like the processes of nature" had "selective values" that could "weed out the unfit and incompetent workers." While the outcomes of competition could be justified by analogizing them to natural selection, the outcomes of such competition depended on a variety of socially constructed means of social selection. For example, industrial enterprises reduced costs by relying on physicians "to weed out the misfits" by screening prospective employees for physical and mental disabilities.[25]

Yet the capitalists' "weeding-out" process, which yielded marvelous flowers in economic terms, also seemed detrimental to the nation's demographics and urban life. Even as the economy created fantastically wealthy people, it also seemed to reduce the number of pure and refined Americans. University of Chicago–trained economist Herbert Davenport argued that the rigors of competition drained the "nervous energy" and moderated "the strength of the reproductive instinct" of the most economically powerful, as well as starved and sickened the poorest workers, thus "weeding out the over-productive and improvident specimens of the race." The urban environments where wealth was concentrated maimed some of the best Americans before they flowered. Robert Hessler reported, "It is well known that the city 'takes it out' of strong and robust men—they soon fail." Charles Davenport, who taught at the University of Chicago before establishing the Eugenics Records Office on Long Island, contended urbanization undermined the reproduction of the United States' dominant classes because cities sterilized "the best" and permitted "the worst to reproduce." Davenport argued that within "hundred mile" radii of the nation's "great cities," poor whites proliferated, while the most vigorous rural citizens with the "best protoplasm" migrated to cities, where they did not marry and found "it inconvenient and expensive to have children." Madison Grant claimed that a powerful "Nordic" man who could thrive in an agricultural community struggled to compete in an industrial environment, where "the cramped factory and crowded city quickly weed him out." Samuel Lindsay observed that the American economy was segregating incapable workers "into the slums of our large cities where society must bear the burden and pay the cost." Whether economic competition was actual natural selection or social selection accomplishing the same end, to eugenic-minded idealists it was not viable evolutionarily because it inhibited the reproduction of the best "stock" and risked "degradation of the race."[26]

Immigration and fertility rates of the poor were the two main reasons why eugenicists like Henry Williams, a disseminator of Burbank's writings, worried that "the human garden is in danger of being choked with human weeds." Huntington and Whitney believed that "genuine human weeds . . . from almost every nation in Europe," as well as the native-born rural poor, were "unfit" for urban life. To Huntington and Whitney, cities were filled with "Weeds! Men out of place; Millions of them!" Huntington and Whitney identified the origins of the problem in their greedy forefathers' decision to import cheap labor. However, that European leaders had used "our sleeping fathers['] . . . 'melting pot' . . . [as] a very convenient garbage pail" proved to Huntington and Whitney that many immigrants had been "out of place in their own native environment" and explained why they "were still worse in America." The environments created by "foreign born" people or people not "of Anglo-Teutonic origin" worried physician Myre Iseman. He considered one section of Atlanta with its "Jewish wretchedness" and "negro wantonness" to be "a plague spot." Iseman claimed the "vileness" of this space was exceeded by mill-district housing, with "unpaved streets, choked gutters, open latrines, and yards overrun with weeds, which serve to conceal the filth and refuse of slovenly homes." Such environmental dangers—environments that threatened to continue to weed out civilized native-born Americans—seemed likely to spread as their populations increased. While the increase of such populations demonstrated to physician Theodore Robie the necessity of finding "a way to stop the undesireables from overrunning us," such conditions may have led Burbank to remark that "weeds breed fast and are intensely hardy."[27]

Eugenicists suspected that their own actions and values allowed human weeds to survive and reproduce. They worried that environmental improvements and charity disproportionately benefited human weeds. Psychologist Granville Hall complained that "hygiene and medicine . . . [kept] alive those with weak lungs, wills, minds, eyes, ears, [and] muscles." Hall argued that "cherish[ing] the weeds and neglect[ing] the good plants" resulted in the "proletarianization of a higher race" that drained "our aristocracies." Huntington and Whitney concluded that "human weeds" did not survive in "primitive and backward societies" because life was so difficult. They believed that "our Christian kindness, our medical skill, and our feeling of social responsibility" were "preserving the weeds and spreading their seed." A sociologist's estimate that nearly 90 percent of "poor human weeds . . . salvaged in childhood by the home-finders" became "good citizens" may have further aggravated eugenicists' anxieties about the future.[28]

Eugenicists believed that increasing their kind's numbers and reducing the populations of human weeds were necessary to strengthen the nation in which future Americans would live. They hoped that by "weeding out" the unfit they could counteract the urban ecological pressures that "weeded out" superior native-born Americans. Among the policies and ideals that eugenicists thought could purify America were the 1924 Immigration Act, exclusionary minimum wages, marriage licenses, knowledge of birth control, the prohibition of alcohol, tobacco-free and "narcotic-free fatherhood," sterilization, and retributive punishment, including the death penalty. If natural selection operating via the economy was dysgenic, an array of policies and regulations could employ social selection to enhance the nation. For example, a posthumously published volume of Burbank's remarks included statements that the crowding in cities resembled "the struggle and the pressing in of a multitude of weeds"; that because such conditions produced "spindling, weak," people, the "growth of cities is unhealthy for a nation"; and that spreading people into "outlying districts" where there were "breathing spaces such as parks and playgrounds" and educational opportunities that allowed "the human plant . . . to meet the competition of the crowd and reach a fuller stature, mentally and physically," was a possible solution. By producing more children from Americans who best fit American ideals and by reducing the numbers of unfit people to sustain American culture, American eugenicists worked to cultivate a sparser human monoculture between the Atlantic and Pacific Oceans.[29]

Eugenicists most concerned with the reproduction of people with undesirable biological inherited traits advocated regulating human reproduction. Physicians thought that "weeding out the weaklings" might be possible by prohibiting the marriage and reproduction of the least physically robust people. Bernard Talmey believed that "degeneracy could be weeded out of human society" and advised fellow physicians that "society has a right and even a duty to weed out these delinquent, defective and dependent classes by the prevention of the procreation of the various defective offspring." Psychiatrist Louis Bisch argued that "ending production of human weeds" with eugenic measures simply modernized marriage-selection practices used by primitives and prescribed in the Bible. However, since marriage laws alone could not purify the population, eugenicists advocated sterilizing problematic procreators. Oregon's sterilization law was applauded by *Sunset* magazine as a measure "Aiding Nature by Sterilizing Human Weeds." When the Supreme Court upheld Virginia's sterilization law, *Helena Independent* editors opined that sterilization was "one of the mild methods of disposing of human weeds." They suggested that eugenic sterilization was a conservation measure in that it advanced "the greatest good for the greatest number, and that is after all the end of all government."[30]

While sterilization was a method of population control imposed by physicians on the unfit, providing women with birth control knowledge could also advance eugenic ends. In addition, eugenic discourse could be used to advance other social changes. Margaret Sanger worked for the "emancipation of working women" by discussing eugenic issues such as population reduction and national vitality. To Sanger, teaching women to use birth control so that their children were "conceived and born in joy . . . never as sinister punishments, never unwanted and unwelcome as transients" was "organically bound up with the biological welfare of the whole community." Poor women begged Sanger for help to limit their family size so that their children could grow into "strong and stalwart Americans." Increasing the "value of child life" with birth control reinforced the notion that the quality of children, not the quantity of them, mattered most. Sanger's radicalism might have seemed tolerable to some conservative eugenicists because it helped achieve their reforms. The thoughtful use of contraception permitted the "cultivation of the better racial elements in our society, and the gradual suppression, elimination and eventual extirpation of defective stocks—those human weeds which threaten the blooming of the finest flowers of American civilization." Moreover, Sanger remarked that as people predisposed to "breeding like weeds" became fewer in number, less would be spent succoring "the dependent and the delinquent." However, patriarchal eugenicists rejected Sanger's feminist commitment to improving the lives of all women. Huntington and Whitney argued that feminists undermined social reform by venturing beyond the domestic sphere. They stated that young feminists could "ensure the future success of their cause" by finding husbands "of their own type" and having "five to ten children apiece." The men proposed that the best role for older, childless, or husbandless women was devoting "all their time and money to assisting the mothers who have many children, or to caring for the brightest and best of the motherless children." All eugenicists may have desired a better future America, but they did not imagine the same future.[31]

Sanger's advocacy of birth control to produce healthier children also diverged from Huntington and Whitney's contention that euthanasiac infanticide was an expression of charity and mercy. The hospitalization of a Chicago prostitute's baby afflicted with a venereal disease and pneumonia prompted the men to ask if the best treatment in such a "hopeless" case was to apply "a little chloroform . . . painlessly over the child's face." They thought that "nature, in a less merciful way, was trying to accomplish just that, but man's ignorance intervened." Huntington and Whitney blended humility with envy in wondering if Americans needed more often to emulate "nature [which] is sometimes cruel, but always kind." Reshaping society with the perceived workings of nature might have been a rather exotic undertaking for people professing a desire for a purer America. More than three decades earlier, New York physician Albert Ashmead sensed that Christians' confidence in contesting "the eliminating operations of nature" was being challenged by "the Oriental spirit . . . to eradicate the human weed, to sweep away all human influences detrimental to mankind, whether they be represented by disease or by crime." Eugenicists concerned with the failure of natural selection attempted to devise methods of social selection that approximated, or even accelerated, the efficiency and righteousness of nature.[32]

While some eugenicists feared that "weaklings" and other biologically deficient people could injure the nation over the long term, an array of order-seeking Americans thought that the most dangerous human weeds around them were criminals. The nation's chief crime fighter, J. Edgar Hoover, argued that fighting crime was like fighting "noxious weeds." To get rid of both, Americans had to "pull them up by the roots." Hoover contended that catching a criminal only cut down "the stalk of the weed." The "roots" of crime were the crooked family members, medical and legal professionals, and politicians who provided such menaces with support and safety. Succeeding in cutting down and uprooting such weeds created new problems. In response to Burbank's remark that the nation was "producing too many human weeds," a Pennsylvania newspaper opined that although "human weeds" were not overrunning the earth, they were proliferating in cities such as Philadelphia, and as a result, courts were becoming "cluttered up." Imprisoning weeds seemed to remove them from the garden of society but transfer them to a hothouse that only nurtured their antisocial natures. *New York Evening Journal* editor Arthur Brisbane worried that the Auburn State Prison in central New York was "a dismal garden set apart for human weeds and in it many a good plant is hopelessly driven into the weed class." He pled for the rehabilitation of "the human weeds in prison," who were as diverse as the faces seen on city streets. One analyst and supposed inmate believed that the nation's criminal justice system did little more than just cultivate a "mammoth bouquet of human weeds." Burbank rejected using harsh environmental conditions to punish the "human weeds . . . we classify as criminals." He thought morals could not be developed without "healthful surroundings . . . of sunshine, fresh air and good food." However, the resources expended on rehabilitation and the tragedies resulting from unreformed, paroled criminals appeared to other Americans to result from "an exaggerated notion of fair play and justice" and result in a nation "overrun with . . . human weeds."[33]

Americans outraged by violent criminals insisted on draconian retribution and liberal use of execution. After Clarence Darrow succeeded in protecting convicted murderers Richard Loeb and Nathan Leopold from the death sentence, criminal law theorist and Northwestern University School of Law dean John Wigmore wrote that the biological and psychological bases of the murderers' behaviors had no bearing on "society's right to eliminate its human weeds." *Helena Independent* editors believed that once "the danger of permitting . . . social weeds . . . to live and increase" was recognized, exterminating them would "become as popular as shooting wolves and crushing the heads of rattlesnakes." Tough-on-crime newspaper editors declared, "Criminals are the weeds of the human race and we should tear them up and cast them aside, as we do the endless variety of weeds which are always choking up our lives." Sterilizing criminals to prevent their reproduction was not sound enough. *Helena Independent* editors wrote, "We do not dispose of enough human weeds." Such men and women had "to be pulled up by the roots and absolutely destroyed before the community can be made safe and healthy." Although such fierce statements made clear a preference for the swift destruction of human weeds, they rarely specified whether all violators of the law, whether perpetrators of felonies, misdemeanors, or infractions, were community-endangering weeds.[34]

Americans who believed that society's stewards were failing to police the nation adequately sometimes resorted to vigilantism to protect their communities, which in turn could result in additional outbreaks of violence. To African Americans, white racists and their racism were the most dangerous weeds in urban life. Racial violence increased with the tremendous growth of cities as white employers and powerbrokers pitted working-class native-born whites, immigrants, and blacks against each other. After World War I, economic opportunities for Southern blacks in Northern factories narrowed, and white racism intensified. *Chicago Defender* editors insisted that the American Legion, which had members who belonged to or supported the Ku Klux Klan, had to "weed out these undesirables [to] . . . save their splendid organization." Frustrated *Defender* writers stated that segregation in Chicago "has grown up like a field of weeds in with the same tenacity as weeds out and choke to death all of the finer elements of our civilization. . . . The only remedy, therefore, for the situation is that which is applied to weeds. . . . They have to be uprooted." *Defender* editors realized that the black community also included problematic people, and they pledged that they were "as anxious to weed out and punish the bad element of our race as any other class of people."[35]

While over the long term population control and execution of criminals could reduce the number of human weeds, eugenics advocates also had to contend with young human weeds. Some elite educators favored isolating human weeds and limiting their education. Granville Hall argued that the "weeding out of the dwarfed stalks" from classrooms required tough standards and needed to be "pretty thorough." Colgate University president George Cutten, an IQ-testing proponent, saw education as an ideal, like democracy, that many Americans had no use for or ruined. Such positions seemed to cultivate one group by applying nurture and to neglect another group by withholding nurture, thereby exaggerating whatever distinctions had likely originated in children's environments. These men's distinguished intellects may have alienated them from how young people developed. Weeding children out did not address what a New York City hospital superintendent believed were the root causes of stunted development, such as criminal environments that produced children who were "human weeds" on which "the flowers of the human mind do not bloom." An "instructor of youth" thought of resilient boys who seemed "to come from nowhere" and seemed to grow "by themselves, without training or care," as "human weeds." Yet despite "the rankness of their growth, their spontaneity, their scorn of cultivated things, [and] the unsavory pungency of their acts and ideas," the instructor perceived the potential in their "vitality." In contrast to reformers who worried that indulging their inferiors nurtured irresponsibility, this educator suggested that lavishing care upon them created vulnerability and dependence. Such a view seemed in line with Burbank's optimism that "human weeds [were] . . . capable of improvement and development" because it was possible with "human plants to develop those habits and traits in which they are deficient and to break up those that harm themselves and others."[36]

Although eugenicists were anxious about the worst human weeds, most Americans probably worried about raising decent children and preventing them from becoming human weeds. The *Seattle Post-Intelligencer* warned that a boy who was not trained in "right living" by his parents "degenerates to a human weed." In her Beatrice Fairfax column, Marie Manning advised parents that by making their homes hospitable environments for their children and their friends, they could prevent their children from associating with undesirables in harmful environments, as well as notice a "human weed" from whom children needed to be protected. At a PTA meeting in Pampa, Texas, a north Texas oil boomtown, a minister emphasized the necessity of parental guidance, warning, "If you want to grow a human weed just let a child grow up." Like neglected children, vindictive parents could be weeds too. Chicago blues musician Joe Williams sang about a vicious stepfather who was "a no-good weed." Some Americans thought creating caring communities might compensate for parental negligence. In Bluefield, West Virginia, a newspaper editor encouraged Kiwanis Club members to create youth programs that channeled young people's energy into "proper forms of self-expression." By cultivating the city's social and cultural environment, they could prevent the emergence "of human weeds and stunted and stifled boys, girls, men and women."[37]

Whether human weed analogies were expressed by vengeful eugenicists like Huntington and Whitney or by sympathetic reformers like Robert Hessler, the expressions implicitly or explicitly referenced ecological relationships and community boundaries. When eugenicists added the environmental context of a garden to the analogy, they did so with cities in which people increasingly lived, growing nations, or the civilized world in mind. According to Wilbur Hall, Burbank likened a neglected garden to "an overcrowded city of inferior individuals, all seeking root-room and breathing space and a place in the sun" because "everything has run riot and weeds and flowers elbow and push and pant and struggle to maintain life." Thinking about human relationships as resembling the relationships of plants in gardens permitted eugenicists to perceive the ecological dynamics they were working with and to propose what ideal ecological dynamics would be. Theodore Robie thought the ecology of the garden showed "how much faster weeds can grow than roses." Myre Iseman believed that how the "weeds of the garden overwhelm the plants" reflected how in human history "prolific inferior races invariably drive out the superior." Margaret Sanger imagined transforming the United States from "a disorderly back lot overrun with human weeds" into "a garden for children." Ambitious and enthusiastic American reformers seemed to miss some of the messiness of the garden analogy proposed by the English scientist who developed the field of eugenics, Francis Galton. Karl Pearson, who produced edited volumes of Galton's work, wrote that Galton thought "the garden of humanity is very full of weeds." Pearson also wrote that Galton believed it was necessary for "space in the garden" to be "freed of weeds," in order for "individuals and races of finer growth to develop with the full bloom possible to their species." Euthenistic-oriented Martha Van Rensselaer better grasped Galton's thinking than eugenicists did when she wrote, "Every garden has weeds, as well as plants that we do not call weeds because they are being fostered." Cultivated plants were not naturally superior; they were temporarily protected from ecological processes that would otherwise "weed [them] out."[38]

Nevertheless, human weed analogies helped their creators propose ways of managing spaces—whether cities or nations—and the relationships of their populations. Their solutions often employed a related analogy of crop development. Feminist Charlotte Gilman wrote, "If you are trying to improve corn you do not wait to bring all the weeds in the garden to the corn level before going on." University of Wisconsin zoologist Michael Guyer wrote, "To get a better crop of human beings, we must, as with other crops, weed out bad strains." These analogies were opportunities to see how things could be—but also to see things as they were not. Gilman's eugenic view saw corn as one species and weeds as another species. Whatever human weeds were and however flawed they were, they were the same species as Gilman. Guyer more accurately saw weeds as weak individuals or deficient strains of the same species. But Guyer's analogy failed to identify a viable agent of improvement; he did not specify what nonhuman selective entity would decide which strains to eliminate. Guyer's "we" either made eugenicists nonhuman or deferred eugenic progress to an unknown entity, which abandoned the eugenic premise of social selection. Fields of tomatoes did not weed themselves of weak tomato plants or nontomato plants. The cosmic process indiscriminately weeded out both fruitless weak plants and vigorous plants with juicy fruits. Despite their analogical inventiveness, eugenic advocates could not create environments in which they simultaneously made the garden flower and were the garden's flowers. While gardeners and nature were both agents that weeded, they were not identical agents operating in unison. These flawed analogies either disregarded the nature of species or confused the nature of agency in the garden with the agency of nature.[39]

As would-be intelligent designers, eugenicists sought responsibility for remaking the population from which humanity would evolve. However, they proposed doing nothing with this responsibility other than identifying the problems of people other than themselves as inborn deficiencies. Since they did not look for what flaw was common to all humans—including themselves—they could not use their analogies to discover what was wrong with human nature. This irresponsibility had two manifestations. First, in misunderstanding the nature of agency by seeking to exercise the agency of nature to perfect the garden, they ignored a basic ecological reality. Although both the gardener and nature produced "finer growths," their processes of weeding out forced the less than fine growths—weeds, as they were—to evolve as well. Weeds complicated creating perfection in the garden, but they belonged in the garden just as much as the fine flowers did. As Brooklyn Boys High School botany teacher Abel Grout pointed out, "Our commonest weeds are the plants most successful in the struggle for existence which is constantly going on in the case of every living thing." Second, eugenicists willingly misread human variation as evidence of their superior "germplasm," rather than simply evidence of exposure to different sources of natural capital for different amounts of time in the distant past, as well as access to different resources in the present. Eugenicists essentialized the flowers and fruits of civilization as a product of inherited traits, rather than a product of tremendous investments of energy and resources invested in their cultivation. They distorted the greatest sources of variation. This perversion of nature was exacerbated by using such social and cultural distinctions to define the human weeds to be controlled. Sociologist Lester Ward argued that the disgust reformers dumped on the impoverished increased intraclass competition, which had selective value. Ward thought that the diffusion of tensions among the so-called finest flowers by denigrating inferiors undermined the improvement of people and American life. Attempting to seize the agency of nature to advance their interests severely limited how many iotas eugenicists advanced the race. Inventing an order in nature that did not exist and misusing it was not a very useful way to improve human nature.[40]

However wildly human weed analogies distorted the nature of American life, they nevertheless reflected past Americans' understanding of change. In 1931, ethnobotanist Melvin Gilmore described *Daucus carota* as a plant that "breaks away from civilized, orderly, useful life in the garden, demands 'freedom of self-expression' and insists upon 'living her own life.'" Gilmore feminized the species—possibly to express exasperation with a jazz-driven flapper daughter. He wrote, "Bent upon expressing her worst instead of her best . . . in living her own life she leads an existence neither glorious nor useful, and [is] quite intolerant toward all others." Gilmore's view of a "rude and rowdy" plant of roadsides and vacant lots seemed derived from theologian Walter Rauschenbusch's warning that the "standards of modesty have declined since women have moved out into freedom. . . . It is not simply an increase in sincerity and freedom of self-expression. It is a loss of control." Gilmore also seemed to share Huntington and Whitney's fear that feminists' persistence in "culling the finest young women from all over the land and indoctrinating them with ideas of self-expression" would "wreck civilization." Gilmore warned that failing to reestablish the plant's domesticity—in his analogy, young women's sense of social responsibility—risked reversion "to the wild forms which their ancestors had . . . many centuries or perhaps thousands of years ago." Gilmore's analogy had little to do with women's inherited traits; he was arguing that without environmental controls—social roles and cultural norms—any individual or group of people's wilder propensities could degenerate the species. John Wigmore emphasized that the problem was young people generally, especially those who worshipped the "self-expression" and "uncontrolled search for complete experience" promoted by John Dewey, who was then a Columbia University faculty member. Wigmore argued that a real sense of terror and "sufficient horror"—such as the punishment of hanging—was necessary for the repression of "nefarious social actions." While such a punishment might deter a *Daucus*-like human, nooses of manila, hemp, or flax were insufficient tools to intimidate all of the weeds that threatened society.[41]

Americans who feared human weeds and desired to weed them out thought that degenerate people degraded urban society and that degraded urban environments produced degenerate people. While human weed analogies made the ecological processes of community-undermining social disorder earthier, in the interwar years authorities reified these dangers by identifying plants that seemed to confirm that circular reasoning was an ecological reality. In the 1920s and 1930s, anxious city dwellers turned cannabis, a long-established and naturalized plant, into a criminal weed and deemed cannabis-using people—immigrants, migrants, and transients flowing into, relocating to, and passing through changing cities—into new urban menaces. Perceptions about the relationships among people, places, and plants compelled social regulators to destroy vegetation and imprison people. They policed the interactions of people and plants in urban spaces to reshape urban ecological dynamics.[42]

The criminalization of cannabis and cannabis consumers simultaneously used people's relationships with the plant to deny such people belonged in cities and used these people's presence to demonstrate the plant's danger. Cannabis's long presence in Philadelphia's vacant lots suggested that the plant's nefarious nature resulted from the city's changing nature as much as from the nature of the plant or its consumers. In 1902, Charles Saunders perceived "groves of Polygonum . . . mixed with an abundant growth of Hemp" as "a veritable tropical jungle." Cannabis's "darker verdure was . . . like the gloom of conifers amid deciduous trees." Saunders considered cannabis a fascinating plant because of its "history full of romantic interest." In 1938, cannabis "scattered pretty profusely along with a variety of other wild growth" along Richmond Street dismayed exaggeration-prone "narcotics expert" Arthur La Roe. La Roe described marijuana as a "killer drug" that caused "a sex urge . . . which can turn to blood lust, sadistic and barbaric." While these divergent perspectives suggest that cannabis became a problem as Philadelphia changed, the plant's criminalizers insisted that its nature had always been the same. One pair of crusaders claimed that the plant's "history for three thousand years has been the same—aberration, abnormality, murder, rape, degradation and horror. In coming to America, marihuana has not changed its nature." Like eugenicists, cannabis foes worried about becoming "degraded to the lowest plane of civilization," although the cause was botanical, not genetic.[43]

Antinarcotic activists, physicians, law-enforcement officials, and order-seeking city dwellers alleged that cannabis compelled insanity, transformed people into dangerous criminals, and threatened city life. They believed the plant damaged the psyche and incited violent crime. New Orleans prosecutors and Louisiana judges agreed that cannabis plants possessed "properties deleterious to health and dangerous to the public safety and morals." Young New Orleanians who used cannabis were thought to lead "loose, irregular and frequently criminal" existences; to lose their "social position"; and to become "a curse to their families and to their communities." African Americans worried marijuana use instigated racial violence. Gary, Indiana, police subdued "quarrelsome" youths in a downtown hotel who had been victims of "a systematic effort to enslave high school students to marijuana, or 'loco weed.'" Chicago police attributed violence among "shack"-dwelling Mexicans to their marijuana use. The *Minneapolis Star* reported that "school children and degenerates of the city" were purchasing "marijuana or hashish weed . . . [and] blossoms" and speculated that use of "the weed" had increased "attacks on women in the city."[44]

The dangers cannabis presented to people's minds may have been no worse than and impossible to distinguish from the dangers posed by other mind-imperiling weeds. Physicians outraged by a New York pediatrician's address to the International Medical Congress chided him for forgetting that "the most despicable weed that grows in the garden of the soul is ingratitude." Robert Schauffler, a writer who preached enthusiasm in life's pursuits, advised that a "poisonous emotion such as fear, envy, hate, worry, remorse, [or] anger" had to be "pull[ed] . . . up like a weed." A popularizer of academic research described human motives as "hidden impulses that so often hurry us to rash actions . . . weeds in our minds that need to be uprooted lest they obtain a fatal dominance over our constructive energies." George James, an English Methodist minister who settled in the sunny American Southwest, wrote, "In the garden of our minds . . . the more fertile the soil the more evil weeds will grow apace if we water and tend them. Our jealous worries are the poisonous weeds of life's garden." Des Moines's "city mother" gave a "heart-to-heart" lecture at the city's Odd Fellow Temple about the "Psychic Forces in a Human Weed Patch." The *Defender*'s astrologer insisted, "You must constantly pull the weeds (doubt, fear, foolishness, disease, ugliness, and worry[)] up by the roots." He also wrote, "An evil disposition, like a weed, needs no cultivating." To "folk psychologists," scientifically trained "new psychologists," and individuals striving to become positive thinkers, it may have seemed that non-psychoactive-substance-altered minds and hearts were as dangerous as the weeds in their cities. Perhaps nothing was as alarming to John Wigmore as the "vicious philosophy of life, spread in our schools for the last twenty-five years by John Dewey and others," which Wigmore claimed caused young people to behave in ways "in special need of repression."[45]

Law-enforcement patterns indicated the danger of cannabis to city dwellers. In 1933, criminologists concluded that "marihuana smoking is at present a common practice among the young people of the city" from the preponderance of cases from New York, Chicago, Los Angeles, New Orleans, Kansas City, and El Paso. "Horrible accounts" and reports of "police [being] harassed in a thousand cities by users who take to petty and more serious crimes" emerged from Philadelphia, Detroit, Baltimore, Boston, Pittsburgh, San Francisco, Washington, D.C., Newark, Louisville, Denver, San Antonio, Richmond, Fort Worth, Nashville, Tulsa, Jacksonville, Wilmington, Asheville, and Colorado Springs. Federal Bureau of Narcotics (FBN) records from 1936 indicated the urban nature of the problem. Eighty of 108 seizures of marijuana in California were in Los Angeles, Sacramento, San Francisco, Fresno, San Diego, Oakland, and Berkeley. In Maryland, 14 of the Bureau's 15 confiscations were in Baltimore. Detroit, Saginaw, and Flint—where agents removed 2,500 plants growing behind houses and billboards—were the sites of 12 of 15 impoundments in Michigan. Thirty-four of 40 busts in Ohio were in Cincinnati, Cleveland, Columbus, and Youngstown.[46]

The problems in cities attributed to cannabis resulted in laws regulating the plant and intensified perceptions of its dangers. Although the Harrison Narcotic Act of 1914 made cocaine, opium, and morphine use outside of particular legal contexts a criminal act, cannabis prohibition was rare before 1920. El Paso's 1915 ban on the sale and possession of marijuana was one of few such laws. It was passed shortly after stories in Baltimore, Los Angeles, and New York newspapers appeared about "insanity weeds" consumed in Mexico, which included the "marihuana weed," tolvache, "a loco weed," and "the weed called de las carreras." Increasing concern about marijuana use in cities between 1920 and 1934 prompted thirty-two states to pass prohibition laws and led law-enforcement officials and newspapers to cultivate marijuana's reputation as the "killer weed." Although Congress classified marijuana as a narcotic in a 1929 law, the FBN, established in 1930, did not extend the Harrison Act to marijuana. The FBN maintained that marijuana use was an isolated problem among Mexican immigrants in Western states and New York City. Nevertheless, FBN officials vilified the plant and urged states and cities to take "vigorous measures for the extinction of this lethal weed." The 1937 Marihuana Tax Act increased the FBN's involvement with marijuana control as officials decided it was a serious health danger and as public fear about the drug increased.[47]

Criminalizing cannabis growth, possession, and use constrained the lives of city dwellers. Americans considered to be criminals because of their association with cannabis included blacks born in the United States and the Caribbean, bohemians, cosmopolitan and thrill-seeking whites deprived of alcohol during Prohibition, "hindoos," jazz artists, people of the Middle Eastern diaspora, people from Spanish-speaking countries, and the transient underclass. Although knowledgeable cannabis foes were aware that Greenwich Village's "sickly aesthetes [used] the weed—as hashish" in the mid-nineteenth century, they combated marijuana use by revising the genesis of the threat. Cannabis use, they contended, was moving beyond the Southwest and New Orleans as migrant Mexicans, blacks, and jazz artists brought the drug east and north into industrial and commercial centers. Cannabis criminalizers thought urban social order depended on destroying a plant that some Mexican immigrants had cultivated in their homeland; that some African Americans artists, dancers, and musicians experimented with or used in social gatherings; and that flourished on roadsides and in vacant lots where jobless and work-seeking Americans congregated and traveled. In criminalizers' minds, these particular groups made the plant dangerous, and the plant made these people dangerous.[48]

As populations of Mexican immigrants increased in cities, Americans who objected to their presence accused Mexicans of endangering cities with cannabis cultivation and marijuana use. One nativist argued that Mexicans were "a menace" and an "undesireable class of residents" who had a "tendency to colonize in urban centers, with evil results." Such xenophobes thought Mexicans degraded Anglo-American purity, caused crime, and spread diseases. During the Great Depression, nativists scapegoated Mexicans for taking jobs away from Americans in cities such as Los Angeles and Chicago and characterized them as dangerous marijuana users. Jointly conducted local and federal raids in 1931 to pressure Mexicans to leave Los Angeles employed marijuana possession as a justification for expulsion. Although marijuana arrests only comprised about 16 percent of the state's narcotics cases, from July 1931 to June 1932, 60 percent of Los Angeles's narcotics cases involved marijuana. In 1935, Sacramento agribusinessman Charles Goethe declared marijuana "the most insidious of our narcotics" and a "direct by-product of unrestricted Mexican immigration." Cannabis grew in Los Angeles's alleys. A sixty-year-old Mexican man cultivated a twenty-five-acre corn and cannabis field on the edge of Dallas, near an incinerator. Large cannabis patches in a Mexican section of Wichita, Kansas, were believed to make Wichita the center of a "dream drug" ring. FBN director Harry Anslinger warned that because Mexicans migrating through the Midwest planted "the seed in secret patches," cannabis was "often found growing wild along the roads and railway tracks." *Minneapolis Star* editors feared "the habit followed the weed wherever it grew." The *Chicago Tribune* dramatized the plant and its users as a threat to the city by publishing a photograph of two Mexicans harvesting their "hasheesh" plants growing on undeveloped South Side land. Criminologists concluded that cannabis growing in New York City's Pennsylvania Railroad yards indicated a dangerous "Mexican influence" and showed how "far north and east of its natural habitat . . . the weed spread under cultivation."[49]

While Mexican immigrants were blamed for spreading cannabis across the nation, African Americans were blamed for popularizing marijuana use. As black laborers from the South, entertainers in New York and Chicago, and Caribbean immigrants changed Northern cities, tolerance-deficient whites perceived them as damaging to unions, neighborhoods, schools, morality, and public health. Efforts to contain African Americans in particular areas imposed deprivations on them but did not prevent new ideas, sounds, and styles from emanating from such neighborhoods or arrest cross-class, intraracial exchanges from shaping youth culture. Some of the creators and participants in jazz culture used cannabis. Pianist Hoagy Carmichael remembered marijuana smoking as enlivening Chicago's South Side jazz scene. Anita Colten, a marijuana-energized lindyhopper from Chicago's Uptown neighborhood, became a jazz vocalist. Saxophonist Mar-

Figure 2.5. In 1929, the *Chicago Tribune* published this photograph of Mexicans purportedly harvesting cannabis in a "southern part of the city" near their "box car homes" to illustrate the necessity of criminalization measures. Photograph courtesy of the Newberry Library, Chicago.

shal Royal thought marijuana use provided an escape for musicians with limited economic opportunities. In Manhattan's Roseland Ballroom, young people such as Boston's Malcolm Little mixed alcohol, marijuana, music, and dancing in their pursuit of independence and freedom. The place of the plant in jazz culture was recognized in songs performed or recorded in the 1930s, such as "Garden of Weed," "Weed Spook," "Weed Smoker's Dream," and "Weed Blossom Blues." The meaning of "weed" was often encoded in the music. One alleged signal a "weed song" about the "love weed" sent to tuned-in listeners was that marijuana could be purchased at a particular location. A song with "weed" in the title did not necessarily indicate that cannabis was an ingredient of a musician's creativity. Don Redmon, a West Virginia native who rejected "any smoke other than good tobacco," reportedly wrote "Chant of the Weed" because the odor of burning cannabis "annoyed him"; at the very least, Redmon's claim hinted at the ordinariness of marijuana smoke in clubs.[50]

Uneasiness with jazz sounds preceded widespread awareness of the use of cannabis by musicians who played jazz. In 1900, cellist Louis Blumenberg dismissed ragtime-style jazz as "a rag-weed of music . . . highly detrimental to the cause of good music." Critics who recognized the significance of jazz's popularity hoped it could be refined. *Etude* editors characterized the rawness of jazz as a "piquancy and originality" that was "not unlike some of the very beautiful wild flowers which we have seen springing up from a manure heap." They wondered if "the weeds of jazz may be Burbanked into orchestral symphonies by leading American composers in another decade." The role of cannabis in some jazz styles confounded cannabis foes. Although the *Science News Letter* reported cannabis did "Not Improve Playing of Swing," according to the *Newark Ledger*, swing could not be played without cannabis because "marihuana smoking . . . [gives] swing musicians exactly the stimulus they need." Law officers attempted to arrest the cultural impact of jazz with marijuana busts. The FBN sought opportunities to apprehend marijuana-using musicians such as Louis Armstrong, Ella Fitzgerald, and Cab Calloway. Police apprehended three Harlem musicians harvesting "contraband vegetation" in a dump in Secaucus, New Jersey. *Defender* editors worried that police intervention weakened the jazz economy. Hotel and club owners seeking to avoid reputation-damaging raids were reluctant to hire a musician whom they suspected of possessing "weeds in his pockets." The most effective regulators of marijuana may have been bandleaders such as Benny Goodman, Artie Shaw, and Tommy Dorsey, all of whom urged or required their players not to use marijuana or alcohol before or during performances to facilitate communication and coordination on bandstands. The persistence of these habits through World War II irritated W. E. B. Du Bois, who encouraged African Americans to abandon "the neglected weeds of cabarets, jitterbugging and drunkenness." For nearly fifty years, Du Bois had issued similar remarks out of concern that the depravities of popular culture hindered advancement of blacks' equality.[51]

Urban Americans also perceived cannabis as dangerous because transient Americans used the plant. Law-enforcement officials and antinarcotic reformers associated drugs with the mobile poor. In 1921, many people arrested for narcotics violations in New York City were "non-residents, found amongst . . . [the] floating population." When Seattle officials decided that drugs, including marijuana, were a problem in the mid-1930s, they prosecuted narcotics users "under the general vagrancy laws." Anslinger warned that "wandering dopesters gather the tops [of cannabis plants] from along the right of way of railroads." New York psychiatrist Walter Bromberger observed that "vagrant youths" became marijuana smokers after "coming into intimate contact with older psycopaths [*sic*]." After a "floater" sheltered at Providence's Federal Transient Bureau found cannabis plants along the waterfront, authorities arrested "twenty tramps in possession of some of the weed" and discovered "nearly a ton of marihuana." A jazz musician arrested by St. Louis police for storing "marijuana weed" in his trumpet case confessed to obtaining it "from a man working with a carnival in South St. Louis." A Memphis man who grew cannabis in his backyard and started "hoboing his way to Chicago on the Illinois Central" was arrested in Cairo, Illinois, with "three large cans of marijuana weeds." The "idle and irresponsible classes" who possessed cannabis were dangerous because they were able to identify cannabis, use cannabis, and transport cannabis around and between cities.[52]

The danger of these people and the need to control them arose not just from their lack of social status and their use of marijuana but also from the changes in and changing perceptions of urban environments. Cannabis users were ubiquitous dangers because cannabis, unlike other psychoactive substances regulated by authorities, grew all over cities. To criminalizers, cannabis leaves, flowers, and vapors were potential and actual causes of disorder and disarray. Antimarijuana crusaders believed that everywhere cannabis grew, marijuana consumers could or did damage urban life. *New York Evening Journal* writer Allen Bernard called cannabis "a virulent weed, growing rapidly, another threat to civilization." In San Francisco, a field of marijuana large enough "to put half of San Francisco asleep for a month" was found in a vacant lot in the Excelsior neighborhood. The half acre of six-foot-tall plants was allegedly one place where "Anna," the city's "Marihuana Queen," harvested what she sold. Americans who wanted to prevent unwanted urban social change had to not only block marijuana trafficking but also eradicate the cannabis growing in and around their cities. Just as some city dwellers cultivated cannabis for personal use or the market, social regulators cultivated cannabis as a growing urban environmental threat.[53]

Ubiquitous cannabis annually blossomed into danger. The *New Orleans Times Picayune*, which called cannabis plants the "weeds of illusion," reported that the plant grew "in almost any kind of soil, along railroad tracks, between corn rows, around city dumps, and in vacant lots." Physicians warned that the Crescent City's fertile soil and semitropical climate allowed "a dangerous plant" to grow "in our very back yards" and claimed that there were "vast fields of this weed" just outside of New Orleans. New York officials' assumption that smugglers imported marijuana obscured a problem sprouting from the city's soil. In 1925, police discovered "twenty acres of a brownish-green weed, growing wild in an old dump alongside the Sunnyside Yards" in Queens, a site described as a "drug field as large as six city blocks, just across the river from the heart of Manhattan." In subsequent years, cannabis grew wilder. A cannabis field bounded by tenements near the Brooklyn Bridge demonstrated the homegrown dimension of the marijuana problem. In 1935, officials located 309 acres of cannabis. "Hasheesh weeds" removed from 361 lots were estimated to weigh nearly three hundred thousand pounds. When officials set fire to a ten-foot-high and fifty-foot-wide heap of the "extremely dangerous weed" uprooted over a six-month period in 1936, the pile burned for nearly three hours. Neighboring New Jersey cities likely produced cannabis consumed in New York. Although Arthur La Roe called Newark "a great big reefer farm" because cannabis was "growing on the roofs of tenement houses . . . [and] in the parks," cannabis also grew in Elizabeth, Hoboken, and Jersey City. Moreover, many a

Figure 2.6. Cannabis growing in Brooklyn in 1934.

"marihuana infested lot" remained "unknown," the *Patterson Morning Call* noted, "except to those who seek such dangerous spots." While cannabis grew in cities' soils, there were innumerable other places to buy the processed plant. On Chicago's South Side, police found a "reefer nest" teeming with "more than 20 pounds of marijuana weeds" and a "weed nest" where drug distributors allegedly used marijuana to introduce young people to other drugs.[54]

Although the Department of Agriculture insisted in 1935 that illegal marijuana came "from small patches grown on city lots or on vacant land in the suburban districts," cannabis that grew beyond cities increased the available supply in cities. The arrests of three "handlers of the weed" in Memphis, where "use of the dope at dances . . . [was] very common," led federal and local agents to confiscate "over two tons of marijuana weed" growing a few dozen miles down the Mississippi River on a Tunica, Mississippi, farm. Investigators uncovered a distribution system that shipped cannabis grown in Iowa and Minnesota fields to Chicago and New York City. Anslinger stated that cannabis was cultivated "in practically every state" and that "almost everyone . . . in rural communities" had seen it. Cannabis, whether growing in the center of a metropolis or in the heart of corn country, became a weed because cities changed. Cannabis growing in cities and fears about city-dwelling cannabis consumers inspired criminalizers to vilify the plant.[55]

Cannabis often belonged in urban spaces where it grew because past Americans had planted the seeds there or processed the plant there or because present city dwellers were actively cultivating it. Cannabis arrived in the Americas along with enterprising British, French, and Spanish colonists, who planted and exported it to sail and rope producers. Jamestown colonists attempted hemp production before embracing tobacco cultivation, and New Plymouth required householders to plant hemp each year. Pioneers sowed cannabis seed across the continent. Farmers and gardeners purchased cannabis seed from nineteenth-century mail-order catalogs. There were hemp plantations in Missouri in 1835 and in California by 1912. Bird producers and owners fed the seed to pigeons to promote feather growth, to hens to lay more eggs, and to canaries to enhance their jovial chirp. Manufacturers used seed oil to produce paints, soap, and linoleum. Physicians and pharmacists employed cannabis extracts to treat Americans' neurofacial pain, bladder spasms, muscular rigidity from tetanus, and skin diseases. On the eve of World War I, the U.S. Department of Agriculture (USDA) collaborated with pharmaceutical companies such as SmithKline to supervise cannabis production. Into the 1930s, USDA Bureau of Plant Industry botanist Lyster Dewey advocated planting hemp as a ground cover to prepare soil for future crops. Suburban gardener Parker Barnes used cannabis for landscaping. The plant, Barnes found, was "excellent for temporary screens and backs of borders," and he advised starting it "indoors in pots." Cannabis was something of an all-American plant in cities that authoritarian city dwellers transformed into a weed.[56]

Since cannabis was desirable, useful, and valuable to cultivators, cannabis despisers had to abuse American English to denigrate the plant as a weed. Criminalizers cast aspersions upon the plant, calling it "killer weed," "nefarious weed," "sinister weed," "weed of madness," and "weed with roots in hell." Anslinger described cannabis as "a big, hardy weed, with serrated, swordlike leaves topped by bunchy small blooms supported upon a thick, stringy stalk." Law-enforcement officials claimed that Mexicans "clandestinely plant[ed] patches of this weed." Newspapers conveyed the value of a "crop" of "the weeds"; described a "flourishing garden of the weed"; claimed cannabis was as easily "cultivated as any rank weed"; and even reported that a "10 foot by 15 foot bed of marijuana weed" found in the Joliet penitentiary "was started by seeds blown there by the wind." In the last instance, the three prisoners caught smoking the cannabis—not the plants or the wind that was allegedly responsible for the mischief—were punished "in solitary confinement." Cannabis foes' wild rhetoric confounded the agency of people and plants and complicated enforcement. La Roe complained that convictions were hard to obtain "because it is difficult to prove whether the weeds are growing wild or under cultivation." In 1938, FBN agents failed to make arrests when they found cannabis on "the property of the Illinois Brick Company" because they did not witness the Mexican workers touching the plants and were apparently reluctant to arrest the property's owners.[57]

Cannabis criminalizers seized upon the ubiquity of cannabis growing in and around American cities as evidence that cannabis was a wild weed, an invasive nuisance, and an omnipresent health, safety, and moral threat. Cannabis appeared to be wild because the plant covered land all over and around cities, all over the country. A textile engineer saw "wild hemp growing in practically every vacant lot in Atlanta," as well as in Knoxville, Bristol, Roanoke, Washington, and New York, "right on Riverside Drive" and "within 50 feet of" the Rockaway Beach boardwalk. Cannabis flourished a few blocks from the New Haven green. Twenty-five acres of cannabis grew on river bottoms "less than a mile" from the Dallas courthouse. In Chicago's streets and vacant lots, cannabis was "locally common"; on the city's outskirts, cannabis grew along the edges of a brick factory's clay pits. Schenectady police found cannabis everywhere "from garbage dumps to some of the city streets." In Providence, cannabis was found "in dumps, along railroad rights-of-way and in vacant lots in the industrial sections," including "in the shadow of a big gasometer, the tanks of one of the large oil companies." Where rope factories around Omaha had abandoned land, cannabis grew "in great masses . . . as a wild weed." Cannabis foes emphasized the plant's growth in degraded, dirty, disturbed, and neglected environments to demean it as a despicable plant like other weeds. New York State's chief of drug control stated that "the largest part of our growth . . . is on poor soil . . . in dumps or soil that has a high content of ashes or cinders. . . . The wild growth seeks that kind of soil." A *Science News Letter* writer insinuated that cannabis was a "weed" by describing it as a "tall, rank-growing plant, able to compete even with the giant ragweed for its place in the sun." Cannabis earned its bad reputation because it could weed out, or outweed, ragweed, a plant that botanist Oren Durham complained "pollute[d] the ground . . . produce[d] nothing of utility or beauty and deserve[d] only one fate—extermination." According to Anslinger, trying to stop cultivation of the plant would be like attempting to "ban dandelions."[58]

Demeaning cannabis as a rampant weed also depended on misrepresenting and ignoring what was known about the plant's biology, history, and ecology. Law-enforcement officials relied on their authority, not their authoritative study of the plant, to malign cannabis as an out-of-place plant that did not belong in American cities. Antinarcotic forces emphasized the plant's biological origins in Central Asia and its historical uses in Asia, India, and Africa to link the plant with "deteriorated" cultures. Anslinger emphasized the foreignness of marijuana use as an "insidious invasion." Kansas criminologists pretended the plant was a native of "the Torrid Zone" and lied that "marihuana was introduced to the United States from Mexico." An American working on the border between Mexico and the United States described the plant as "a weed of the Mexican desert." Perhaps the most bizarre charge against cannabis blended the plant into a cosmopolitan threat: a California official claimed that when Mexicans sprinkled the "jazz weed" on a baking tortilla, "the lowly delicacy vibrates and . . . sends forth weird tunes not unlike those seeping over the walls of a Sultan's harem rendezvous." Linking cannabis with so many diverse and distant cultures to malign it seemed to blind most criminalizers to an uncomfortable reality—the plant belonged all over the globe. In 1931, physician Frank Young remarked, "If there is a plant indigenous to every part of the world, I suppose this one comes as near filling it as any." FBN officials and their allies also ignored the knowledge of scientists who worked with the plant to construct cannabis as a weed. In the early 1930s, USDA botanist Lyster Dewey, a turn-of-the-century advocate of weed eradication in cities, reported, "Hemp Assists in Killing Weeds," especially quack grass and Canada thistle. Dewey did not invent this use of cannabis; one common nineteenth-century name for cannabis was "choke-weed." Nearing the end of a forty-five-year career in government service, Dewey told the House of Representatives Ways and Means Committee at hearings for the Marihuana Tax Act that of the "thousands" of inquiries he had received about weeds, "only four" were about "hemp as a weed." Dewey stated, "Although I was looking for weeds and all plants that might be troublesome as weeds, I never found the hemp plant to be really a troublesome weed." Nevertheless, FBN officials decided that the plant's ability to grow in the North, South, East, and West made it a dangerous marauding invader.[59]

Anticannabis crusaders' willingness to hastily communicate inaccurate knowledge and their reluctance to find out what was not known about the plant were also essential to cultivating Americans' animosity toward the plant. After passage of the Marihuana Tax Act in 1937, FBN chemist Hans Wollner asserted that "virtually nothing" was known about marijuana's "narcotic principle." A USDA chemist believed that tall cannabis plants grown in the United States produced good fiber, but less oil and less potent chemicals. Cannabis plants varied with cultivators' efforts to produce tougher fiber, more seed oil, or a more potent drug. However, FBN policy makers clung to a 1935 League of Nation's Opium Advisory Committee report that stated, "The American form of Indian hemp is more virulent in effect than that found anywhere else in the world." Anslinger warned the director of New York's State Police Laboratory that "all cannabis is potentially dangerous and should be destroyed unless it is being grown under proper regulations." Although FBN officials had finally determined for themselves that different plant strains had different levels of "the active narcotic principle" in 1940, they were still alarmed by this warning and asked for an investigation of a Public Health

Service official's alleged remark that the "weed in Kentucky and in most other states . . . [was] practically harmless." Closely guarded facts and widely disseminated lies manufactured misinformed cannabis foes, who often could not accurately identify the plants or its dangers. One police chief was unaware that "a large patch of strange weed about six feet high growing in his garden" was cannabis. A *Defender* writer attempted to malign cannabis and its psychoactive effects by writing, "Nicotine from the weed affects the mentality and changes the user into a state of depravity, often leading him to commit acts of violence. It is the favorite intoxicant of sex fiends." Although such a diatribe may have been inspired by or confused the legacy of Chicagoan Lucy Gaston's battle against tobacco and cigarettes, it did not lead the FBN to scrutinize the multimillion-dollar tobacco industry.[60]

Changing cities made it seem necessary to cannabis foes to rewrite the plant's history and distort perceptions about the plant's ecology. Denigrating cannabis as a noxious, dangerous weed asserted their power over urban environments. Policing the plant permitted policing the boundaries of society and culture and delineating the bounds of urban life. With increased concerns about the drug's use in cities, the roots of cannabis's past in American soil suddenly disappeared, and just as "suddenly, almost overnight," marijuana crusaders created a new threat to American youth and a national menace. However, as cannabis was well rooted in the American past, it was well adapted to changing cities. After World War II, the plant continued to be cultivated and remained wild in cities. Chicago police arrested two African American men "picking marijuana weed" near railroad tracks. An "Alert Citizen" found a Chicago vacant lot "overflowing with marijuana" and wondered what the "crop of iniquity [would] produce." New Yorkers informed city officials and police when they spied cannabis growing in neighbors' yards or witnessed "mysterious personages cutting the weed and carting it off." In 1951, the New York City Department of Sanitation investigated 2,116 properties for cannabis and found plants the "size of Christmas trees." Workers spent four days destroying the labors of "daring marijuana farmers" who had "virtually taken over the vacant lots of Brooklyn and Queens." In Pittsburgh, a five-hundred-square-foot "reefer 'garden'" hidden by an "unchecked growth and spread of weeds" aroused the suspicion of residents because policemen seen inspecting the area did nothing about it.[61]

If cannabis was a weed—a plant with a bad reputation—this reputation revealed more about the plant's detractors than about the plant. Cannabis criminalizers who believed cannabis was definitely dangerous because of its potential dangers were like turn-of-the-century Americans who thought all plants of certain heights or all plants at certain dates were per se public nuisances. In using cannabis as a weed to cultivate fear and tighten social order, guardians of society proved they were as capable of misusing cannabis as the Mexicans, African Americans, transients, and an array of other city dwellers whom such guardians regarded as nature-abusing outcasts. Imagining cannabis was always deadly made the plant incredibly useful to antinarcotic reformers. Whatever the nature of cannabis actually was, the plant was certainly not a useless weed to the moral entrepreneurs and ardent foes who used it as a tool of social regulation.[62]

In the first decades of the twentieth century, urban Americans' hostility toward weeds did not evolve, even as their cities continued to change. Just plants remained in cities and influenced the ecology of urban life. In-between places with in-search-of-place people and happenstance plants were sources of possible urban growth and probable change. Mixes of new peoples and new plants in transitioning areas formed novel rudimentary communities that sustained urban growth. From New Haven to Sacramento, Americans who perceived urban environmental change as disorder often developed ecological interpretations of change that denied the belonging of certain people and plants. Even people in nascent urban settlements in the Rocky Mountains worked to establish community boundaries with ecological observations. Iowa-born Blanche Soth, a resident of Blackfoot, Idaho, monitored plants and people in Pocatello, Idaho, about twenty-four miles to the south. She reported that *Atriplex laciniata,* which she thought had not yet been seen in the United States, had appeared on vacant lots throughout Pocatello's east side. Soth noted that Greeks and Italians had been living in this area for many years, "so the connection is easy to establish." Similarly, *Lancisia coronopifolia,* an exotic plant common on the Pacific Coast, was spreading in ditches north of Pocatello. Soth's vigilance suggested which plants and people—although established—did not belong in the region. At best, such ecological analogies could have been early indicators of future overpopulation problems. In effect, they did little more than stereotype the nature of the city into an environmental problem for American society.[63]

Some films and songs from the 1930s handled transient people and plants more sympathetically—albeit ambiguously—than scornful nativists did. Metropolitan photosynthesizers and the spaces that they inhabited occasionally permitted hope, playfulness, and endurance. In Charlie Chaplin's *Modern Times,* the little tramp and the gamine briefly relax on a vacant lot and dream of domestic bliss and material comfort before being awakened and displaced by a beat cop. Failed by a faulty society, they eventually take to the road and leave Los Angeles behind. Thistles on a vacant lot near a Disney studio make cameos in *Fantasia.* Three Disney storyboard artists gathered the plants and employed them as models from which they drew Russian Cossack dancers to illustrate the changing seasons in the *Nutcracker Suite* segment. Although Walt Disney was impressed with the thistle-inspired figures, he later dispatched a team to a nature preserve—rather than another vacant lot—to recruit more botanical extras. Lively animated happenstance plants, among other creatures, were unable to rouse many American moviegoers during the Depression's protracted economic turmoil. Americans looking for work and for homes may not have imagined vacant land as much more than a place to rest on the way to somewhere else. One American who saw no future in Oklahoma and Texas boomtowns full of fortuitous flora and quickly built and rapidly abandoned homes was Woody Guthrie. Amid happenstance plants on his way to Tucson and to Los Angeles and in Redding, California, Guthrie joined rail-riding men who made coffee and found ephemeral migrant communities hidden from the road. His song "Pastures of Plenty" expressed the travails of people traveling on hot roads, passing through cold mountains, and sleeping on the moonlit edges of cities. Like the seeds of happenstance plants, these people arrived "with the dust" and departed "with the wind." The optimism exuded by architects and boosters of Sunbelt urbanization was not intended for nor extended to all. Most displaced Plains people lacked the capital and the cleanliness to be welcomed in the sunny Southwest. They were often turned away or forced violently to move on. Some likely made their way to the federal government's temporary shelter at the Weedpatch Camp near Arvin, California, a shelter for migrant laborers named for an otherwise dry area once filled with plants invigorated by irrigation waters.[64]

Botanomorphizing people as weeds accomplished little more than naturalizing contempt for them. Essentializing people's problems as genetic in nature did not solve the nation's problems. Decades of disparaging weeds and human weeds did little to prepare for or prevent their production during the Great Depression. Assistant Secretary of State Francis Sayre criticized proposals to create an isolated national economy independent of foreign trade, since many laborers depended on the revenue produced by exporting agricultural surpluses. Taking 40 million acres of average farmland out of production would jeopardize the livelihoods of 3.2 million rural people. Sayre asked, "What shall be done with these human beings? Are they to become human weeds?" Some may have. The *Helena Independent* had little sympathy for many struggling people. The newspaper guessed that "about 5,0000,000 of those listed as 'unemployed' are human weeds. They will not work steadily." The editors called for a presidential commission that could discover "the exact conditions under which people who will later become weeds are born; or the circumstances under which they develop in later life." Devising a simple system for "remedying the evil" would be a great gift "to our economic system, to all mankind." But the newspaper seemed skeptical that any commission could achieve this because the nature of such weeds could not be changed: "human weeds will be ever present like weeds in garden and field, they do nothing of the slightest value to the rest of mankind and cultivation does not improve them." Yet even as these Americans believed the nature of weeds was unchangeable, Frank Thone mustered up Emersonian and Burbankian optimism in writing that people with "scientific curiosity and patience" might be able to extract "wealth from weeds." Thone imagined that radiation, colchicine, and new plant hormones—"new genetical magics"—would make possible a new era of plant alchemy in which even the most "useless" and "repulsive" weeds might become "profitable."[65]

CHAPTER THREE

Creating Ragweed Frontiers

On sunny summer days in East New York, garments gently tugged on clotheslines and elevated trains rolled along above Livonia Avenue. However, on 17 July 1953, an outburst of warfare upset the Brooklyn neighborhood's ordinary feel. Curious onlookers watched from window perches as four young municipal employees attacked herbaceous enemies of public health that had amassed on vacant land. Leonard Moriarty hoisted and aimed a hose. Another soldier turned on the pump. The 2,4-D held in their truck's spray tank pulsed through the hose, exited the nozzle, cut through the air, and soaked the vegetation. Four days later, the plants had withered. The litter that had accumulated in the lot, which often remained concealed until plants died in fall frosts, was no longer hidden. Wherever the forces of Operation Ragweed engaged their enemy, they may have produced such patches of land.

Destroying ragweed with 2,4-D was a public health measure to liberate city dwellers from seasonal ragweed hay fever. In order to help hay fever–afflicted New Yorkers, who may have numbered around four hundred thousand in 1955, New York City's Operation Ragweed (NYOR) had to defeat a heavily armed enemy. A single ragweed plant could release a billion pollen grains each year.

Figure 3.1. Herbicide spraying in East Flatbush, near its border with Brownsville, in 1948. Courtesy of NYC Municipal Archives.

NYOR leaders were not deterred by the failures of urban Americans throughout the first half of the twentieth century to eradicate weeds from their cities. These postwar New Yorkers believed chemical manufacturers' claims that phenoxyacetic herbicides promised a new era of weed control. The chemicals inspired an ecological dream that killing ragweed with herbicides would produce new vegetation, cleaner air, and healthier people. While NYOR was possibly the most intense battle against ragweed, other local, county, and state officials around the region launched wars against the plant too.[1]

New York City, neighboring municipalities, and surrounding counties consisted of an array of places with a variety of patches of happenstance plants, all of which were changing at different rates. Diverse Americans' pursuits of wealth, comfort, health, and recreation produced sprawling urban environments. From the 1930s through the 1960s, an ecology of metropolitan growth generated changing populations of just plants that changed the land in unanticipated and unfamiliar ways. Like sunflowers in St. Louis and thistles in Chicago, ragweeds were among the New York City area's most notable plants. Ragweed inhabited and connected multiple frontiers around the metropolis—an unsuppressed frontier on the city's edge, a long frontier along the region's roads, and an emerging frontier of postponed growth inside of cities. Yet seeing places around the region as drastically different spaces—as no longer valuable urban environments, as wasted urban fringes, as delightful escapes from cities—encouraged unecological vegetation management that made them more similar.

Like other fortuitous flora that city dwellers fought in preceding decades, ragweed could not be eliminated from evolving metropolitan areas because the plants grew with and belonged in changing cities. In the New York City region, where so much land was in transition, falling out of use, and waiting to be used or reused, ecological dynamics were more complex and more encompassing than urban weed-control advocates initially realized. Unlike rigidly drawn municipal boundaries, ragweed pollen did not remain fixed in space but drifted and shifted across borders. Herbicides killed plants, but they did not diminish the boundless nature of pollen pollution, did not remake real estate markets, and did not transform the ecological dynamics that perpetuated metropolitan photosynthesizers.[2]

NYOR was the flowering of hostilities toward ragweed cultivated in the first decades of the twentieth century. Nineteenth-century physicians commonly believed hay fever was—and sometimes dismissed hay fever as—a neurotic disease. Scientific measurement of pollen in the air that correlated with hay fever symptoms gradually shifted medical opinion. A biologist's 1922 statement that "the pollen of ragweed is believed to be a cause of hayfever" concealed nuance and ambiguity in passive objectivity. Such a remark did not address how significant ragweed pollen was among a number of hay fever causes. The mounting evidence that ragweed pollen sickened people led a Minnesota doctor to argue that ragweed should be controlled "by law just as strictly as is marijuana . . . [because] the misery caused hay fever and asthma patients by ragweed was more widespread than ills caused by the use of marijuana as a drug."[3]

New York City was one of the first cities to declare ragweed a public health threat. In 1915, the city prohibited property owners from allowing "poison ivy, rag weed, or other poisonous weeds to grow . . . [or to] extend upon, overhang, or border upon any public place" and prohibited them from allowing "seed, pollen, or other poisonous particles or emanations . . . to be carried through the air into any public place." Unlike most late nineteenth- and early twentieth-century weed nuisance laws, New York's law was novel in identifying two particular plants suspected to cause health problems. However, the vagueness of "other poisonous weeds" expressed an uncertain suspicion that perhaps every plant might be harmful in some way. Unlike laws in St. Louis and Washington, New York's law applied to all land in the city since the wind could carry pollen from almost anywhere in the city into streets and public spaces. However, since the law did not eliminate ragweed growth, during the 1920s and 1930s New Yorkers complained publicly about their hay fever, which in the words of one New York physician was a "vile disease." In the *New York Times*, one sufferer described hay fever as "an affliction . . . beyond description and impossible of understanding by one free from it."[4]

Ragweed's danger became more material as scientists discovered how much pollen filled the air. William Scheppegrell, a New Orleans physician and the president of the American Hay Fever Prevention Association, worked in the 1910s to measure pollen levels in order to estimate how much pollen people were exposed to. Oren Durham, a botanist who worked for pharmaceutical manufacturers such as Swan-Myers and Abbott Laboratories, increased the accuracy of pollen counting by improving the instruments and procedures used to obtain data. By establishing a standard method for pollen sampling, Durham encouraged the production of data with which different environments could be compared. He also created a pollen count network. Whereas only eight American cities had conducted pollen counts by 1928, Durham was collaborating with allergists and U.S. Weather Bureau station meteorologists in one hundred locales by 1935. Durham's pollen index—an admittedly "arbitrary rating"—was based on a place's average total number of days with pollen in the air, the highest pollen count recorded, and the average annual total count. This scientific research informed people about what point in the year the air became filled with different plant pollens, as well as to which places with pockets of ragweed-free air they might escape.[5]

Hay fever sufferers were inundated with pollen and information about pollen each season. Seasonal indices, and later daily pollen counts, informed them of the potential danger in the air they breathed. The *New York Times* published stories about pollen counting during the 1930s and 1940s, and the paper printed pollen counts on its weather pages in the mid-1950s. The novelty of pollen data required experimenting with their meaning and significance since researchers had not determined how counts correlated with probability or intensity of symptoms. New York City's 1927 daily average pollen count of 26 grains per cubic yard led Durham to conclude that Manhattan's "pollen situation . . . is not very serious." The statement reflected the fact that the pollen count in Oklahoma City was 614 grains per cubic yard. To highly sensitive, easily irritated individuals, this comparative perspective provided little relief. As the seasonal patterns of pollen levels became better understood, data anomalies could surprise New Yorkers. When the air became saturated with pollen earlier than anticipated in 1938, the Central Park meteorologist insisted the pollen had "no right to arrive before August 15."[6]

Figure 3.2. East Flatbush, near its borders with Crown Heights and Brownsville, in 1937. Brooklyn Public Library, Brooklyn Collection.

Anthropomorphizing ragweed as polluter emphasized its detestable nature. Ragweed plants have separate pollen-producing staminate heads and seed-producing pistillate heads adapted for wind pollination. When individual plants grew in isolation in the Pleistocene, this anatomy aided pollen dissemination in the wind so the species could survive. Their ancient biology also made the species out of place in modern New York. Health officials estimated that each square mile of New York City produced about 101 pounds of pollen annually. The Hay Fever Prevention Society claimed that a square mile of ragweed produced "SIXTEEN TONS of ragweed pollen." To ragweed hay fever sufferers, no plants, whether powered by the sun or by ConEd, may have been as polluting as *Ambrosia artemisiifolia* and *Ambrosia trifida*. One New Yorker pleaded for city departments and landowners to destroy them before they could send "poisonous pollen forth to pollute the air." Another griped that indifferent—and presumably unallergic—city officials lacked sympathy for "victims of ragweed pollution." Although Oren Durham naturalized the hazard of pollen as a seasonal phenomenon when he claimed that in 1928 the Eastern United States, including New York, had been hit by an "invisible pollen storm . . . a summer blizzard . . . [of] billions of toxic particles," he denaturalized the plant by classifying its pollen as toxic particles in the air at a time when air pollution was commonly associated with coal soot and industrial emissions. The ease with which pollen was described as pollution suggests that the concept of pollution remained malleable enough to apply to natural processes.[7]

Vilification of ragweed as an unredeemable polluter often disregarded the environments in which the plant thrived. The consolidation of Manhattan, the Bronx, Brooklyn, Queens, and Richmond in 1898 created a nearly three-hundred-square-mile city. In the following decades, developers, engineers, and laborers razed and rebuilt structures, as well as expanded transportation networks throughout and beyond the metropolis. Manhattan grew upward with skyscrapers, while people moved to outer boroughs connected via rapid transit lines and parkways, as well as to settlements west of the Hudson River and New York Bay linked by tunnels and bridges. Between 1920 and 1930, the population of Manhattan declined, but Brooklyn, the Bronx, and Queens each gained more than 500,000 people. New residential neighborhoods in these boroughs provided opportunities for lower-density living. In Brooklyn, for example, middle-class residents migrated out of industrial, noisy, and dirty areas to Sheepshead Bay and Canarsie, where roads were unimproved and there were no sewers. With 28 to 105 people per acre, the city's "subway suburbs" and nascent neighborhoods were four to fifteen times more dense than the municipalities outside New York City, which in 1930 averaged 7 people per acre. Populations outside New York City were growing by tens of thousands of people as well. From the 1920s to the 1960s, ecological conditions in old slums and new neighborhoods alike continued to change, albeit at different rates of population growth and land development over time. Yet just about every area of the metropolis had places in which ragweed could grow.[8]

Surrounded by, running through, or marking the edges of this city building were marshes, beaches, riverbanks, hills, woods, and meadows. Spider lilies, pickerelweeds, white hellebore, heartweed, and joe-pye weed covered marshy, damp areas. Dry roadsides were lined with clovers, cow vetch, dogbane, ox-eye daisies, and king devil. Queen Anne's lace, burdock, sunflowers, coltsfoot, goat's beard, and chicory colored emerging neighborhoods. In cemeteries, poison ivy evaded or returned after occasional mowings and left visitors with rashes. Chickweed, milkwort, and lance-leaved tickseed broke through gravely paths and old sidewalks. In the early 1920s, journalist Lewis Gannett marveled at a "whole hillside" of asters and goldenrods behind the Tiffany Building near the New York Central Railroad's tracks. Gannett was also impressed by the ability of the ailanthus tree "to endure New York's burden of smoke and soot," which allowed it to become known as "the back-yard tree." Japanese knotweed in the Trinity Church Cemetery in upper Manhattan had "taken possession and defie[d] dislodgement" from the rocky soil. The young Alfred Kazin stood "in a daze" and stared so long at the harsh-to-his-hands "fibrous stalks" of goldenrod that he failed to do what he set out to in Brownsville. Kazin walked by Canarsie's "infinite weedy lots" to reach Jamaica Bay. On Staten Island, Donald Peattie encountered "survivors of old gardens" and "wandering escapes" such as dame's rocket, hyssop, and apple of Peru. After World War II, botanist Harry Ahles searched the southeastern Bronx for plants growing among ballast and rubbish. He spotted silkybent grass and hoary alyssum on the ballast dumped in Castle Hill, an area tucked between two creeks. Sweet sagewort, late-flowering thoroughwort, and hawkweed oxtongue remained in the vicinity of the Bronx River Parkway. Ahles observed annual wallrocket and curlytop knotweed on the future site of a section of the Cross Bronx Expressway. When visiting the city in the 1940s, Norwegian writer Sigrid Undset observed that magenta petunias "thrive[d] like weeds in all the small front patches by Brooklyn houses."[9]

In New York City, as well as many other emerging metropolitan areas, ragweed often ruled changing land. The plant's reputation as a twentieth-century nuisance obscured its ancient roots. *Ambrosia artemisiifolia* (common ragweed) and *Ambrosia trifida* (giant ragweed) flourished in the northern and eastern areas of what would become the United States from 1.8 million to eight hundred thousand years ago. Around 1000 B.C.E., some indigenous peoples encouraged ragweed growth on a small scale for seed food. Although ragweed was a "native American," few sixteenth- and seventeenth-century European explorers encountered the plant. However, as European immigrants became Americans who turned grasslands and forests into crop fields, widened paths into roads, and bored into the ground with mining and canal construction, ragweed began to spread again. Botanist Roger Wodehouse believed this period of "wasteful exploitation of natural resources" that accompanied the "coming of civilization" generated a "ragweed boom." The making of America was making hay fever more widespread. Modernizing the continent created ecological conditions that resembled prehistoric ones in which ragweed thrived. Ragweed, which in paleobotanical terms was an ancient American, had become, according to Detroit sanitary engineer John Ruskin, "an urban plant" and a public health problem. To malign the plant, ragweed's despisers referenced bygone days and modern ills. The *Brooklyn Eagle* conjured up the nineteenth century, asserting that "ragweed blooming in an open lot . . . ranks with pigs running wild in the streets." Herbicide-spraying contractor Thomas McMahon imagined that ragweed was "a vandal" because it harmed other plants.[10]

Ragweed not only returned with "civilization" but survived its discontents during the Great Depression. New Yorkers found it in Inwood and St. Nichols parks, as well as along stretches of Highbridge and Riverside drives. While "excessive smoke" produced "withered and dwarfed" trees and shrubs in Central

Park, the park's ragweed endured in "magnificent clumps." In the Bronx, six-foot-tall ragweed and its associates hid the sidewalks. Ragweed grew among the rubble of odd-sized lots and in the shadows of the elevated trains. In Brooklyn, there were "jungles of weeds from six to eight feet high" in Flatbush and along Ocean Avenue, leading to Sheepshead Bay. In Chicago, some children called these "ragweed jungles" prairies; later in their lives, they cherished the "spacious freedom" they had enjoyed there. The *New York Times* suggested that surviving ragweed hay fever was a tribulation of progress common in cities, which had lost nearly all of their "spontaneous vegetation." The editorial writer tried to console hay fever sufferers by reminding them that the pioneers who settled the frontier endured "the fever and shakes." In this analogy, just as Midwesterners had established farms, real estate developers, public works crews, and construction businesses would build denser and better cities; as concrete replaced ragweed, hay fever would disappear. This anticipation of progress saw New York City's urbanization as a finite process, rather than recognizing that metropolitanization was possibly a perpetual process with undefined, distant boundaries.[11]

Aggravated ragweed hay fever sufferers sought more immediate relief than a hypothetical future stage of urbanization. During the 1920s and 1930s, some New Yorkers saw physicians who injected them with small doses of the allergen to desensitize them or to give them immunity to pollen. The city sponsored allergy clinics in thirty-eight hospitals for poorer people. In 1930, New Yorkers made some seventy-five thousand visits to such clinics. However, physicians were not confident that medicine could eliminate ragweed hay fever. One otolarynologist envisioned "an organized movement to uproot all rag weeds." He proposed relying on scouting organizations and other civic groups to provide the labor, and he hoped a private foundation could "further the propaganda" necessary for the formation of a public-private partnership to "actively destroy or remove this nuisance." Hay fever sufferers and leaders of their organizations wished their city would confront the plants. An exuberant Queens resident who believed hay fever was "not a malady that can be cured . . . when ragweed is . . . allowed to grow and exist all over" was shocked that "THE GREATEST CITY . . . IN THE WORLD" was "HELPLESS TO COMBAT A WEED THAT CAN EASELY [*sic*] BE DESTROYED." One minister acknowledged the city's generosity in providing free hay fever clinics but counseled, "Prevention would be more economical and a larger number of people would be benefited." A Hebrew Hay Fever Relief Association representative maintained that destroying ragweed would not only be effective but "seems to us practical." Many ragweed hay fever sufferers believed that preventing the environmental cause of the disease was the best policy and their best hope.[12]

Energetic advocates in hay fever organizations urged municipal governments to combat ragweed. Julia Ellsworth Ford, a wealthy hay fever sufferer, presided over the formation of the Society for the Prevention of Hay Fever after the demise of the Hay Fever Association, a group started in 1861 by Henry Ward Beecher. Ford, who lived in Rye, New York, about ten miles north of the Bronx, believed that "victims of ragweed . . . should bestir themselves to take a militant part in an eradication program." Since about half of Rye's parcels were vacant in the early 1930s, ragweed was probably common. When officials opted to cut grass "for beauty's sake" rather than to fight ragweed, Ford hired workers to harvest the plant and dispose of it on "the Town Hall porch." When her work with Governor Franklin Roosevelt to pass a statewide ragweed destruction law did not succeed, Ford turned to persuading local boards of health within twenty-five miles of New York City to advance an "urban district law" banning ragweed growth. Inside New York City, the Society for the Elimination of Ragweed proposed educating public schoolchildren to fight the plant. Public parks, streets, school yards, backyards, and vacant lots "overrun with ragweed" led Queens residents to form the Anti-Ragweed Society. Louis Fucci, who organized the Bronx's Ragweed Eradication Committee in 1932, joined with Ford to establish the locally and nationally active Hay Fever Prevention Society. To realize "freedom from noxious ragweed pollen," its members destroyed ragweed, notified health officials about violations, pressed New York to sponsor a "Ragweed Eradication Week," and wrote to national leaders. Thomas Hughes, the society's president, claimed, "It is a waste of Government and public funds to seek to find a cure when the cause and its prevention is scientifically known."[13]

As the number of hay fever sufferers in New York increased, as medicines remained unsatisfactory treatments, and as New Yorkers became confident that ragweed could be destroyed, the city worked harder to decimate the plant. Over time, New Yorkers fought ragweed with better-organized forces and more powerful weapons. However, people's conviction that ragweed was dangerous and undesirable was stronger than the city's capacity to banish the ubiquitous plant and its invisible pollen. Without control over the dynamics of metropolitan growth, battling ragweed evolved into a seemingly endless conflict and perpetuated ecological conditions that allowed it to return year after year.

In the 1920s and early 1930s, ragweed elimination in New York was much like weed control in turn-of-the-century cities. Despite the 1915 law prohibiting ragweed growth, New York's health department sporadically assaulted the plant. In 1925, the city compelled Queens landowners to clear 101 premises of ragweed. "The first time" that the department made a real effort to eliminate "ragweed and other noxious weeds" was in 1929. In Queens, crews removed twenty acres from the Corona Meadow Dump and thirty acres from Flushing Heights, Jamaica,

Little Neck, and Douglaston. Although the city cleared 1,800 land parcels in 1931—mostly in Manhattan—this was 700 fewer properties than planned. In 1932, the city cleared equal acreages in Queens and the Bronx. The effectiveness of this work may also have been limited by the imprecision of missions targeting "ragweed, poison ivy, and other similar plants." Commissioner Shirley Wynne admitted the city could not discover every violation of the law. Nor did Wynne's pledge on WNYC radio that "any complaint will receive prompt investigation" mean that problem properties would be quickly cleared. The department struggled to track down and secure the cooperation of property owners. An experiment with a title company's data increased the cost per search without substantially improving the accuracy of the results. Securing and assigning seasonal workers complicated efforts. Competition for authority among centralized health department officials, borough presidents, and their sanitary superintendents inhibited development of a citywide strategy and coordinated use of aggregate resources. Each borough carried out the work as its leaders believed necessary. Although New York officials struggled to adapt their bureaucracies to the environmental and ecological challenges of controlling weeds, they were probably no more disorganized than officials in other cities where the work was "loosely attached to the health, or street or public works department."[14]

During the Great Depression, multitudes of jobless city dwellers were available to fight ragweed, but funds to employ them were not always available. In 1932, a hay fever sufferer wrote to the *Times* that deploying the unemployed to uproot ragweed would be both "a godsend to many of the destitute" and "highly humanitarian work." In 1932, an "army of otherwise unemployed men" uprooted more ragweed than in previous years. New Deal programs to provide work relief for New Yorkers permitted a new scale of weed control. New York received more than one billion dollars in federal funds, and it was the only city to have its own Works Progress Administration (WPA) unit. In 1934 and the first half of 1935, the health department reported that relief workers used scythes, shovels, spades, hoes, and wheelbarrows to clear nearly 2,300 acres of "noxious weeds" on 112,000 lots. The 1935 decline in acreage cleared resulted partly from a "preoccupation with Hasheesh elimination." In 1936, workers cleared more than 3,044 acres. In 1937, men eliminated ragweed from 1,200 acres and removed poison ivy from 72 acres of land. Critics of work relief who suffered from hay fever supported such projects. A resident who believed many WPA projects wasted "funds and energy on an apathetic community" claimed that eradicating plants "growing in rank profusion would not only be of tremendous benefit but would be met with widespread approval." In Chicago, "prolific" ragweed compelled Good Will Industries and the Chicago Women's Clubs to collaborate on employing 1,350 men in city shelters to fight ragweed. The Illinois Emergency Relief Commission assigned up to 750 laborers a day to work clearing weeds. Field Museum botanist Paul Standley disparaged this work. He acknowledged that "local weeds" menaced nearby residents but concluded that cutting ragweed was "of little value." Standley warned that "there is no hope that relief may ever be obtained for people with ragweed hay fever by suppression of the source of the affliction."[15]

Although the WPA "campaign to clear . . . ragweed . . . in all five boroughs" attempted to better utilize new funding and existing resources than past efforts did, established divisions and funding limits hindered progress. Colonel Brehon Somervell, New York City's West Point–educated WPA administrator, intended to dampen ragweed hay fever sufferers' sneezing and its "devastating crescendo . . . when the ragweed brings up its heavy artillery and starts spraying the air with pollen in all directions." However, Somervell's troops went into battle without the ballyhooed "electric 'pistol'[s]" that when touched to weeds "literally cook[ed] the heart of the plant." Pollen counts from four stations in each borough assisted clinicians' efforts to match patient symptoms with types and levels of pollen in the air and provided "a check on the efficiency of the eradication work." Somervell's battle plan was set back when regional backing could not be secured, when allies squabbled, and when funding was suspended. Health Commissioner John Rice met with health officers from Westchester and Nassau counties in 1937 to negotiate taking the battle against ragweed beyond city borders but could not persuade them to open this new front without assuring them federal funding. A dispute between Rice and Somervell over the health department's role in supervising the work led Somervell to reconsider funding for it. Although Somervell claimed he would "approve an application for the destruction of noxious weeds, marajuana [*sic*], hashish, etc.," Somervell failed to procure the funds. Washington administrators insisted funds were only for "public improvements" and that managing vegetation—"work . . . recurrent in nature"—was "not permissible." Another WPA official claimed it was "undesirable to assign . . . [men to] weed eradication . . . who can be more usefully employed on construction work." New York's experiences may have deterred the USDA from supporting such programs elsewhere in the nation. USDA officials maintained that the department's "inadequate information and experience" prevented it from administering weed-control work-relief projects and that weed control was an individual, local, or state responsibility. Secretary Henry Wallace stated that the "practicability of ragweed eradication campaigns" depended "entirely upon local circumstances."[16]

Ragweed control in New York City was not cleaning up a finished city but addressing temporary environmental conditions of a city that was continuing to develop. During the Depression, federal relief funds supported infrastructure construction such as the Triborough Bridge, tunnels underneath the East and Hudson rivers, dozens of miles of expressways, and sewage treatment plants, as well as adding thousands of acres of parks and playgrounds and establishing or repairing health clinics, libraries, schools, fire stations, and police stations. However harmful ragweed pollen was to hay fever sufferers, its public health threat was useful for stimulating demand for labor. The work was simple, but it helped sustain working men's sense of utility and assured hay fever sufferers that officials cared about them. However, unlike better-conceived, better-funded, and better-organized Civilian Conservation Corps projects, ragweed eradication did not transform the laborers or the land that they worked. Removing ragweed ephemerally changed the land and marginally improved air quality. Although allergist Robert Cooke criticized "the prevention of hay fever by the destruction of ragweed alone as a stupendous, expensive and impractical plan" after the WPA terminated operations, the work established the city's role in and even responsibility for protecting the public from hay fever.[17]

The short WPA campaign against ragweed and the long depression that undermined urban development perpetuated fields of happenstance plants and resulted in bureaucratic turf battles. In 1941, Bronx borough president James Lyons claimed that "expeditious removal of ragweed . . . is of even greater importance than snow removal." Lyons and Bronx councilman Joseph Kinsley drafted legislation giving borough presidents the power to authorize cutting weeds on properties that owners did not cut, and their bill received the support of hay fever sufferers; the New York Association of Biology Teachers; and civic groups such as the Elmhurst Manor Community Council, the Flushing Garden Club, and the Good Citizenship League. Mayor Fiorello LaGuardia signed a revised bill in 1942 that the *Times* reported "doom[ed the] ragweed plant." Ineffective implementation that summer compelled ragweed despisers to attempt to amend the ordinance by requiring vegetation exceeding twelve inches in height to be removed every three weeks between 15 May and 15 September. However, as the demands of World War II had already led the health department to reduce many nuisance elimination programs, the commissioner argued that the proposed changes were not "practical" and would not be "effective." The *Times*, which a year before had cheered ragweed's imminent doom, now opined that "uprooting [ragweed] would do little good. . . . It is easier to exterminate a wild animal than a weed." In 1943, the department issued 1,163 weed notices.[18]

After World War II ended in Europe, New Yorkers' confidence in ragweed eradication surged. The Coordinating Council of the Medical Societies of Greater New York planned a ragweed elimination campaign to be carried out by the Boy Scouts with the herbicide 2,4-D and invited several city departments to cooperate. City officials reviewed the plan and informed the council that it lacked the proper equipment and that the boys lacked the capability to carry out the work. The Department of Health instead devised battle plans for a larger campaign—Operation

Ragweed. Officials envisioned "the complete elimination of ragweed" by spraying 2,4-D on thousands of acres of ragweed-covered land each summer. They intended to persuade dozens of surrounding communities to participate as well. These plans were inspired by the potential of the herbicides, not the emergence of unanimous scientific opinion about preventing hay fever. While the *Journal of the American Medical Association* stated that ragweed control was a "waste of time," a Cincinnati professor of medicine deemed NYOR a "modern and forward looking approach to the control of the environment." New York officials agreed with botanist Roger Wodehouse's view that "the only way" hay fever could possibly be cured was "by treating the environment."[19]

Mass production of phenoxyacetic herbicides made the city's plan possible. Early twentieth-century "contact herbicides" killed plants by poisoning or burning plant tissues. These inorganic chemicals were expensive, toxic, and flammable; they killed vegetation indiscriminately, and they corroded spraying equipment. District of Columbia officials attempted to kill weeds in alleys with sodium arsenite in the early 1920s. New York's health department and the WPA had only used them experimentally. In 1942, Du Pont sent advertisements to the Parks Department about its ammonium sulfamate weed killer. A parks official determined that the chemical was "not a weed killer in the true sense of the term as normally used, as it would kill grass, trees, and anything else that it is sprayed onto." Research conducted by state health officials also indicated that such chemicals were too costly and killed too much vegetation. They insisted that "removal of the [rag] weed from the ground is the most satisfactory method." During the war years, scientists also discovered the hormones that regulated how plants rooted, ripened, or seeded and learned that applying too much of these chemicals destroyed plant tissues. These substances were used to develop herbicides like 2,4-D. By 1944, researchers had demonstrated that 2,4-D killed dandelions in lawns. A chemical that killed broadleaf plants but spared grasses was heralded as a powerful weapon in the "War on Weeds." Commercial production of 2,4-D expanded from less than one million pounds in 1945 to nearly thirty-one million pounds in 1952; the price fell from two dollars per pound to fifty cents per pound. Alfred Fletcher, who served in NYOR, wrote in 1950 that the "elimination of the sources of pollen pollution appeared hopeless" until the combination of "engineering intelligence" and 2,4-D made possible a new "mass public health measure."[20]

The herbicide promised to accomplish what cities had been unable to do by cutting and pulling—to reengineer nature in urban environments. Chemical manufacturers and scientists encouraged city officials and city dwellers to wield this weapon. Dow Chemical's film *Death to Weeds* promoted the use of the herbicide to manage "parks, highways, playgrounds, golf courses, and gardens." The company marketed 2-4 Dow Weed Killer as the means for "improving appearance of vacant lots, [and] roadsides," "controlling . . . undesirable . . . plants along . . . rights of way," and "eradicating . . . undesireable vegetation in home and public recreation areas." The Brooklyn Botanical Garden's George Avery marveled that 2,4-D "kills ragweed like magic." In a radio drama about a physician with allergy patients, the doctor explained to a patient that "ragweed is the big thing—and we hope we can knock it out. . . . The public can buy [2,4-D] . . . to kill ragweed in their yards. . . . Just ask for 2-4-d." Dow pointed out that ragweed was just one of the one hundred odd plants that 2,4-D could destroy. In the *Brooklyn Eagle*, Frank Thone wrote that unless the unwanted plants were near flowers or vegetables, "you can let the spray swish where it will."[21]

Several city departments and organizations collaborated in NYOR. The City Council granted the health department the legal "authority to spray ragweed wherever found, even when on private property." Health officials coordinated programs, trained workers, and obtained supplies. The police commissioner appointed "every patrolman a 'spotter' of the hay-fever producing plant." Their reconnaissance indicated that 5 percent of the city's land was covered with ragweed. The sanitation department and the borough presidents provided street-flushing trucks and drivers. The health department sprayed plants on privately owned land, while the boroughs sprayed public properties and roadsides. The parks department handled ragweed on its properties and permitted the Brooklyn Botanical Garden to conduct experiments on them. The health department and the Brooklyn Botanical Garden attracted two hundred individuals to the conference Control of Plants Harmful and Annoying to Man. The assembled botanists, horticulturists, and gardeners discussed weed-control techniques and viewed *Death to Weeds*. George Avery encouraged attendees to "go back to your communities" and start weed-control programs. In 1947, the health department held evening seminars on "Control of Plants Detrimental to Public Health" for metropolitan area health officers, hay fever society members, weed-control committees of women's clubs, garden clubs, and general community organizations. Officials scheduled outreach events at venues like the Jewish Community House of Bensonhurst and the Sunset Park Health Center. Health department flyers recruited New Yorkers to join the "All Out War on Ragweed" by buying 2,4-D. Boy Scouts distributed posters and circulars. New Yorkers saw giant ragweed and common ragweed on posters in buses, streetcars, and in public places.[22]

When NYOR began in 1946, the department decided the campaign would be "prosecuted vigorously." Health Commissioner Israel Weinstein claimed New York was "setting the example for the rest of the nation . . . [and winning] the heartfelt appreciation of several hundred thousand hay-fever sufferers." Philip Gorlin, the coordinator of NYOR, planned to launch attacks from the city's 5,700 miles of streets. Spraying three thousand of the city's ten thousand acres of ragweed with 2,4-D cost about $33,000—including $17,500 for labor and $14,000 for 2,4-D. A crew sprayed between two hundred to three hundred gallons of 2,4-D per acre and treated about two and a half acres each day. Officials predicted that it would take workers three years to kill the city's ragweed and that a "small routine maintenance program" would prevent new growth. They attempted to procure additional funding from the U.S. Army and the Federal Security Agency. Citizens offered tactics for the health department's consideration. The Bronx Heights Improvement Association recommended that workers not spray from the sidewalk or road, but penetrate properties so that ragweed could "be conquered." A Gravesend, Brooklyn, woman urged the department to do the work "in the spring before the plant was in full bloom." Along with such suggestions, health officials received "letters of congratulations, commendations, and gratitude," rather than "thousands of complaints about noxious weeds . . . the usual experience since 1915."[23]

The promotion of its initial successes to public health and public works professionals made New York a model for other cities. Chicago's Board of Health asked Commissioner Weinstein to share his methods. Philadelphia's Bureau of Municipal Research requested technical information about equipment, program costs, and securing the support of neighboring municipalities. Detroit health officials studied New York's program and decided that ragweed elimination was a "justifiable use of public funds" because efficient programs lowered pollen concentration in the air. St. Louis Street Department officials analyzed NYOR to improve a weed-control program that had become "the unwanted stepchild of city operations." Officials in Worcester, Pittsburgh, and Houston also inquired about New York's war.[24]

Training and planning improved NYOR in 1947, but the program remained imperfect. Two health inspectors from each borough were to enroll in seminars to improve execution of their work. When police department surveys revealed railroad properties harbored more than 130 acres of ragweed, Weinstein encouraged these corporations' representatives to attend the training seminars as well so they would be able to support the program. That summer, NYOR drenched almost forty-eight hundred acres of ragweed with more than one million gallons of herbicide. About one-third of this land was in the Queens neighborhoods of Flushing, Bayside, Richmond Hill, Astoria, Woodside, Forest Hills, and Whitestone. Although crews could not kill every plant, officials hoped the spraying would reduce the density of future ragweed populations. Some citizens were disappointed when they found out NYOR was targeting "dense growths in heavily populated districts" and leaving other patches for future summers. A Bronxite invited politicians who wanted "to go slumming" to drive along Thwaites Place and "admire our ugly ragweed." Cost overruns forced modifications of the program. Manhattan's borough president agreed to divert his resources to the outer boroughs.[25]

Although the mission of NYOR was to reduce "pollution of the air" by eliminating the source of "toxic pollen," officials' reliance on 2,4-D permitted many anemophilious plants to remain and become more common. Attacking ragweed indirectly determined what plants would grow in the slain enemies' place. Although the sanitary code prohibited plants that released "poisonous particles," the city did not ban the grasses and trees that caused allergies. A ragweedless New York would not be a hay fever–less city. People who hoped the war be would widened to destroy other allergy-causing plants were disappointed. A Queens resident stated that the ragweed elimination made his grass allergies worse and requested that the city mix 2,4-D with ammonium sulfate to kill the grasses that flourished in ragweed's absence. The health department responded that it could not control the "six kinds of wild grasses" that caused allergies. Nor did the department concede that NYOR increased the population of allergy-causing grasses. NYOR leaders possessed conservationist pragmatism. Attacking ragweed removed what did the greatest harm to the greatest number of allergy sufferers. Less offensive allergenic plants remained part of the city's nature. Only paving vacant lots would have worked as well—or better—if it could have been done more cheaply.[26]

Despite ongoing assaults, NYOR became a war of attrition. In 1949, the department estimated that the city's ragweed had been reduced to four thousand acres. The department began respraying resurgent growth. However, as ragweed continued to grow in some of the same places each year, officials realized that additional reductions would be difficult. Rainy springs assured "lamentably abundant" crops that compromised work from previous years. Estimating "the remaining amount of ragweed" became harder because small patches in built-up neighborhoods were hard to locate. Some varieties of ragweed found refuge among other plants. In early summer, giant ragweed was an easy target. Common ragweed, which looked "just like a dozen other weeds," eluded eradication teams and went on to mature. "A chronic griper" who implored health officials to take action mistook "mustard weed, pigweed, and a number of other weeds . . . as ragweed." A botanist observed that many happenstance plants other than ragweed were being sprayed in Central Park. In addition, eliminating ragweed involved not only killing existing plants but preventing growth of new plants that could emerge from ever-shifting land uses. Two critics who thought the region favored "profuse ragweed growth" claimed "the weeds have returned almost as rapidly as they were 'eliminated.'" A former health official who thought weeds near his home indicated that NYOR was limited by ordinary bureaucratic woes asked the commissioner to "light a firecracker under the seats of some of your underlings" to improve the effort. In 1952, the health commissioner conceded his foes' tremendous forces remained strong and lamented that the obliteration of the city's ragweed would not prevent hay fever because of the amount of pollen that the wind carried into New York. A 1956 study indicated that NYOR had produced "no consistent decline" in the pollen index. The researchers—two allergists—concluded that "anything less than complete success in the battle against ragweed is a doubtful gain." Herbicides with the power to kill ragweed did not imbue New Yorkers with the power to conquer hay fever.[27]

In 1932, an irate New Yorker informed the health commissioner and the *New York American* that "ragweed is being cultivated. . . . Careful investigation would reveal a man made cause for the profuse growth." On streetcar rides through Queens from College Point to Jamaica, Javier Adrianzen saw ragweed crops planted to exploit hay fever sufferers. Although offending property owners ranged from real estate concerns to the police department to gardeners, the key conspirators were the manufacturers of "QUACK treatments" and the newspapers that sold them advertising space. Yet in constructing this hypothesis, Adrianzen omitted one group of ragweed cultivators—its would-be eradicators. Men who trudged across land and ripped the plants out also positioned seeds to germinate. Giant ragweed seedlings often emerged in soils with varying levels of moisture in March, despite cool temperatures. They began claiming space before the seedlings of other plants that needed slightly warmer temperatures and moderate soil moisture appeared. With this head start and its rapid development, giant ragweed claimed water and nutrients and prevented other plants from obtaining much sunlight. Similarly ambitious but less dominant plants included smartweed, lamb's quarter, common ragweed, and velvetleaf. Although Adrianzen had little evidence of a conspiracy, he was nevertheless correct that ragweed patches had "man made" causes. The laws of economics, the laws of New Yorkers, and the laws of ecology helped cultivate fortuitous flora and contributed to their distribution, variety, and density. Cutting ragweed did not get to the root of the problem—a real estate market that generated externalities city officials could not control. The city's struggle to subdue ragweed frontiers in dense neighborhoods and on wide-open borough land before World War II helped sustain them in the postwar years.[28]

Elements of past, present, and future real estate markets were apparent in New York's 1916 zoning law. Although it was primarily intended to regulate building heights, property values, and land use in Manhattan, the law also attempted to shape development in the Bronx, Brooklyn, and Queens by assigning types and proximities of future land use—industrial locations, areas for large apartment complexes, and spaces for detached single-family homes. Where these zones did not match property owners' plans, or when property owners sought to change designations, patches of happenstance plants worked to color the environment. For example, some developers acquired low-density residential land and then petitioned for approval to build apartments. When they were successful, adjacent property owners who had planned to build houses could become sellers of vacant parcels. Similarly, when home owners protected their neighborhoods by working to have nearby land rezoned as low-density residential land to prevent construction of apartments or establishment of industries, they delayed development. City building and city upkeep were also diffused with the extension of infrastructure and services into outlying areas. Beneficiaries of these investments often showed little concern for conditions in established sections of their boroughs. As a result, older commercial and apartment buildings became shabbier if their owners failed to earn enough to pay their mortgages and taxes. Banks that financed development of new neighborhoods and denied loans to owners of older properties deterred new building or reinvestment in older neighborhoods. Nevertheless, there remained land so isolated that development was irrational or impossible until automobile-driven lifestyles became more common. For nearly five decades—until 1961, when a new zone plan was passed—these interactions of speculators, property owners, and city officials permitted metropolitan photosynthesizers to inhabit vacant lots in older neighborhoods and large undeveloped tracts in newer neighborhoods. Such urban environments could seem unappealing to potential home builders and home owners, who then continued to search for property outside the city.[29]

Although zoning could influence when land came into use, changes in land value resulting from investments in infrastructure and construction on adjacent land, from property owners' decision making, and from the passage of time itself were more important factors in land development. William Loucks argued that owners who invested capital in sewers, paved streets, and sidewalks added real value to the land and could expect their properties to rise in value. In contrast, Richard Ely stated that simply purchasing land that was being or had been used for agriculture or resource extraction for future urban real estate development enhanced its value. The costs of owning property—assessments, taxes, and loss of interest on capital that might have been invested otherwise—were the bases of Ely's "Law of Ripening Costs." Property owners who incurred these costs were justified in selling their land for the highest price—prices critics thought were unjustifiably inflated. When speculators succeeded in harvesting much wealth from property in which they had invested little, they could further delay when the land would be built on and used. Some economists concluded that many landowners overestimated how long it took land to ripen and thus never enjoyed the fruits of real estate investment. Ripe land could quickly rot; its ripeness was a fleeting "five years or less." In unpredictable urban real estate markets, lucky, highly sophisticated, or well-capitalized investors most often profited from their investments. When ordinary Americans bought land and held onto it for a decade or longer, they lost their savings. Speculators also profited by persuading optimistic and naive people to buy properties as investments or potential home sites. Such people could not afford the taxes and special assessments for infrastructure extension. When vacant tracts remained between built-up areas and their investment, many buyers, including speculators, lost their land. Tight credit markets reduced the number of investors who could produce subdivisions and Americans who could buy homes. Ownership of the land reverted to banks and metropolitan photosynthesizers.[30]

On the eve of the Depression, economists estimated that more than 20 percent of the land area was vacant in cities such as Altoona, Davenport, Flint, St. Louis, Salt Lake City, and San Francisco. While Midwestern cities had some of the highest proportions of vacant land (64.1 percent of Duluth), there was also vacant land in the country's oldest cities, such as Boston (16.9 percent) and Norfolk (9.5 percent). Population growth had slowed, and the number of people able to afford homes was declining. In the early 1930s, for every lot that had been built on in Chicago, Cleveland, Detroit, Milwaukee, and Birmingham, there was a vacant one. New York's populations remained clustered in the 1930s. Nearly 90 percent of the people living in Manhattan, Brooklyn, Queens, and the Bronx lived within a half mile of a mass transit line. They lived on less than 40 percent of these boroughs' total area. There were vacant lots in their neighborhoods, although more than 80 percent of vacant parcels were in outlying areas where few people lived. Both near and beyond densely developed areas—in Central Park, on the west side of Manhattan, in Red Hook near the Gowanus Canal, and off of dirt trails in Queens—squatters lived in shacks and huts next to patches of urban herbs.[31]

To hay fever sufferers, land that was purportedly ripening in the hands of real estate speculators was rotting when the ragweed on it proliferated and polluted the air. Ely's "law of ripening costs" seemed primarily to conceal decisions "to assign an extravagant value" to land in cities because of the great wealth that could be harvested from renting out a square foot of land. However, New York economic analyst Philip Cornick noted that "vacant lots [in cities] produce nothing beyond weeds and headaches, and both of those products are unsalable at any price." Ernest Fisher described cities' undeveloped land as "a great belt of . . . idle land, grown up in weeds, delinquent in taxes, and derelict." Cornick observed subdivided land in New York where surveyor's stakes had "been lost in the underbrush." Even after World War II ended, real estate developer Joseph Laronge observed that "the weeds of failure are still found growing down the middle of many such streets and between the cinders of many such sidewalks."[32]

In the absence of an effective and affordable ragweed- and ragweed-seed-obliterating technology or method, city officials' basic strategy was to identify violators and secure their cooperation. Although they knew that speculators and property owners financed ragweed's transgressions, city officials were unable to develop a viable economic penalty to break this alliance and persuade property owners to fight the plant. Health Commissioner Wynne considered fines "far superior" to liens on properties, ostensibly because fines brought "faster results."

However, the market economy continued to produce both undeveloped land inhabited by happenstance plants and many economic agents who disregarded the economic consequences of their illegal inaction. An optimistic Richmond Hill, Queens, resident imagined that the right "tax" could make "the weeds . . . disappear in twenty four hours and hay fever with it." The termination of the WPA program and the failure of fines led ragweed hay fever sufferers like Councilman Kinsley to secure passage of a 1942 law that assessed liens against properties for weed-removal work. Kinsley may have considered liens to be superior to fines since a lien could reduce an owner's profit when the property was sold. When the health department continued past practices without completing the property research for liens, Kinsley charged that its staff had "devitalized" the law "because if the city simply hauled ragweed away, there was no incentive for property owners to act." Even had liens increased compliance with the law, health officers insisted that they lacked the resources to inspect land and to identify all noncompliant property owners. The health commissioner opposed proposed revisions to weed-nuisance laws in 1943 because his underfinanced staff would have been overwhelmed by more properties violating the law more frequently.[33]

While resource constraints inhibited penalization of ragweed cultivators, political and bureaucratic competition also complicated enforcement. Department heads and borough leaders alike sought to shape weed-control programs. Frustrations with the health department's efforts and the centralization of power in city hall in the 1930s created opportunities for borough leaders to show they were more responsive to addressing local needs such as fighting weeds. The Bronx's Kinsley thought borough leaders and the right legislation could properly assign the "responsibility" for and "financial burden" of ragweed control to landowners. In contrast, Mayor Fiorello LaGuardia believed that citywide and regional perspectives were necessary and that experts in municipal departments could better manage such operations. These politics surfaced in LaGuardia's rejection of attempts to give borough presidents more authority in ragweed control because they violated the city charter and his insistence that only health and sanitation officials could order ragweed removal. However, after LaGuardia secured his preferred division of labor, he did not insure departments fulfilled their roles. The sanitation department resisted lending staff to inspect roughly twenty thousand land parcels and to search nearly ten thousand property records. This reluctance to exercise responsibility may not have troubled LaGuardia, since like the city's property owners, he opposed more stringent enforcement. Although LaGuardia argued that a centralized approach could more effectively handle problems with ragweed, he subverted expectations that the bureaucracy could reduce hay fever suffering by stating that statewide efforts and regional cooperation were necessary to remedy the hay fever suffering caused by "obnoxious weeds." Emphasizing the regional dimension of pollen-filled air was used to justify minimizing the city's expenditures to kill the plant.[34]

Limited capabilities occasionally led officials to find ways to limit their responsibility for landscape management and extend it to city residents. Officials refused to till ragweed with tractors or horse-drawn plows because "the city would be held responsible for the destruction of private property." However, falling short of New Yorkers' expectations could harm the health department's reputation and cost money to repair it. In 1930, Francis Jones, an *Irish World* magazine editor, complained that "the existing local conditions" that caused his wife's hay fever problems risked bringing "the city administration into disrepute" in his Long Island City neighborhood. The department tried to appease Jones the next summer by surveying the area and mailing out warning letters to the owners of ninety-six vacant lots. A health official who suggested holding a "neighborhood get together" to clear a lot of weeds without explaining "just why a group of taxpayers should assume the responsibility of another taxpayer" incensed a Bay Ridge man. Even NYOR left some residents with the impression that the city had not assumed enough responsibility for ragweed control. When officials solicited citizen participation, a man who lived about a mile east of Prospect Park objected that it was "not the duty of the ordinary citizen to search out ragweed fields."[35]

Irresponsible property owners and real estate speculators reduced their "ripening costs" by externalizing the costs of vegetation control to the city and the costs of pollen pollution to ragweed hay fever sufferers, who lost time to illness and spent money on medicines and visits to physicians. Even in an idealized real estate market that imposed no land-use regulations—and thus no costs—on landowners, the ecological time that transpired while owners waited and worked to realize their desired sale price permitted urban herbs to grow. To laissez-faire idealists who opposed regulation, requiring property owners to manage vegetation—and enforcing their cooperation—was unjustifiably costly. Doing nothing was best. Avoiding intervention would expedite growth, raise property values, maximize the profits of developers, intensify the happiness of new home owners, minimize the cost of government, and increase municipal revenues from taxes. In New York, such wealth was paid for with the "suffering and torture" of people on days when the air was "saturated with the pollen."[36]

Before World War II, New York City's ability to moderate the public health consequences of a dysfunctional real estate market was undermined by the acreage of the problem and tempered by the awareness that enforcement could cost more than revenue generated. City officials' rationalizations of inaction suggested that irritating and disappointing allergy sufferers was preferable to the alternatives: depleting budgets by enforcing the law against every property owner, paying for damages to property that could arise from making land conform to the laws, exacerbating real estate market problems with liens, or possibly even upsetting powerful property owners who opposed the measure. Enforcing ragweed removal could not rework the real estate market so that properties would be sold and developed faster. City officials may also have worried about where ever-stricter enforcement could lead. If liens on properties were excessive, the city would become the last-resort owner, at which point the city would receive no tax revenue from a property but instead be responsible for the cost of maintaining it.

While property owners and city officials both worked to minimize the cost of their responsibilities, ragweeds went about their business of preventing erosion, even if they did not do so particularly gracefully. Deaf to the aspersions cast upon them and unable to defend themselves from verbal abuse, ragweeds could not evade what faults were dumped on them. The wastelands ragweeds inhabited were dumps created by New Yorkers' evasions of their responsibilities. Ragweeds were not useless. They were useful scapeweeds for New Yorkers unable to figure out how to distribute and manage land. While externalizing responsibility for hay fever to ragweeds was convenient, such attitudes did not change the mismanagement of land and the miscalculation of the value of metropolitan photosynthesizers. The bursts of violence against ragweed often reinjured the land and yielded more ragweed. The ingrained cultural revulsion to weeds reinforced by mounting concerns over ragweed pollen as a public health threat also reinforced the unecological economics sustaining New York City's ragweed frontiers.

The desire to improve public health and the resignation that the acreages of land and complexities of the market compromised effective action may have persisted for years had World War II not renewed city officials' confidence in the power of well-run organizations and new technologies. NYOR officials thought 2,4-D might be powerful enough to realign city departments to obliterate ragweed. Strategic use of 2,4-D seemed to make it feasible for the city to internalize some of the costs externalized by real estate owners. Officials attempted to use 2,4-D to lower the labor cost of destroying vegetation and waste less time and energy attempting to secure property owners' cooperation. For example, fifteen hundred WPA workers reportedly cleared roughly 4,000 acres of land from part of 1936 through 1937. In 1947, fifty-two municipal workers sprayed about 4,800 acres; in subsequent years, fourteen crews of three men each sprayed upward of 3,100 acres. When the Hay Fever Prevention Society's Louis Fucci proposed serving summons to property owners who violated the law as a public health education measure in 1954, the health department stated its determination to avoid this less effective "negative approach." Before World War II, even if New York had had the resources to clear every lot of ragweed, the plants would have returned because workers cleared land by "cutting, pulling, [and] grubbing," which cultivated future fields of ragweed. NYOR bombarded ragweed in June and July with herbicide to prevent pollination and seed production, as well as to avoid disturbing the ground and encouraging additional ragweed growth.[37]

Throughout the ragweed wars, officials opted for destruction rather than cultivation. In 1935, health officials discussed—but decided against—removing weeds and planting "the area with grass . . . [to] keep the weed growth out." Hay fever sufferers recommended enlisting vigorous plants to fight ragweed. Some imagined engineering a "botanical war to the death" between ragweeds and sunflowers. One New Yorker observed how sunflowers were able to spread and subdue "all other growths," including a tall, thick patch of "pestiferous ragweed." A Bronxite saw this ecological tactic as an economic way to lower the costs of the city's battle. During NYOR, city officials recognized that establishing new vegetation was potentially a "permanent solution." They considered planting clover or Japanese honeysuckle to smother and keep out ragweed. A parks department official commented that honeysuckle was the "best bet [because it would] engulf everything." Although parks officials may have been wary of introducing plants that could harm ornamental shrubs and small trees, they did not oppose them on their geographic origins. One pragmatic official rejected experimenting with "so-called competitive plants" because the city lacked the funds to buy the plants and to pay for the labor to establish them. Parks officials conceded it was "cheaper and easier" to spray 2,4-D annually than to establish preferred plants. The city lacked the resources to invest in reworking land that officials assumed would someday soon be developed.[38]

New York officials seemed to hope that it was possible to have more healthy citizens and fewer nuisances at less cost simply by killing "the weeds without injuring the grasses." They also seemed to mistake destruction for cultivation. The herbicide only temporarily stopped ragweed growth, and it did not establish a grassland ecology. The future of a sprayed space depended on the seeds in the soil and the seeds that continued to arrive. Although NYOR commanders embraced Roger Wodehouse's contention that eradicating hay fever required annihilating ragweed, they may not have grasped his claim that modern Americans only appreciated "two types of landscapes"—wild ones and ones "brought completely under control." Wodehouse thought Americans had no interest in "anything in between." However, the city could afford neither to force owners to manage land properly nor to manage the land properly for them. Killing happenstance plants left urban spaces perpetually in between. Delaying and scrambling ecological time planted a perpetual feel of disarray and prevented the emergence of loftier vegetation. City officials' campaigns against ragweed unintentionally cultivated ambiguous spaces that could be misinterpreted as evidencing neglect or decline.[39]

The assumption that vacant land only needed to be temporarily managed also suggested that officials were not fully aware of metropolitan changes. In 1949, land

Figure 3.3. Vacant lot before herbicide spraying in eastern Brooklyn (Vermont and Livonia avenues), July 1953. Courtesy of NYC Municipal Archives.

Figure 3.4. Vacant lot after herbicide spraying, June 1954. Courtesy of NYC Municipal Archives.

economist Frederick Aschman argued that land "prematurely" and "badly" subdivided in the 1920s had become "chronically tax delinquent or abandoned vacant land." He stated, "Always unsightly and weed-producing, 'dead' lots, when scattered about a partially built-up area, prevent the injection of 'new life' into the neighborhood." Many large Eastern cities, including New York, possessed such "dead land." Residents of Jackson Heights, Queens, were displeased with the "many empty lots" strewn with garbage and the sidewalks that were hidden by "veritable forests of ragweed during the summer time" in their neighborhood of "beautiful garden apartments and private homes." In Brooklyn, "extremely high weeds" were rumored to be "lairs for potential rapists and thugs." At the turn of the twentieth century, such areas were anticipated sites of growth; after mid-century, such spaces seemed to lack any future. Happenstance plants lived in these areas at the same time that city neighborhoods were changing demographically. As African Americans from Harlem and Puerto Rican immigrants from the Upper West Side made homes in Brooklyn neighborhoods such as Bedford-Stuyvesant, Brownsville, and East New York throughout the 1950s, whites moved away. On some streets, whites quickly sold to blockbusting realtors; in many cases, absentee owners and banks ceased investing in properties there. As neighborhoods changed and as workers continued to obliterate weeds throughout the city, few inhabitants and fewer outsiders may have been able to discern where vacant lots had long been empty and how long plants had been there. City dwellers could more easily misread weeds as evidence of an area's flawed social nature rather than as the consequences of past unecological economics.[40]

While ragweed frontiers remained in New York City, some new ones may have emerged on sites of future neighborhoods on Long Island and in New Jersey. To metropolitan planners, New York City was the 300-square-mile center of a 5,528-square-mile tristate region. The city's roads and rail lines led to suburbs and towns where job and population growth exceeded the city's. Accelerating urbanization of the metropolis slowed development of city land. Yet despite renewed speculation and urbanization through the production of suburbs after World War II, botanists learned that pollen levels were lower than in the past. Ragweed pollen levels perhaps surged and declined drastically in the region from 1890 to 1930, when people in Bergen County, New Jersey, abandoned their farms and moved into cities. In this demographic shift, large acreages of land became covered with plantain and ragweed and then with red maple and gray birch. After World War II, this reforesting land quickly became roads and lawns rather than enduring ragweed frontiers. From 1932 to 1967, ragweed pollen levels—and pollens of other herbs too, in both New York City and its most developed neighbors—remained steady and declined as open land became developed. Despite the limits of NYOR, postwar hay fever sufferers may have been somewhat more comfortable than earlier generations. Locally at least, pre-Depression urbanization patterns had limited future pollen production. To the extent that metropolitan-grown ragweed irritated hay fever sufferers, these problematic plants were spread throughout already urbanized areas home to a shifting metropolitan population.[41]

Ragweed inhabited geographically separate but historically inseparable spaces: vacant lots in changing neighborhoods, undeveloped land at the edges of New York City, developing suburbs, vacation sites, and distant farms. Although NYOR commanders, soldiers, and allies obliterated thousands of ragweed plants on thousands of acres of land, they could not arrest the metropolitan changes that permitted ragweed to persist. NYOR tacticians thought it was possible to develop "considerable" interest in ragweed eradication within a one-hundred-mile radius because officials from Nassau and Westchester counties promised to "do everything possible to control the ragweed in their respective communities." Jersey City's entry into the alliance was certain, given that "the mayor's secretary happen[ed] to be a ragweed addict." Although NYOR partially subjugated—temporarily—the city's ragweed frontiers, city pollen levels depended not only on "parcels of solid ragweed" in the city or the metropolitan region but also on land uses in places beyond. Allergists argued that cleaning the air required "thorough and continuous eradication" of ragweed on land within "a few hundred miles" of the city. While such a program might have seemed unimaginable, in 1964 a Crown Heights man wrote to the city's health department to express hope that Connecticut, Rhode Island, Massachusetts, New Jersey, New York, Pennsylvania, and Maryland could work together so that "ragweed could once and for all be controlled."[42]

Pollen drifting to and fro in the breezes demonstrated to metropolitan Americans the shared atmosphere of environments conventionally distinguished as large cities, suburbs, and surrounding countryside. On the ground, people's economic, political, social, and cultural differences made life in city neighborhoods, their metropolitan neighbors, and the wider region different. Yet these unequal environments had commonalities besides pollen-laden air. They hosted or were near rights-of-way for infrastructure that joined people and economies near and far. In 1928, planner and wilderness enthusiast Benton Mackaye described the outward spread of cities along transportation corridors as a "metropolitan invasion." However, it was not only roads and rails but also utility lines and pipelines that crossed many different landscapes, connected urban nodes to each other, and integrated in-between points. After World War II, American metropolitan areas and modern urban life consumed more land as people used new expressways and highways to live farther away from where they worked in large tracts of mass-produced suburban-style homes. Over time—rather than at any single point in time—these transportation corridors supported the development of denser, larger metropolitan regions with the decentralization of cities and the urbanization of suburbs. Managing the infrastructure connecting these places extended urban economic development and reworked adjacent landscapes to serve metropolitan Americans' safety, health, and lifestyles. Even as rights-of-way spaces allowed people to travel to urban, suburban, and rural places, they created similarities in and blurred distinctions between these spaces.[43]

Rights-of-way managers controlled vegetation to protect infrastructure networks and increase their efficiency. Railroads eliminated vegetation to reduce ground moisture that hastened the decay of ballast and ties, to increase the safety of track work, and to reduce the likelihood that fires along tracks would spread to nearby properties. Although sodium arsenite was used to kill plants in the 1880s, its expense limited its use. Enthusiasm about electricity led to experiments with railroad cars that brushed weeds with fine wires to deliver "a deadly current" that "traverse[d] the roots to their very tips" and killed plants "outright." Specialized cars that burned oil on their undersides to destroy vegetation were used where there was no risk of starting fires. By the 1920s, some railroads were applying chemicals to sterilize the ground, and many more did so with the proliferation of herbicides after World War II. Power companies maintained one-hundred-foot-wide rights-of-way to prevent burning trees from touching conductors and landing on lines. The expense of hiring men to clear some of a region's "roughest and toughest land" with axes meant work was often only scheduled when vegetation was within feet of the lines. After World War II, cost-cutting managers used herbicides to reduce this cost from $161 per mile to $91 per mile. Water suppliers used herbicides to eliminate trees and shrubs that interfered with valve and pipe operation and that shed leaves that affected water color and taste. A Du Pont salesman promoted chemical control of vegetation bordering public water supplies as a conservation measure "to insure the country's continued growth." Sewage treatment facilities used herbicides to eliminate the peppergrass and smartweed that grew on and reduced the efficiency of sludge-drying beds. Governments mowed and sprayed cheap diesel oil on roadside plants that could hide signs and guardrails and cause accidents at intersections. By the 1950s, new construction standards to accommodate faster traffic required additional vegetation control. Killing plants prolonged the life of road shoulders, and sterilizing the soil on which roads were built could prevent plants from cracking through them. Highway officials contended that reducing vegetation-maintenance costs released money and labor to expand highway systems, thus increasing the rate and range of urbanization. One vegetation-management contractor argued that highway departments able to persuade farmers, allergy sufferers, and hotel operators of the benefits of chemical ragweed control programs could win their political support for tax increases.[44]

By the 1950s, maintaining rights-of-way for these networks shaped tens of millions of acres of land. This land became more visible as metropolises grew and people traveled them more widely. Controlling rights-of-way vegetation between cities and beyond city lines to protect urban people's safety and health turned more plants into weeds, made more weeds, and added new ragweed frontiers. Although rights-of-way and vacant city lots could be geographically distant, spraying herbicides on these two noncrop landscapes and justifying this spraying with the same rhetoric produced spaces with similar vegetation and ecological dynamics. The condemnation of more plants as out of place because they interfered with infrastructure generated conflicts among ecologists, conservationists, herbicide manufacturers and applicators, and public works engineers.[45]

While infrastructure was used to deliver resources into cities that made urban life comfortable, convenient, efficient, and safe, infrastructure also made it possible for more urban people to spend time in places well beyond their homes. For New Yorkers who desired and could afford to travel to places with reportedly ragweed-pollen-free air, transportation networks allowed them to reach them. However, where ragweed frontiers ended and the frontiers of ragweed-pollen-free air began was blurry. Over time, some attempts to reduce the temporal distance between allergy sufferers and cleaner air increased the geographical distance between them. Both maintaining faster, safer roads and making ragweed-free places more appealing to city dwellers could increase ragweed growth. Corridors that made it possible to escape ragweed complicated and interfered with the pursuit of allergy relief.

In the second half of the nineteenth century, hay fever resorts established in the White Mountains of New Hampshire, the Adirondacks of New York, the shores of the Great Lakes, and the plateaus of Colorado were refuges for wealthy city dwellers from their cities and the urban poor during hot and humid months. Reputation—not rigorously constructed knowledge—made these places into hay fever havens. In 1931, New York City's health commissioner advised allergy sufferers to "avoid the ragweed as you would the plague," and he advocated traveling to a pollen-free place as the "best way to avoid hay fever." Joy Wheeler Dow, who had plenty of the "entrancingly happy-looking, crisp Treasury notes" necessary to spend four months in Kennebunkport, Maine, gloated that the destination had "such pathetic, forlorn, little weeds, you wouldn't have the heart to uproot them." Dow's experience suggested that without accumulated capital, rails and roads alone could not deliver much hay fever relief. After World War II, allergy sufferers could consult "Hay Fever Holiday." In 1948, Abbott Laboratories worked with the Carrier Corporation, a manufacturer of air conditioners, and the Chicago Automobile Club to distribute this index of high and low pollen environments to travel agencies, health officials, medical schools, and physicians. Abbott printed twenty-eight thousand copies of this pamphlet from 1952 to 1954 and nearly eighteen thousand copies in 1966. The index appeared in magazines like *Consumer Reports*, *McCall's*, and *Science Digest*. The Allergy Foundation of America referenced this information in its brochures. The most promising low-pollen or pollen-free environments were southern Florida, Arizona, southern California, Nevada, Utah, and Canada's Atlantic and Pacific coasts. Allergists and aerobiologists also contended that places "protected by mountains or by wide stretches of water" and able to eradicate the ragweed within their boundaries were likely to be spaces where ragweed hay fever sufferers could escape allergies.[46]

Costs and time made New England and northern New York the simplest places for many people living in New York City to escape ragweed pollen. New York State's Department of Health monitored places "free from the evil." The department certified areas as havens for hay fever sufferers based on knowledge that ragweed was not naturally found there or that "through systematic, organized effort the trouble-making weeds [had] been rooted out." For example, in 1943, state officials reported that "the entire central Adirondack area was free of ragweed . . . [except] only two small patches . . . found in more than 150 miles of highway." Pollen indices showed that air quality beyond the New York metropolitan area could be—but was not always—better. The indices for New York's five boroughs ranged from 24 to 35. In Sullivan County, Liberty had a pollen index of 13, while South Fallsburg, about a dozen miles away, totaled 54. Given these numbers, a New Yorker of limited means might try to guess when pollen pollution would be at its worst locally and get away to Sullivan County's comparatively accessible and affordable hay fever resorts.[47]

Sullivan County comprised the southwestern edges of the Catskills, a region of tanneries, quarries, farms, and streams that shipped leather, bluestone, eggs, and water to metropolitan New York. The county's economy evolved as city dwellers traveled there and created environments to suit their needs. Discontent with one urban nature guided efforts to seek and shape another urban nature. In the nineteenth century, the area's clean air and seemingly wild hills helped it become a refuge from city summers and yellow fever, Asiatic cholera, and hay fever. Tuberculosis patients traveled there to recover. Boardinghouses proliferated in Sullivan County at the end of the nineteenth century to serve the middle classes and some people of modest means. In the first half of the twentieth century, Jewish New Yorkers also began traveling to Catskills resorts. Wealthier families spent long summer vacations there, while poorer New Yorkers moved seasonally between city and Catskill economies as workers. Sullivan County was a temporary alternative to and a seasonal extension and variation upon urban life. When metropolitan Americans who dwelled in high-rise apartments and split-level homes for most of the year traveled to Catskills hotels, resorts, and bungalows, the countryside became more urban. Between 1920 and 1950, Sullivan County's urban population increased from 0 to 20 percent. By midcentury, Sullivan County's forty thousand permanent residents were joined in the summer by two million vacationers. The county's resources also continued to be poured into New York City. In 1954, its Neversink Reservoir became connected to the system of aqueducts constructed in the Catskills to provide New Yorkers with water.[48]

Transportation improvements eased city dwellers' travels to Sullivan County. The county had few improved roads until execution of the state engineer's 1906 master plan resulted in improvement of existing roads outside of cities and roads linking important cities, as well as construction of routes into scenic lands such as the Catskills and the Adirondacks. By the 1920s, Route 4, which ran from New York City to Buffalo through the Sullivan County towns of Liberty, Middletown, and Monticello, swelled with bumper-to-bumper traffic on summer weekends. In the postwar years, resort owners sought additional improvements to increase the area's accessibility to vacationers from East Coast cities like Boston. The highway was expanded into the four-lane "Quickway." Promoting the area's resorts as sanctuaries for hay fever was another way to attract more tourists. Although the county's hills looked different from Suffolk Street in the Lower East Side, the distance between them in space was bridged in part by the roads accelerating exchanges between entwined, mutually dependent economies.[49]

The state's investments in roads included management of highway rights-of-way. Road construction created bare roadsides and hillsides; revegetating these areas limited soil erosion. In the early 1950s, state public works officials collaborated with scientists in agricultural colleges and the federal government to advance the production and maintenance of roadside vegetation. Although researchers learned that simply mowing the fortuitous flora that appeared on roadsides could produce a dense, persistent cover "difficult to surpass," officials recommended planting dual-grass cultures. They also developed guidelines for herbicide use, especially in areas that were difficult and expensive to mow. Researchers also evaluated controlling poison ivy with "chemical mowing" in Buffalo; in Tompkins, Dutchess, Putnam, and Westchester counties; and on Long Island. Although managing rights-of-way to improve public health was not a priority of or mandate for state or county road crews, researchers assessed how roadside vegetation management could enhance public health. That rights-of-way vegetation could affect public health was acknowledged in a 1947 state health department report that indicated that ragweed pollen levels seemed to be increasing in places "previously . . . free from such contamination," in part because of "a slackening in efforts to eliminate the weed by mowing strips adjacent to highways." Such findings were reinforced by the New Jersey Department of Health, which claimed that roadsides harbored "up to 65%" of the Northeast's ragweed.[50]

Even as Catskills resort owners hoped better roads would bring more tourists, established tourist areas also had to work harder to retain old visitors and attract new ones, since automobiles and airplanes were expanding the range of New Yorkers' vacation destinations. Pine Hill, in Ulster County in the eastern Catskills, fought ragweed on roadsides, farms, and three miles of railroad rights-of-way between 1946 and 1950, reducing its pollen index from 6 to 2. Aggressive publicizing of this air quality won community support and increased how long hay fever sufferers remained. It also may have inspired the Sullivan County Hotel Association to propose a countywide ragweed-spraying program. Although the county's resort owners relied primarily on family traditions and celebrity entertainers to bring summer visitors, they recognized that ragweed eradication could help retain and attract New Yorkers. Advocates of ragweed spraying contended that pollen cost the "hotel and restaurant industries considerable business," while a ragweed-free environment could "attract" some, or more, of the estimated eight hundred thousand metropolitan New Yorkers who suffered from hay fever.[51]

In September 1953, McMahon Brothers, a rights-of-way maintenance contractor and distributor of Dow Chemical products, met with the county's Board of Supervisors. After a June 1954 meeting of Raymond McMahon and Robert McMahon, the supervisors, and an executive from the Sullivan County Hotel Association, county officials decided a ragweed elimination program was in "the best interest and health" of county residents and vacationers. McMahon Brothers submitted the only bid for the project, which would cost the county one dollar in 1954, twenty-five thousand dollars in 1955, and twenty-five thousand dollars in 1956. The supervisors approved the contract with the right to terminate the program after the second year. The McMahons were dedicated to eliminating the plants that inhabited "practically every foot of our roadsides everywhere" and that gave "practically every road in our land . . . a disagreeable, weedy appearance." Raymond McMahon claimed weeds were dangerous because they obscured "every curve, crossroad, driveway, culvert, guide rail, warning sign and marking post . . . [making] the highway . . . an obstacle course of well-concealed hazards." The company considered herbicide spraying to be low-cost roadside vegetation maintenance. Better public health was just "a bonus," a bonus that McMahon told highway officials "looking for the justification of chemical spray" they could use to their advantage.[52]

Sullivan County's ragweed elimination program was to be "the largest of its kind ever attempted." The Hay Fever Prevention Society declared 28 July 1954 "Ragweed Day . . . a godsend." Crews sprayed herbicide along both sides of more than seventeen hundred miles of state and county roads, and the state health department managed pollen-sampling stations. The Hotel Association encouraged its members to assist. Townships organized volunteer groups and Boy Scouts for community sprayings. The county established free 2,4-D distribution stations at town halls, stores, and gas stations for residents. After the first year, pollen counts reportedly fell by 60 to 70 percent. Although program monitors found that spraying was a "complete success" on roadsides, they reported it had been a failure around driveways, tennis courts, outhouses, barns, garages, and vacant lots because of "apathy at the community level." Louis Fucci thought the county was "still dirty with weeds." Nevertheless, the *Times* ran a story entitled, "Ragweed Reported Abolished in County."[53]

Although state officials claimed that such programs required "at least three consecutive years" to determine their effectiveness, county supervisors decided in 1956 to abandon the program before its completion. McMahon Brothers presented them with a bill for $33,333.33, but the supervisors formally rejected it in 1957. They might have determined that continuing the service at a cost of $25,000 annually or more would be fiscally imprudent. The supervisors may also have doubted the program's efficacy after the *Liberty Register* and the *New York Times* printed plant ecologist Frank Egler's criticism that spraying would expose bare soil, "providing an even better seedbed for future crops of ragweed." Egler estimated that highly disturbed areas like roadsides could require fifty years of spraying to prevent ragweed growth. The ecologist compared "spraying ragweed to kill it" to "treating an external sore that is due to a fundamental systemic disease." To eradicate ragweed, Egler recommended improving the soil "to allow other plants to grow which in turn would crowd out the ragweed, and make future reinvasions impossible." Elsewhere in Sullivan County, Egler attacked the Rockland Light and Power Company's maintenance of a utility line that ran parallel to a state highway as "a flagrant violation of common sense and good taste" that demonstrated "either a blind spot, or else a delusion" about appropriate management techniques. Egler argued this spraying contributed to the county's ragweed problem, noting that "the ammate treatment of an earlier year has rootkilled the grass, and ragweed is beginning to grow luxuriantly in these places." Egler faulted DuPont, the manufacturer of ammate, and the utility for failing to establish a test area. The effect of herbicidal attacks on ragweed populations was disputed territory.[54]

The battles against ragweed in Sullivan County became a battleground over ecologically sound vegetation management when Egler decided to challenge the business of ragweed destruction. After World War II, Egler, who had earned a Ph.D. in ecology at Yale University in 1936, consulted on forestry practices and conducted experimental work in vegetation management for infrastructure sites such as radio transmitter towers. In 1950, Egler helped form the R/W Maintenance Corporation within the Osmose Company to develop a rights-of-way maintenance program for utility companies. In 1951, Egler established a research project at the Boy Scouts' Ten Mile River Camp in the Sullivan County town of Narrowsburg. After R/W Maintenance Corporation mismanaged a project for AT&T, Egler began distancing himself from the company in 1952, but he did not extricate himself completely until the spring of 1954. In these ventures and experiments, Egler worked to develop a method of removing specific plants to establish low-growing vegetation that would need only occasional attention. Dow scientist Lawrence Southwick respected Egler's work, but he found Egler's methods less practical for corporations' needs. He thought the variability in plant cover complicated establishing stable vegetation over large amounts of land. Like many scientists and engineers, Southwick considered grass-covered rights-of-way simple and economical to obtain and maintain. The failure of corporations and markets to develop sound rights-of-way maintenance practices compelled Egler to enter conservation politics to disseminate his ideas. Egler used his association with the American Museum of Natural History and various conservation organizations to advocate the need for conservation-minded herbicide use.

Egler was not opposed to herbicides, but he criticized how herbicides were typically used. In 1951, Egler had accepted an appointment to the American Museum of Natural History as a nonsalaried research associate. He established a fund to support his rights-of-way research with contributions from family; friends; organizations such as the National Audubon Society; businesses including Osmose, Beechwood Packing Company, Dunning Sand and Gravel Company; and herbicide manufacturer Du Pont. Egler's technique was a stark contrast to blanket spraying, which covered all rights-of-way plants with herbicides. Egler worried that if chemical manufacturers persuaded rights-of-way managers that blanket spraying was less expensive than mowing, they would have little incentive to develop less chemically intensive methods. Egler thought that the low costs of blanket spraying disregarded the costs of habitat destruction. In 1952, Egler asked the museum's Conservation Department head, Dick Pough, if they should "ring a gong of alarm." Egler tried to convince industry scientists that although herbicides did not poison animals, improper use of the chemicals did destroy their habitats. He drafted manuscripts for potential publication in outdoor magazines and sent them to chemical manufacturers in hopes they might curtail commercial herbicide sprayers' destruction of game habitat, which was "infinitely worse than death to the animals themselves." In "Du Pont Desert," Egler charged that the company's attempts to persuade outdoor groups that herbicides benefited animals were made possible by ignoring the basic facts of game management. He criticized Dow for refusing to consult experts in wildlife management and plant ecology. Dow's public relations director asked whether Egler's views represented the museum's official view. Although museum director Albert Parr was concerned by Egler's advocacy, he defended Egler in a letter explaining the museum's

Figure 3.5. Blanket spraying of vegetation on a telephone line right-of-way between Danbury and New Milford, Connecticut, 1949. UC1-0699, Archives & Special Collections at the Thomas J. Dodd Research Center, University of Connecticut Libraries.

commitment to conservation, the study of natural history, and Egler's reputation as an accomplished ecologist. Parr called for a meeting between company officials and the museum's conservationists to address their disagreements.[55]

Egler's criticism of ragweed elimination in Sullivan County renewed these tensions. At the 1955 Northeastern Weed Control Conference, Egler's antagonists attempted to embarrass him. NYOR coordinator Philip Gorlin asked Egler to speak out against a resolution in a public health session. When McMahon introduced a resolution to commend Sullivan County for its spraying program, an audience member asked if there had been any opposition to the program. McMahon noted that the county had failed to allocate funds to continue it after "someone from the 'American Society of Natural History' had objected." At this point, Egler spoke about the ecology of ragweed and inquired whether an individual who possessed "ecological awareness" had been involved in the program's design. McMahon named a Dow scientist. Although Egler reported that a large minority abstained from voting, only he voted against the resolution. McMahon went on to claim the resolution passed unanimously.[56]

The attack on Egler was not surprising, given the domination of the conference by agricultural scientists and chemical manufacturers. The organization's meetings advanced knowledge about killing weeds, rather than fostering scientific inquiry. Its members likely also disregarded the perspective of botanist Oren Durham—the scientist whose pollen counts inspired ragweed-control efforts—who at the 1955 conference warned that preventing hay fever by destroying ragweed was difficult since even "highly effective herbicides" were "no match for our present agricultural practices, for our neglect of waste land in urban, suburban, and rural areas, and for our thoroughly weed-seed-saturated soil." A much more reassuring message was delivered in 1959 by former New York City commissioner of health Israel Weinstein, who remained confident that "ragweed, and with it hay fever" could be "entirely eliminated from this part of the world" with carefully conceived spraying programs.[57]

Whether, as Egler wondered, the Northeastern Weed Control Conference truly advanced scientific knowledge, the McMahon Brothers had succeeded in using it to manufacture a professional consensus to undermine Egler's credibility. Robert McMahon complained to the museum's directors that Egler had upset their clients by telling them that their work was based on poor science. Pough defended Egler by pointing out that roadside ragweed control would not reduce hay fever allergies, given the plant's abundance on fallow fields. Pough also argued that indiscriminate blanket herbicide spraying annihilated spring violets, summer's orange butterfly weeds, and fall's wildflowers, ivies, and shrubs, all of which made New England roadsides beautiful. Parr, however, decided to avoid further conflict with corporations and their clients rather than perpetuate scientific controversy. In

February 1955, Pough asked Egler for his resignation from the museum because his advocacy conflicted with the museum's desire to be seen as objective. Soon after, the museum disbanded the Conservation Department as well.

The professional politics and political expedience that resulted in Egler's resignation did not end conflict over rights-of-way management practices. The environmental, economic, and ecological issues Egler had identified could not be resolved by either routine commercial practices or weakly conceived weed-science research. Agricultural and engineering authorities were developing a hybrid natural and technological system in which some plants did not belong. However, whether such changes were beneficial and acceptable depended not just on cost calculations, but on how Americans experienced these environments. While management techniques seemed to be an obscure technical domain, they were not just technical trade-offs arising from technical-professional backgrounds, but divergent ideas about how America would look and feel in the future.

As the array of herbicides proliferated, spraying contractors and highway officials experimented with weed killers, soil sterilizers, and growth regulators to control roadside vegetation with "chemical mowing." A landscape architect in Ohio's highway department announced that these chemicals "established a new front line . . . in our endless war against weeds." Roadside sprayers believed that blanket chemical spraying was superior to highway-mowing programs because herbicides killed broadleaf plants and lowered maintenance costs. Many sprayers used trucks with pumps that could deliver 25 gallons of herbicide per minute and apply up to 250 gallons of herbicide per acre. A few companies used airplanes or helicopters to release herbicides. The chemicals destroyed the leaves and sometimes killed the stems of the broadleaf plants that absorbed them. On roadsides, McMahon Brothers recommended three sprayings per season in the first three years, and after grasses dominated, two sprayings per season. Blanket sprayers were dissatisfied with their work when they damaged plants on private property or failed "to kill what is supposed to be killed." However much herbicide was necessary to achieve the latter was never excessive—presumably as long as it stayed within estimated costs—because the result was "one of universal acceptability—no weeds!" If turn-of-the-century Americans thought weeds did not belong in their cities, at midcentury, the McMahon Brothers and their allies were conceiving of a far greater metropolitan area in which weeds no longer belonged.[58]

Blanket sprayers sold vegetation destruction with their enthusiasm for conventional aesthetics. Highway managers held the aesthetic view that grassy roadsides were the best "from the standpoint of downright beauty." The McMahons specialized in converting "weed-infested" corridors of "dandelion, plantain, chicory . . . to turf like this . . . beautiful lawn." Raymond McMahon promised the highway manager "a billboard as long as his highway system. . . . By this 'billboard,' I mean endless grassy roadsides." Sprayers' celebration of the magnificence of grass was reinforced by simplistic and vague denunciations of weeds as enemies of grasses and people. Agricultural scientist Erhardt Sylwester viewed unkempt rights-of-way as "unproductive [land], abandoned to the ravages of weeds." Herbicides protected an investment that too often "weeds have stolen from us." Where and when "nature rushes back with its profusion of weeds and bushes to frustrate the man who will cut them," McMahon Brothers vowed to end the problem. Their solution, however, created new environmental problems. In addition to the destruction of game habitat, blanket spraying was often indiscriminate. Operators not only aimed the chemicals along the roadsides but also shot them twenty to twenty-five feet beyond the shoulder, producing masses of brown, dead vegetation.[59]

While blanket-spraying operators used herbicides to produce grass, Egler employed selective basal herbicide spraying to produce dense plant communities. Selective herbicide spraying to "*root*kill the unwanted plants" could preserve slow-growing plants. Egler believed that over the long term "the cheapest control for many plant weeds is—not chemicals,—but *other plants*." Per gallon, Egler's solution used more herbicide than the solution utilized by foliage sprayers. Egler thought the problem with rights-of-way management was that aggressive vegetation destruction created ecological dynamics that favored unwanted plants. On roadsides, Egler contended that "the ragweed problem is essentially a site problem. The occurrence of ragweed is merely a 'symptom' of the 'disease.' If the *site-disease is eliminated, the ragweed-symptom disappears*."[60]

Although both Egler and the McMahons relied on herbicides to manage rights-of-way, they employed plants differently to demonstrate the capabilities of the technology. Egler cultivated vegetation to show that targeted herbicide applications reduced costs. Blanket sprayers used dead herbs and new grasses to demonstrate the power of chemicals. Egler saw landscape destruction by fervent "herbicidists" as efforts to manufacture "spectacular results, sufficient to impress the most lethargic executives." Egler also worried that blanket sprayers' worship of grass turned all other plants into "weeds," including brush, which Egler noted had once simply referred to a mixture of shrubs and young trees but had "become a derogatory term, comparable to weed." The McMahons' redefinition of all plants on the roadside except grass as weeds was almost Emersonian entrepreneurship. They discovered ragweed's irritating persistence was a virtue; the plant's biology made it a perennial crop that returned year after year to generate profits by being destroyed. Dangerous or useless weeds became valuable when they were objects that could be profitably destroyed.[61]

Whether covered with grasses or with changing vegetation, the roadside environment colored Americans' journeys from cities to resplendent natures. Automobiles facilitated and complicated the development of Americans' environmental values. Dozens, hundreds, or thousands of miles distanced nature-seeking metropolitan denizens from state and national parks. Although cars carried people to these wonderful places, they also transported noise and pollution into forests and mountains. As conflicts over vegetation management demonstrated, accommodating the vehicles also altered travel corridors connecting cities, the countryside, and the in-between. At some point within the right-of-way, the natural and built environments converged. Egler called the plants in this transitional space "seminatural roadside vegetation." He contended that this landscape was "seen by more people more often in the USA than any other type of land." Egler realized that the obliteration of the nature that existed in between wild spaces and urbanized spaces altered the experience of the distance between the places people were leaving from and heading to. Taken to the extreme in the New York metropolitan area, although Manhattan's concrete and steel and Catskill State Park's forests were separated by factories, suburban capes and ranches, and old barns, efficiency-oriented rights-of-way management would produce uniform vegetation for the length of the entire trip. Through the windshield of a speeding car, a right-of-way became a green stripe that did not change even as the traveler moved through time and space.[62]

Egler's challenge to indiscriminate herbicide spraying also challenged where the appreciation and protection of nature began. Unlike national park administrators and landscapers who labored to provide automobile travelers with a "windshield wilderness" of magnificent vistas, Egler conceived of bringing attention to simple natural glories on the county and state roads that Americans most frequently traveled. Scenic splendor did not have to be relegated to hidden corners of national parks. Egler believed that just as the public chose superhighways over country lanes, the public could decide "whether it wants grass, shrubs, or cockleshells along the roadsides." The technocratic preference for grass as the lowest-cost soil cover risked further narrowing Americans' perceptions of desirable vegetation. Pough thought that Egler was struggling to limit how completely the "drab sameness" of modern life reworked Americans' environments. Although Pough did not oppose "modern efficiency," he thought it could not improve "our birthright of roadside beauty." Pough suggested that if "drab sameness" was what blighted city slums and mass-produced suburban homes had in common, chemically mowed vegetation spread it over an ever-greater area. Pough and Egler were among the ecologically informed and aesthetically minded Americans who believed traces of the wild could and should be kept closer to cities and the countryside.[63]

While many New Yorkers traveling to Sullivan County may not have cared about what plants grew next to the highway, the rights-of-way that they passed were shaping Americans' environmental values. In *Silent Spring*, Rachel Carson alerted the nation to the environmental destruction wrought by profligate pesticide use, including blanket spraying. The outraged McMahons called Carson's words "harsh, uncharitable, and untrue" and complained that her source was "one of our frustrated competitors." They challenged the terminally ill Carson and Egler to have the divergent methods used on Delaware highways evaluated by federal agencies and studied by private-sector accountants. Carson ignored the letter and refused to indulge the firm and its industrial backers in a public-relations ploy. When the *Hartford Times* interviewed Egler about the matter, he trivialized the challenge by claiming that history and ecology had already proved Carson right. He informed the reporter that new projects were not needed; scientific evidence already existed in sprayed areas that permitted "going backward in time." Egler devised a counterchallenge in which the McMahons would permit their operations to be scrutinized by scientists and the results would be turned over to the Ecological Society of America and the President's Science Advisory Committee. While the brothers admitted that they were businessmen and not scientists, they grasped for the mantle of scientific authority. They attacked Egler and Carson as Luddites. Egler and Carson were responsible for "fear, aggravated by gross exaggerations and partial reporting. . . . In this new witchcraft, science has become the witch and the witch doctors have supplanted the progressing doctors." According to the McMahons, environmentalists were harbingers of the destruction of civilization.[64]

The managers remaking rights-of-way did not necessarily make these spaces ecologically uniform. Variations in budget, staff, and equipment, as well as myriad environmental factors, ensured that one environment differed from another. While these "herbicidists" did not achieve the production of self-sustaining grasslands, they did continue to dominate policy making in highway departments and to intimidate those who entertained incompatible views of roadside vegetation management. In 1975, landscape enthusiast May Watts, who had studied with Henry Cowles at the University of Chicago, described interstate highway medians as "vacant" and characterless. When she asked a conservation magazine about advocating a "'Strip of Reality'" that would not be mowed or sprayed so that Americans driving east to west would see pitch pines on the Atlantic turn "to oak woods, to mixed-mesophytic forest, to beech-maple forest, to tallgrass prairie, to shortgrass prairie, to plains, to sagebrush, to mountain evergreens, to redwoods," magazine editors decided not to print her piece for fear of having "the highway men at our necks." To those who would regulate ecological dynamics with chemicals, progress involved fully extracting the promise of old investments rather than using new scientific knowledge to shape the future.[65]

In 1956, the New York City Department of Health scaled back NYOR to evaluate how the program influenced pollen counts and hay fever morbidity. The department concentrated ragweed control in the Bronx and suspended spraying elsewhere in order to permit comparison of pollen counts, ragweed hay fever incidence, and sufferers' symptoms in areas sprayed and not sprayed. Complete ragweed spraying in the Bronx did not result in consistently lower pollen counts there, compared to other boroughs without ragweed control. In addition, Bronx ragweed hay fever sufferers did not feel better or have fewer symptoms than in the previous years or in comparison with ragweed hay fever sufferers in other boroughs. Officials also found that nine thousand acres of the city's ten thousand acres of ragweed remained. In 1958, the health department withdrew without victory.[66]

The years spent battling ragweed to advance public health put the health department in an awkward position in subsequent years, when the plant was seen trespassing in the city. A Queens physician described the ragweed abutting Forest Hills High School as "more grandiose and terrifying than ever," and she volunteered to clear the weeds from "no-man's lands." Although officials in New York and elsewhere once cultivated such public-spiritedness, they advised her not to proceed without a property owner's permission. Hay Fever Prevention Society members continued to debase ragweed as subnatural, protesting in 1966 that "ragweed is litter of the dirtiest kind." Upset by the happenstance plants that grew year after year in their neighborhoods, a group of Bronx residents complained, "Where ever we go, the weeds are still growing like weeds." The department assured concerned New Yorkers that "it continues to be the prerogative of a citizen to submit a complaint." When officials encountered properties suffering serious neglect, pest control bureau staff were dispatched to abate the nuisance. Like turn-of-the-twentieth-century cities, New York kept the authority to prohibit weeds but lacked the resources to eliminate them.[67]

From the 1920s through the 1960s, enemies of ragweed hunted and blasted the plant from Corona to the Catskills. Weed warriors killed many plants, but ragweed hay fever sufferers continued to sneeze, and ragweed continued to grow. Attempts to remake metropolitan nature after World War II relied on using chemicals created from scientific research but ignoring scientific knowledge, whether applying ecology or rigorously evaluating efforts. At best, ragweed warriors reduced plant populations that seemed to interfere with urbanization. At worst, they wasted money, destroyed colorful vegetation, and narrowed perceptions of the plants that belonged in cities.

The ecologies of New York City's metropolitan area were neither extremely malleable nor easily reworked. Commissioner Israel Weinstein believed that NYOR would succeed because "the greatest concentration of this noxious weed" was in "urban and suburban communities. . . . Ragweed growth is related to land use. It is usually the first plant to grow on freshly disturbed or abandoned lots." The powerful chemicals used in NYOR unfortunately kept such land suspended on the verge of abandonment, rather than allowing a succession of plants to replace ragweed. The multiple fronts on which New Yorkers battled ragweed—

and thought ragweed should be battled—demonstrated ragweed's metropolitan geography. Just plants remained and continued to evolve with a changing New York City and its changing region. Enormous metropolitan areas divided into ever-smaller pieces of land—home to shifting populations and subject to various land uses—produced ecological changes that could be countered but could not be controlled. Confusion over the nature of midcentury growth—varying kinds and densities of people across an urbanizing metropolis—could make happenstance plants that had been there for many years seem to be agents of decline.[68]

In working to influence the management of rights-of-way, ecologist Frank Egler recognized that the pressures on rights-of-way land reflected this metropolitan growth. The expansion and management of rights-of-way networks were among the ways that cities transformed landscapes beyond their borders. Destruction of "seminatural" vegetation on these corridors produced increasingly uniform vegetation. Egler wrote, "The cities—from slums, through the bedroom towns to the exurbanite areas—are in the process of exploding and spattering themselves over the open lands. It might be said that open lands exist *for* the cities." Egler considered slum, suburb, and exurb—places that sociologists distinguished with demographics, geographers distinguished with measurements of density, and ecologists distinguished with specific site impacts such as the differential effects of sewers and septic tanks—all to be parts of expanding metropolitan entities. Indeed, aggressive herbicide sprayers such as NYOR warriors and the McMahon Brothers were attempting to turn New York's vacant lots and Sullivan County roadsides into similar spaces. While Egler, like Shelford and other Chicago-trained ecologists such as William Skinner Cooper, with whom Egler briefly studied, doubted whether what he saw as ever less diverse metropolitan vegetation retained enough wonder to be worth seeing, other Americans imagined even the most homogenous places as evidence of nature's potential to inspire.[69]

CHAPTER FOUR

Weed Capitals of the World

In his 1938 essay "Conservation Esthetic," forester and wildlife ecologist Aldo Leopold wrote, "The weeds in the city lot convey the same lesson as the redwoods." Ordinary herbs and stunning conifers alike revealed the beauty of evolution and ecology. Advanced academic training was unnecessary to sense this wonder. To Leopold, perceiving and appreciating evolution and ecology were the purest and most sustainable form of recreation—"infinitely" renewable and divisible. Leopold's observation that the intricacies of glorified distant places also existed inside of cities challenged the assumption that seeking inspiring nature required escaping from everyday environments. The pressure that Americans and their cars were putting on wild animals, wildflowers, and wild places may have driven Leopold to annihilate the distance separating ordinary people from splendid nature. This recreational activity was not much different from the tramping of naturalists and scientists on the edges of cities in decades past. However, Leopold hoped that repurposing such outings and expanding their appeal might help protect the natural world from what his fellow wilderness advocate Benton Mackaye called the "metropolitan invasion" of areas beyond cities.[1]

While Leopold contemplated the uses of ecological thinking, during World

Figure 4.1. Pittsburgh Courier illustrator Wilbert Holloway drew these "ruins of a great metropolis" in the wake of Detroit's 1943 race riots. Courtesy of the *Pittsburgh Courier.*

War II, African Americans questioned whether cities could evolve rapidly enough to accommodate all Americans who sought space in them. They employed weeds to symbolize the racial animosity undermining urban life and the dangers of failing to overcome this racial animosity. When violence erupted in Detroit in 1943 over housing shortages and employment discrimination in factories, a *Chicago Defender* cartoonist drew Uncle Sam looking into his victory garden—the cities producing the people, machines, and materials to win the war—where "race riots" and "discrimination" had become wild weeds that he "should have stopped . . . before they got along that far." The riots led *Pittsburgh Courier* writers to ask whether American cities, like once great ancient cities, would soon deteriorate into "weed-grown temples" inhabited by goats. They warned readers that "SELFISHNESS, greed, injustice, stupidity, social imbalance, declining statesmanship, class and race prejudice, moral decay . . . RUIN cities and empires, returning them to the weeds, the dust and the reptiles."[2]

When both Leopold and the *Courier*'s writers thought about the futures of American cities, they measured how great cities' distance from nature should be with urban herbs. Leopold hoped that the wonders of nature could be so close that cars were not required to experience them. The *Courier* contended that cities infiltrated with wild nature might soon disappear. Taken together, their concerns identified dynamics that were creating metropolitan environments that seemed to be "giant urban wildernesses" to historian Sam Bass Warner. In the second half of the twentieth century, the array of rates of and extents of urbanization created more spaces for happenstance plants to color cities from coast to coast. Fortuitous flora emerged on the edges of cities, whether those edges were near wild places or near downtowns. The geographic expansion and fragmentation of metropolitan America coincided with ongoing social and cultural changes that sustained many Americans' fears of weeds as agents of disturbance and degradation that compromised safety and damaged refined spaces. The conventional perception that troubled cities were overrun with disorderly weeds, which prevented most city dwellers from discovering the virtues of the plants, coexisted with the common declaration that cities were devoid of nature. Some city dwellers hoped that metropolitan photosynthesizers could be used to improve urban environments. Nevertheless, municipal officials continued to view weeds as nuisances and to use police power to eliminate them. Their weed ordinances remained inadequate tools to address the challenges of persistent, ubiquitous ecological time.[3]

Americans' problems with weeds diversified and diverged after World War II as Americans dispersed into sprawling metropolitan areas. Nostalgia for tradition and simpler times, Cold War fears, and environmental amenities led some people to seek homes on larger lots outside of the largest cities. New housing in smaller municipalities became increasingly available with technological changes in transportation and construction, while aging buildings and infrastructure in large cities made existing homes less attractive. Government housing and education policies complicated city life, while immigration, racism, violence, and crime seemed to make it more dangerous. Cities' municipal finances weakened and the services they provided suffered when their populations became poorer, corporations relocated, and industries shut down. Although many of these problems had emerged by the 1930s, they were used to define postwar cities. Social unrest during the 1960s increased awareness of these problems, and cycles of blight and population shifts produced "extensive areas of dereliction" over and over again. Postwar metropolitan growth also created marvelous new residential subdivisions and small, exclusive municipalities. California Tomorrow founders Alfred Heller and Samuel Wood, however, regarded some newly built areas as "slurbs—sloppy, sleazy, slovenly, slipshod semi-cities." When Americans discussed weeds to emphasize the diverging circumstances and futures of places around a region and of cities in different regions, they often obscured different places' ecological resemblances. Deteriorating inner-city areas and expanding urban fringes were not necessarily disconnected competitors, but two edges of their evolving regions. For example, when in the spring of 1966 weeds in the St. Louis area grew "faster, higher, and tougher than ever before," officials in the City of St. Louis and surrounding St. Louis County alike struggled to assert control over the "jungle-like weeds."[4]

Urban decentralization, disinvestment in older urban areas, and speculation well beyond these areas allowed fortuitous flora to flourish. Creating new built environments brought more land into urban real estate markets, yielded leftover parcels that waited to be developed, and redirected investment away from older areas. Geographers, economists, planners, and conservationists, among others, described postwar urban growth as sprawl—the spread of random, discontinuous, "almost strictly urban" settlements throughout a region. Analysts and experts found weeds throughout sprawling metropolitan areas. Small parcels evolved into "weed infested, bottle-strewn lots" in both longtime slums and "the best residential neighborhoods." The "wild expansion or 'explosion' of the cities" produced "miles of sidewalks running through weed-covered vacant land." While speculators waited for land buyers, ecological time transformed pastures into "field[s] of weeds" and covered hillsides with "second growth and poison ivy." Looking down on cities from airplanes, Americans saw "large patches of . . . weeds or weedy, barely used structures." Homes on large lots designed to exclude undesirable people from exclusive subdivisions had peculiar costs. A commission observed, "A homeowner can't do much more with a one-acre lot than he does with a half-acre except to spend more time mowing grass and pulling weeds." To delay ecological time, such home owners also spent more money on fuel for mowers and pesticides. In 1966, a little more than thirty years before the *Atlanta Journal-Constitution* opined, "Urban Atlanta is spreading like kudzu," U.S. Department of Housing and Urban Development secretary Robert Weaver stated, "Our metropolitan areas have grown more like weeds than well-tended land."[5]

Many of the nation's rapid-growing metropolises were in Southern and Western states. Through combinations of annexation, migration, and immigration, over time some Sunbelt cities surpassed Eastern and Midwestern cities in terms of land area and total population. Whether these cities experienced moderate, steady, or rapid growth, they remained as vulnerable to the whims and winds of ecological time as cities east of the Mississippi River. Before their most intense periods of growth, these cities had expanses of happenstance plants. *Santa Fe New Mexican* editors attempted to combat "unrestrained and uncurtailed weed production" in 1923 by proposing that the Russian thistle growing throughout the city was "a fortune waiting . . . for the right man," who could harvest and market it as stock feed. Despite San Antonio's 1929 weed law that made the "common vacant lot variety of weeds" as illegal as "the marihuana weed" on any property that was not a homestead, one of the city's block-size "mosquito weed pastures" led a neighborhood to form a search party to rescue a lost ten-year-old. Pollen-filled air ruined the concentration of sneezing students and worried officials that tourists would avoid San Antonio. "Tall weeds" on corner lots increasingly created dangerous intersections as more people traveled in cars. Houston's intermittent and casually designed weed

Figure 4.2. Two young women stroll through a "jungle" in southeastern San Antonio, Texas, 1949. San Antonio Light Collection, UTSA's Institute of Texas Cultures.

destruction operations used horse-drawn mowers to clear dense vegetation, sometimes with tragic results. In 1940, officials dispatched a mower to Ralston Street, in an area three miles northeast of downtown where there were few homes and no sidewalks. A young girl who was either scared by the horse-drawn mower or walking on a path made through the plants was hit by the machine. She survived, but both of her legs were amputated beneath her knees.[6]

As postwar San Antonio, Houston, and Phoenix, and their neighbors, grew larger, fortuitous flora abounded in them. Four miles southeast of downtown San Antonio, three-year-olds could get "lost" in "forest[s] . . . of weeds and Johnson grass." San Antonio health officials mailed more than three thousand letters to owners of properties with "overgrown weeds" in 1958. After city officials decided that mosquitoes did not reproduce in weeds and that they did not want to cut "everybody's vacant lot in town," they concentrated on removing weeds that created blind intersections. However, a failure to perform the work because "the ground was too wet" led housewives to chop down weeds themselves. Even as "a hundred thousand tons of air-conditioning" helped Houston become a "white-towered megalopolis of the world of tomorrow" with "asphalt prairies" and "concrete canyons," the city's center and fringes provided homes to happenstance plants. Newspaperman Sig Byrd described life in an "arrogant slum" in a "weedy bluff" a few blocks north from city hall across the bayou. Walking down 75th Street toward the ship canal, Byrd passed "waterfront dives, honkytonks, [and] vacant lots grown into ragweed jungles." Seven miles south of downtown, Byrd visited a woman made rich by oil-lease royalties but who lived in an old house surrounded by a "thicket of brush," "untended rosebushes," and "weedy petunias growing out of rubbish heaps." Ordinary vegetation surrounded the tens of thousands of Americans who migrated to Arizona's metropolitan areas. In 1961, Mesa's business leaders encouraged owners of vacant properties to control weeds because such "unsightly" lots were "definite fire and health hazard[s]." Chandler was among the first municipalities to take advantage of a state law permitting municipalities to assess the costs of destroying weeds against properties. In Tucson, rolling tumbleweeds sometimes caused nighttime car accidents. City officials issued burn permits and provided "sprinkling equipment" so residents could safely destroy vegetation. Urban herbs persisted in smaller cities, too. In 1959, the *Santa Fe New Mexican* criticized the city for failing to enforce the weed law. The paper, which regularly reminded its readers of the importance of tourism in the local economy, thought that "rascally weeds" gave "the City Different an atmosphere of genuine messiness." After outlining a strategy to eliminate weeds, they concluded, "There's no reason why Santa Fe must hold the title of Weed Capital of the World." When postatomic weeds continued "to obliterate the charm of Santa Fe," they complained, "you can't see Santa Fe for the weeds."[7]

While metropolitan photosynthesizers remained in Sunbelt cities being built up, they also inhabited Eastern and Midwestern cities where land had never been built on, where buildings had been razed, and where languishing renewal efforts left areas waiting to be rebuilt. In the early 1950s, parts of Chicago consisted of "cracked sidewalks . . . unimproved street right-of-way overgrown with weeds, and unused utility poles." Although in 1963 almost 75 percent of Chicago's vacant land consisted of parcels at least two acres in size, nearly two-thirds of the city's 12,100 vacant lots were less than one-fifth of an acre. The removal of deteriorated housing added to the number of the small lots, although there were also many properties that had long been vacant and home to urban herbs. The *Chicago Defender* criticized renewal as a "scorched earth policy" that left "a residue of blocks of weeds in gaping empty lots." In Pittsburgh, houses razed for a university's expansion project became a "reefer 'garden'" hidden by vigorous vegetation that the city had not "chopped down." While the renewal process often upset a neighborhood's inhabitants who had to relocate, distant critics were unimpressed with completed projects. The mayor of Mercedes, Texas, a small city near Mexico and far from the nation's largest cities, disparaged this spending as "the planting of rosebushes in a weedpatch." When Buffalo approved redevelopment plans for 160 acres but then delayed construction for thirteen years, ecological time turned an "overgrown lot" into a field of trees. Although many areas of abandoned Detroit became covered in prairie grasses, one area was called "ragweed acres." Frustration with slow-going renewal projects was exacerbated by absentee landlords who failed to maintain their properties and subcontracted their destruction to arsonists. As fire-ravaged buildings fell apart, happenstance plants grew up around them. In New York City, "city-owned wastelands" populated only by "ragweed and the occasional ailanthus" emerged when contractors "too lazy" to drive "to legal landfills" dumped debris on demolition sites.[8]

Cities and their inhabitants worked to reverse or suspend ecological time on public spaces, undeveloped lots, and suddenly vacant parcels, but they could not maintain all of these spaces. When Cleveland closed a bridge damaged by a freighter and delayed repairing it, happenstance plants began growing on the bridge deck. New York City sanitation and education officials collaborated with the Citizens Committee to Keep New York Clean to tidy up city-owned vacant lots and convert them into small gardens tended by schoolchildren. The program managers hoped to inspire city residents to buy and landscape vacant lots rather than ignore them. In Pittsburgh, a block club used "rakes, picks, shovels, brooms and . . . eager little hands" to remove a "jungle-like growth of weeds" and a "mass of debris" from a parcel before covering it with gravel and building a playground. In 1965, Chicago's Department of Streets and Sanitation was able to "refurbish about 30 per cent of the vacant areas in the city most in need of cleaning" by supervising and assisting 686 "underprivileged" men in the Neighborhood Youth Corps, who "cleaned, raked, and cut back weeds in more than 1,000 vacant lots." Like ragweed elimination programs in previous decades, such efforts helped young men become responsible for maintaining urban environments, but they did not create ecological conditions that would inhibit the return of rapid-growing vegetation. Weeds helped turn Morningside Park, on the eastern edge of Columbia University's campus, into an "abyss of decay" and a "raw no-man's land between a ghetto and affluence." The *New York Times* questioned the city's reduction of its parks maintenance force in the early 1970s because so "many of the city's islands of greenery [were] facing serious deterioration." Without adequate budgets and without continual labor, cities could not banish motley vegetation from these spaces. When cities were able to regularly clear land, vegetation never matured. When and where mowing was irregular, happenstance plants changed with ecological time.[9]

While weeds were indicators that cities and businesses were neglecting and disinvesting in African American neighborhoods, weeds also became pernicious symbols of the inequalities facing black residents and the deprivations of the poor. In 1963, Charles Gaines, chair of the Chatham Avalon Park Community Council in Chicago, observed that white property owners often were awarded zoning variances to construct multiunit apartment buildings in black neighborhoods

Figure 4.3. Watts, California, 1967. Los Angeles Times Photographic Archive, Department of Special Collections, Charles E. Young Research Library, UCLA.

consisting primarily of single-family homes and duplexes. Gaines contended that when such neighborhoods later became overcrowded and seemingly blighted, the city reduced services such as "clearing vacant lots of weeds." Two summers after the 1965 riots in Watts, California, a business reporter wrote that "a vacant lot filled with weeds and a sign" were all that remained of a chain supermarket that would not be rebuilt. In 1968, Sterling Tucker, executive director of the Washington, D.C., Urban League, described the ghetto environment as houses and tenement buildings surrounded by "cold, grey concrete." Tucker pointed out there were no trees or grass because they were not budgeted for by authorities and poor residents lacked the income to procure these luxuries. "Straggly weeds" were the only green things Tucker saw. Yet in some areas, there were green things among the "ghetto grass" that some city dwellers gathered for their meals. Into the early 1960s, just five miles northwest of Chicago's Loop on the south edge of the North Center community area, Greek immigrants and Mississippi-born African Americans plucked dandelion greens in their neighborhoods.[10]

Removing the metaphorical weeds from and pointing out the actual weeds surrounding public housing units were among the ways Americans discussed the problems of such places. While social workers believed public housing helped preserve families who struggled to afford shelter and could help them move out of poverty, they debated how to prevent unruly inhabitants from harming and repelling families in need, as well as from undermining public support for such facilities. James Fuerst, who worked at the Chicago Housing Authority before joining Loyola University's faculty, argued that "weeding out the worst families"—those unable to "adjust satisfactorily to project living"—was necessary for improving the social environments of housing projects. Richard Scobie, a Unitarian Universalist reformer, criticized Fuerst for not being straightforward enough about "whom he plans to weed out." Scobie argued that reformers who were "determined to 'weed out the garden'" needed sound methods for identifying problematic people if they were to be denied housing. Americans who criticized public housing to sustain their prejudices against people who inhabited it were unprepared for the consequences of its demise. Marginalizing public housing tenants exacerbated the difficulties of integrating them into urban life when and where city officials decided to empty buildings. Metropolitan photosynthesizers encircled projects that would eventually be razed. Happenstance plants took root between the edges of concrete pads around the Pruitt-Igoe complex in St. Louis. By the end of the 1970s, Boston's Columbia Point, a fifty-acre superblock of public housing, was a "dismal landscape of dirt and dying grass, cracked asphalt, invasive weeds and wind-blown litter."[11]

Although establishing and maintaining weed-free landscapes might have distinguished Sunbelt cities from Eastern and Midwestern cities with rundown environments, fortuitous flora persisted. San Antonio officials' tolerance of weed nuisances compelled one newspaper to use its Auntie Litter column to criticize such properties, such as a lot "totally overgrown with weeds and grass" on the northwestern edge of downtown. Latinos and African Americans formed Communities Organized for Public Services to press officials for environmental improvements on the south and west sides. In 1977, the group urged the city council to increase the fines for violations of the city's weed ordinance. When, in 1972, "a record crop of weeds" was becoming "head high" in and around Las Cruces, New Mexico, a resident recommended making the heraldic crest of Dona Ana County "Russian thistle and ragweed rampant on a field of three crosses." The *Albuquerque Tribune* reported in 1965 that "no segment of the city is without its share of the unsightly weeds." Weeds obstructed sidewalks both near to and far from downtown, and urban herbs repopulated city-vacated alleys. Phoenix's *Arizona Republic* used its Citysore Box Score to pressure property owners and city officials to clean up nuisance properties ranging from burned-out buildings, to "scenic and ethnic garbage piles, junk yards and rubbish heaps," to lots with weeds. In 1968, the *Republic* noted that "most complaints are about weeds." A weed-covered corner lot about two and a half miles north of downtown led writers to crack, "All this jungle needs is Tarzan." About two miles farther north was a home surrounded by "hip-high weeds"; to emphasize the fire hazard, writers commented, "If this was a national forest, Smokey would be worried." A city official explained that Phoenix was "so spread out" and had "so many vacant holes" that unsightly "Citysores" were "much more acute than in the Midwest." In Tucson, a man who claimed that the city's west side was "the weed capital" criticized both the city's poor enforcement of the weed law as well as calls for new parks in the El Rio neighborhood, where people did little to control "unsightly weeds." Uncontrolled vegetation seemed to permit illicit activity north of San Diego. In the mid-1980s, a stretch of Highway 101 in Leucadia was being frequented by prostitutes, "undocumented aliens . . . [who were] harassing people," and "transients . . . seen lounging among the weeds and trash" along nearby railroad tracks.[12]

Despite the shifting fortunes of cities across the country, weeds were an environmental problem that they all shared. In 1956, the street superintendent of Fort Worth, Texas, considered vacant lots with weeds among the possible "forerunners leading to the eventual deterioration of a clean and progressive neighborhood." For "the benefit of the city," his department mowed such land "without cost to the property owner" with an allocation from the city's general fund. While many cities may have liked to follow Fort Worth's example, most cities in the Sunbelt or beyond did not or could not afford to maintain vacant lots indefinitely. In 1973, a Texas newspaper columnist declared that Lubbock was the "Weed Capital of the World" because of the city's failure to mow weeds. By the mid-1980s, it was hardly changing Pocatello, Idaho, that was "fast becoming 'the weed capital of the world.'" In the city's commercial areas, happenstance plants poked through pavement cracks and curbs, while five-foot-high plants lined Yellowstone Avenue, which headed north away from downtown. When Pocatellans submitted eleven hundred complaints about weeds in the summer of 1985, the city manager admitted that weeds were an "unsightly mess" that did not "enhance the community." Regardless of the size, population, or age of a municipality, weeds irritated urban Americans across the nation.[13]

While ordinary vegetation often covered land where city dwellers anticipated development, metropolitan photosynthesizers also flourished on sites burdened with toxic chemicals and isolated from economic activity. After passage of the Comprehensive Environmental Response, Compensation, and Liability Act (Superfund), some property owners, including banks, litigated rather than remediated land or kept properties out of the market to prevent environmental assessments. In the Chicago-Joliet corridor, abandoned gas stations became "weed infested targets for vandalism and dumping." Urban herbs colored a fifty-eight-acre Superfund site just southeast of Houston off of the Gulf Coast freeway. A forty-acre railroad switching yard hemmed in by Los Angeles's freeways was a "weed-filled expanse, strewn with building debris and rusting metal and makeshift encampments for the homeless." Environmental Protection Agency administrator Carol Browner noted that many cities had brownfields in the form of "Old parking lots, cracked and choking with weeds. Chain link fences with signs warning, danger, keep out, toxic waste." The sight of fortuitous flora year after year led many municipal officials to question the expenses imposed by regulations regarding site remediation. Columbus officials were astonished when preliminary work to pave ground next to a maintenance garage exposed chemicals in "a patch of weeds and mud" that would cost $2 million to clean up. Columbus's mayor joined 114 mayors from forty-nine states to petition President Bill Clinton to review remediation rules. Yet vigorous vegetation did not always indicate a site's indefinite toxicity. Cleansing brownfields permitted plants to grow with little restraint as they extracted heavy metals from the soil. For example, India mustard pulled lead out of ground once used by a Trenton, New Jersey, battery manufacturer.[14]

Although weeds grew all over American metropolises all over the country, in the last decades of the century, many writers and analysts used weeds primarily as symbols of the most destitute urban environments and, perhaps, of urban vitality that had been irretrievably lost. The *Chicago Sun-Times* printed headlines such as "Group Weeds Out 'Slum' Image" and "Lawndale Tries to Weed Out Eyesores" for stories about North Lawndale's "Slum Busters," a group that controlled vegetation on vacant lots, boarded up buildings, and sought repairs of crumbling infrastructure. In Chicago's Englewood neighborhood, one reporter felt that "despair seems as invasive as the rag weed and elephant ears that push up through cracked concrete." Academics employed weeds as photographic evidence of "decimated blocks" and "once prosperous" parts of Chicago's East Garfield Park and Woodlawn. Urbanist James Kunstler described a section of Cleveland as a "weed-filled wasteland." When Camilo Vergara was photographing "the process of decline" in Baltimore, Camden, Chicago, Detroit, Gary, Los Angeles, Newark, and Philadelphia, he encountered a longtime Camden resident who pointed to "overgrown lots where some form of paradise had once existed." Apartment towers surrounded with plants cracking up asphalt compelled psychiatrist Mindy Fullilove to ask, "What is a prairie doing in the middle of the city?" To Fullilove, such landscapes were outcomes of a flawed social system that perpetuated "the existence of well-tended places side by side with neglected places." Many of these statements dramatized urban decline but also obscured how long such plants had been in cities and how many prairies there might be in cities had past Americans not so regularly resisted the workings of ecological time. As weeds, plants were evidence of troubling environmental changes and destruction; as longtime inhabitants of metropolitan areas, they belonged wherever they grew.[15]

When many postwar Americans remarked upon weeds around them, they identified environmental problems that could usually only be temporarily resolved by removing plants. However, since happenstance plants remained inhabitants throughout metropolitan areas, curious environmentalists sought better understandings of this vegetation and the ecological dynamics of cities. There were also city dwellers who genuinely appreciated fortuitous flora, some of whom looked for spaces to introduce these plants. Not all nature-loving city dwellers were enthralled with motley vegetation. By the last decades of the century, admirers of native plants labored to destroy hazardous transformer species in and around cities to protect nature throughout the nation. Studying, working with, and fighting plants were not the only ways to improve the nature of urban environments. Americans who feared that cities were perpetually troubled because too many immoral and dangerous people inhabited them sought to identify and punish human weeds. In contrast, reform-minded city dwellers discussed the plight of human weeds to address the conditions from which they emerged. Understanding and disseminating urban herbs as well as removing or reforming human weeds both appeared necessary to improve metropolitan America.

Midcentury St. Louis contained numerous places where botanist Edgar Anderson could discern ecological dynamics. Anderson, who taught at Washington University and was affiliated with the Missouri Botanical Garden, was uniquely qualified to assess the belonging of happenstance plants, given his interest in unraveling the "tangled history of weeds and man." He joked that his 1952 book *Plants, Man and Life* was not an "orderly history of weeds," since it did not progress through discussions of "Iron Age weeds," "Holy Roman weeds," and "Elizabethan weeds." Anderson was probably reluctant to write a Eurocentric history because much of his "detective" work focused on maize and took place in herbaria around the country and croplands in Mexico, Guatemala, and Honduras. *Plants, Man and Life* also lacked, as Anderson quipped, "a few final paragraphs about post-atomic weeds," because he was spending so much time snooping around St. Louis's vacant lots for more clues about them.[16]

St. Louis's evolution in the first half of the century had not transformed the urban landscape into a weedless one. Vitality measured in terms of population growth—575,238 people in 1900 and 856,796 people in 1950—was not mirrored by the city's physical condition. In the 1910s and 1920s, civic leaders became concerned with poor residents' deteriorating housing that surrounded downtown. While old structures fell apart, some city blocks had not been fully developed, and nearly 30 percent of the city's land was vacant. Isaac Lionberger, an attorney who lived in a fine Westmoreland Place home near the city's western edge, lamented the machine shops and vacant lots a couple of blocks away from his private street. A couple of miles to the south of Lionberger's home was the Cheltenham neighborhood, where manufacturers mined clay and baked it into bricks and sewer pipes. These operations produced smoke and dust and attracted poor laborers, all of which interfered with speculators' attempts to profit from their properties. During the Depression, geographer Lewis Thomas observed that the part of Cheltenham from which clay was extracted "lies idle, is covered with weeds, or is occupied by scattered squatters." By the 1940s, home construction in and business relocation to the edges of St. Louis and surrounding municipalities added to the worries of St. Louis's civic and business leaders, who were troubled by the persistent roughness of the city's central corridor, where they had invested much capital.[17]

St. Louis's myriad changing environments provided Anderson and his Washington University students with places to study plants he once labeled "peregrinators" because of their capacity to move or to spread widely. Anderson thought young men and women needed to become more knowledgeable about landscapes because automobiles were divorcing them from their surroundings. Although Anderson brought them to the Botanical Garden's arboretum thirty miles west of the city, gas rationing during World War II also influenced his efforts to find easily accessible spaces to study. Anderson focused his students' field research on documenting sunflowers' habitats in St. Louis. When they explored the residential areas between their campus and downtown, they traveled into neighborhoods that once had been inhabited by the elite, as well as neighborhoods of modest houses inhabited by people who had put tenement life behind them.[18]

Anderson's students reported that sunflowers, like henbit, clover, and bindweed, tended to occupy particular kinds of habitats. His 1941–42 field botany class found sunflowers growing in "cramped, hostile conditions." One sunflower was "jutting from a brick wall," while another was wedged in the gaps of a stone retaining wall. Sunflowers inhabited industrial spaces such as ash pits, coal yards, and railroad yards. The plants congregated near billboards and telephone poles. Another class described waste heaps and smoky rail yard air as their typical "humble" homes. This group hinted that sunflowers belonged in cities by offering a sharp juxtaposition: in country fields, the wild sunflower was as "rare as a neon sign." Students did not see sunflowers everywhere in St. Louis. A neighborhood near their university had "clean . . . stores" and a few vacant lots that had been planted with irises. An area built up with sophisticated commercial establishments and grand hotels featured many "neat green" lawns. In a neighborhood of single-family homes, kosher delis, and "second rate hotels," a student perceived that "every bit of space" was being used and saw well-kept grass. The neighborhood lacked the "raw earth" or "trash heaps" where sunflowers could flourish. Anderson's students ordered the "unkempt" spectrum of sunflowered spaces as industrial areas with the most sunflowers, to slum areas, to commercial areas, to middle-class and elite neighborhoods, where there were few, if any, sunflowers. Although they accurately reported that sunflowers were absent both from places where people obliterated all nature and from those where they controlled every green thing, they did not seem to notice this or explain the implications of the fact that intensely cultivated beautiful places devoid of sunflowers had something in common with the environments that St. Louisans so severely abused that "not even a sunflower can sprout."[19]

In her effort to comprehend the sunflower's ecology, Elois Fay tried to correlate areas' environmental conditions with children's manners. When Fay told young people about her search for sunflowers, some chatted inquisitively, some stared at her blankly, and one said that she was "crazy er somptin'." Fay reported an inverse relationship between sunflower populations and youths' dispositions: "In the dirty regions the sunflowers grow well, even if the small thugs do not. . . . The better sections display a few handsome sunflowers and many pleasant youngsters." Like the physician Robert Hessler's studies of Logansport, Indiana, Fay's work included impressions of plant populations and people's home environments. In a walk through "a small crowded region of tiny box-like frame houses [that were] dirty, unsubstantial, inadequate, and unfit for the number of persons living in them," Fay observed yards "allowed to go to weeds." When she met a friend in a neighborhood in which whites tried to segregate African Americans, Fay noticed slovenly conditions. She faulted slumlords' exploitation of renters for the latter's "I don't care" attitudes, arguing that if people owned their own homes, they "would weed out the [sun]flowers." While Fay did not consider such sunflower-filled neighborhoods to be ideal, they were more humane than one "terrible region" that seemed to be "too crowded for people" and where the few sunflowers she saw had taken root in building and sidewalk cracks.[20]

Anderson's students tried to link the quality of social space with the nature of the plants. One group of students claimed not only that sunflowers preferred areas with rubbish heaps but also that sunflowers did not "grow in better environments because the plant [disliked] competition." This analysis mistook human environmental manipulations for the interactions between plants. In places where the elite lived, cultivated plants did not overpower other plants; people eliminated their competitors. If the sunflower appeared throughout St. Louis—a city where the workings of nature were shaped by human competition—the students had apparently misjudged the plant's competitive nature. Where St. Louis was changing or not yet carefully controlled, sunflowers seemed perfectly adapted. Sunflowers were stronger than plants that could not withstand people's sudden, intense, or rough environmental transformations. To the extent that sunflowers' changes to their environments allowed other plants to grow, the plants that succeeded sunflowers did not outcompete them so much as partly owe their existence to them. While the presence or absence of sunflowers were results of changes over ecological time, Anderson's students essentialized the outcomes of changes that had occurred as evidence of the unchanging nature of organisms.[21]

Anderson shared the lessons that might be learned by studying sunflowers with everyone he could. He drove his out-of-town visitors around St. Louis to see the "slums and factories and lots of sunflowers." Year after year Anderson brought his botany classes to dump heaps and alleys because he thought the ecology of such spaces was "more rewarding for the time spent on it" than theorizing about the "grasslands in the Great Plains . . . which one only halfway understands." Anderson not only affirmed Leopold's proposition that fortuitous flora in cities conveyed the same lessons as the redwoods and prairies but also contended that plants in cities conveyed the lessons more clearly because in cities ecological dynamics and people's fingerprints were more easily detected. The lessons to be learned in revered, majestic environments were far from clear because people's influences were guessed at, ignored, or denied. In contrast, cities' happenstance plants allowed Anderson to wonder about nature and to speculate about how Americans needed to evolve. In 1956, Anderson wrote that learning about "weeds of all sorts" and nurturing "a fellow feeling for these organisms with which we live"—rather than eradicating them—was essential to reworking cities "into the kinds of communities in which a gregarious animal like man can be increasingly effective."[22]

Anderson's call to make postwar metropolitan communities more inclusive was complicated by troubled individuals needing to be reformed and the conflicts between people to be addressed. Physicians, humorists, and ministers employed the human weed metaphor to discuss these problems. In his widely syndicated health column, William Brady insisted that a proper breakfast consisted of "PLAIN WHEAT," fruit, "nitrogenous material, of animal origin," and, for adults, coffee. Brady asserted that too many Americans were breakfasting on sugar-laden cereals and sodas, and their consumption of "extraordinary amounts of fuel" gave them an "excess of flabby fat." The exasperated Brady complained that "human weeds" who ate this way had "nothing to do" and devoted "their lives to doing it." In a syndicated column that appeared in small cities across the country—Newport, Rhode Island; Beckley, West Virginia; Hope, Arkansas; Mason City, Iowa; Denton, Texas; Pocatello, Idaho; Bakersfield, California; and Fairbanks, Alaska—Hal Boyle explained what to do when faced with "a human weed and neither you or he can escape the fact he is a weed." Instead of becoming mired in contempt, Boyle humorously recommended planting a flower in a pot and naming it after the antagonist. Tucson's Council of Churches placed a prayer in a newspaper that declared that without Jesus, individuals' lives were "ugly, mean, poor and worthless." The prayer's reader asked Jesus to be employed in the ministry of "transforming human weeds into flowers and fruit." A humorist who authored "comic dictionary" entries defined a sponger as "a species of human weed that is always showing up where he is not wanted."[23]

As broad and as arbitrary as these bases for identifying postwar human weeds were, they were not inclusive enough to identify all human weeds. Americans who perceived other people as human weeds might not have been able to recognize what they possessed in common with those whom they disparaged. Some academic scientists made up for this inherent bias by proposing that all humans were human weeds. The concern of open-space environmentalists with the destruction of places where "nature predominates" prompted University of California–San Francisco tropical medicine professor J. Ralph Audy to examine why such places were disappearing. At the 1963 Sierra Club conference, Audy explained that many creatures, including people, searched for edges where they could access "two or more different kinds of habitat." People possessed a "preadaptation" to seek and spread into such environments. With time, edges became simplified ecosystems where "troublesome . . . introduced species"—weeds—tended to survive and dominate. According to Audy, degradation of edges and loss of open spaces continued because "man himself is now in nearly all places an animal weed." The implications of Audy's remarks in a period of metropolitan expansion was that some of the most "troublesome" weeds were the middle-class Americans migrating to the new edges of cities and the edges of new cities. Three years later, Oklahoma State University botanist Jan de Wet, a South African who earned his Ph.D. at the University of California–Berkeley, wrote that *Homo sapiens* had become "the ultimate of all weeds." These scientists' views that each and every human was a weed abandoned earlier distinctions between people as fine flowers and human weeds. Nevertheless, the well-established human weed metaphor remained an expression that Americans used to assert other people's uselessness or harmfulness.[24]

African Americans striving for civil rights and debating how to advance the Civil Rights Movement considered their obstacles to progress human weeds. At a 1961 Nation of Islam meeting in Los Angeles, Malcolm X expressed his belief in the need for revolutionary change in leadership of and by African Americans by warning black Christian ministers that "the time is coming when we will weed the world just like you weed a garden." After African American juveniles attacked white Brooklyn subway riders and Staten Island ferry commuters in June 1964, black leaders discussed the origins of such violence and how the violence injured the push for equality. The Congress of Racial Equality's James Farmer identified the root causes of the episode as segregated slums and unemployment. However, *Chicago Defender* editors maintained that even if a "sick American society" produced such criminals, the violence perpetrated by "social aberrations" could not be tolerated. The *Defender* urged black leaders to "put their hands on the plough and uproot the human weeds that impede the march of progress." In 1966, Martin Luther King Jr. lived in Chicago's Lawndale neighborhood and participated in the efforts of groups trying to break down the barriers that confined poor African Americans to ghettos and "substandard slums." Mobs assaulted marchers with bottles, bricks, and firecrackers. King wrote, "Swastikas bloomed in Chicago parks like misbegotten weeds.[25]

Americans compared cities with gardens to emphasize the care that neglected cities needed and that neglected city dwellers needed if orderly people were to be cultivated. The *Wall Street Journal*'s Mitchell Gordon asked whether it was possible to "sterilize urban ground against the seed that grows the slum." Arthur Goldberg, who served as President John F. Kennedy's secretary of labor, a Supreme Court justice, and a United Nations ambassador, contended that the "deep-rooted evils" of poverty and racial discrimination warped some juveniles into criminals. He argued that filling prison cells with "poor, semi-literate youth" did not change the fact that "in the back lots of city slums, we are reaping the weeds that have germinated during generations of complacent neglect." When the House Committee on Un-American Activities examined cities for the influence of "subversive organizations" in 1967, knockout king and former light-heavyweight boxing champion Archie Moore appeared before the committee to denounce rioting. Since America had changed dramatically during his lifetime, he believed that African Americans should not mistake past injustices for current circumstances. Moore stated that the nation could become a "beautiful garden" if individuals did not "choke" it with "weeds of hate." The leader in "youth guidance" announced he was "ready to start 'Operation Gardener.'" However, University of Oklahoma education professor Franklin Parker insisted that ongoing, virulent discrimination persisted with the denial of "social, economic, and educational opportunities" to blacks. Parker saw season after season of reaping the "seeds of frustration" and letting "grow wild the weeds of hatred," culminating in "explosions as happened in . . . Watts." In Parker's view, if the nature of American society could not be changed, both weeds and human weeds would continue to proliferate in cities.[26]

The association of degraded physical environments and oppressive social environments indicated the persistence of moral environmentalist thought. In 1955, Bauhaus architect Ludwig Hilberseimer, who spent much of his career in Chicago, argued for the significance of urban space by writing that it was possible that "each man has within him seeds of evil and of good and that his environment may determine which seeds take root and grow." As at the turn of the century, Americans worried that weed-filled places attracted and sheltered malicious people who committed crimes. A geographer thought that cities' "delinquent vacant properties," when "grown over with weeds," constituted "hiding places for criminals and hazards to the community." A Brooklyn councilman complained that "fifteen blocks of vacant land . . . [with] extremely high weeds" in the Bath Beach vicinity made possible "attacks on girls and women," as well as "many muggings of men." In a neighborhood adjacent to Pittsburgh's downtown business district, "tall weeds on a hillside" permitted a man to attack a girl when she tried to take a shortcut to school.

In 1965, the widest missing-person hunt in Texas's history ended when a survey crew found two murdered sorority sisters in "a weed-grown vacant lot" on the edge of Austin. Sacramento's xenophobic conservationist Charles Goethe took advantage of Americans' worries about crime to argue that the worst criminals were immigrant "human weeds" such as the Italian American gangster Al Capone and the German-born spy Klaus Fuchs. Goethe, a self-proclaimed "garden-philosopher devoted to the elimination of human weeds" such as "vendors of narcotics" and gamblers, attributed the nation's booming "marihuana trade" to Mexicans. However, the geographic origins of such weeds may have been less significant than the care exercised by cultivators, especially cultivators of children. When the *Chicago Defender* informed readers how to prevent their children from using drugs, it advised parents to show their children "love and affection" and stressed the necessity of being vigilant because "there's nothing that pollutes a shrinking violet like the jimson weed in the next yard." Since criminals could operate anywhere, both inside large cities and on the edges of future metropolises, weeds remained associated with crime and violence.[27]

While a range of Americans used the human weed metaphor to express the disorder of cities, Isaac Rehert proposed the nature of the sunflower as a metaphor for the evolution of American cities. In 1969, Rehert excerpted dozens of passages of Edgar Anderson's *Plants, Man and Life* in the *Baltimore Sun*. Rehert adapted Anderson's statement that "two of the main races of sunflowers" were the "Great Plains *annuus*" and the "camp-follower *annuus*" as "there are two main races . . . the cultivated type and the city weed." While Anderson's "camp-follower" could be found in cities, it was not exclusively Rehert's "city weed" since it existed throughout metropolitan areas and beyond. Yet by fusing city with weed, Rehert was able to state that it was "appropriate" that a "uniquely native offspring" of the United States was able to "thrive so vigorously in the inhospitable concrete, asphalt, and cinders that a modern city offers it." Like Anderson's students, Rehert thought that city sunflowers were unable to handle "competition from other kinds of plants." The uncompetitive nature of the plants was an indicator of the degradation of their environments. Rehert concluded that the inhospitable cities that had emerged from American history had evolved into perfect environments for uncompetitive weeds. To the extent Rehert was thinking about people as well as (or rather than) plants, by distinguishing the geographically-unspecified "cultivated type" and city sunflowers as races, Rehert suggested that the ongoing segregation of people into different environments, as in Baltimore, was a natural outcome of American history. Yet Rehert's evolutionary metaphor faltered by imagining an unnaturally obvious distance between cultivated plants and weeds, as well as cultivated people and weeds. Anderson acknowledged that there was "an indefinite boundary between weeds and cultivated plants," and Charles Heiser, who had

Figure 4.4. Sunflowers in East St. Louis, Illinois. This undated photograph was likely taken by Charles Heiser sometime after he began teaching at Indiana University in 1949. © 1976 University of Oklahoma Press.

studied sunflowers with Anderson in St. Louis, wrote that weeds were plants that grew in "largely manmade" environments. Since weeds grew where people lived because of the environments people created, people and weeds could not be segregated. Any separation was an illusion of ecological and historical time.[28]

While Americans like Rehert thought weeds were evidence that cities were unlivable, other Americans thought appreciating happenstance plants would make cities more livable. Ecologically aware city dwellers wrote poetry, children's books, and plant guides that nurtured cities' connections with nature by portraying happenstance plants as impressive organisms striving to survive in hostile urban environments. They recognized that maintaining and improving environmental quality—whether in cities, rural areas, or wild places—depended in part on appreciating the nature in the places that most Americans live. In a Loren Eiseley poem celebrating the resistance of beggar's tick, foxtail, and ragweed to the dominion of industrial cities where "nothing" was planted or supposed to grow, the poem's narrator declared that the "undaunted" plants were more memorable than New York City's towers and lights. Schoolteacher Ruth Howell encouraged children to look at green leaves in pavement cracks so they could see "a quiet factory" making food from the air, water, and sun. After botanical illustrator Anne Ophelia Dowden's editor Elizabeth Riley suggested creating "a book about urban weeds" for "city children [who] would probably never see anything of nature except weed patches," Dowden walked around Manhattan looking for the plants around parking lots, the World Trade Center construction site, West Side railroad yards, and under the Brooklyn Bridge. A West Village demolition site was home to "head-high" mugwort, "tall hedges" of lamb's quarter, Mexican tea, and bladder campion. Dowden's exquisite paintings showed dandelion leaves obscuring a broken green glass bottle, short beggar's tick growing out a cracked sidewalk, clovers covering matches and rusty nails, sunflowers reaching above chain-link fences, and plantains towering over pull-top tabs. Dowden told an interviewer, "It's nice to think the natural world has that power of battling all the pollution and depravations that man has imposed on it." Maida Silverman, who described herself as an indigenous city person and a mediocre gardener, was awed by and grateful for the "wild" plants that were able to overcome her city's "sterile mantle of asphalt, concrete, and glass." Nancy Page and Richard Weaver published a wildflower guide that featured the plants that thrived in urban environments and questioned the accuracy and adequacy of labeling them as weeds. Although they recognized that land with metropolitan photosynthesizers was often neglected, they disapproved of the behavior of property owners rather than the activities of the plants. They believed that just about any plant that grew was preferable to bare rubble. Poet Amy Clampitt expressed delight that pokeweed grew from seeds that a "vagrant bird" deposited on a vacant lot. These writers treated urban herbs as full-fledged natural denizens of cities that were working to improve urban environments. In contrast to naturalistic landscapes and beautiful ornamental vegetation in venerable parks, such plants were a seemingly independent and fleeting wildness amid relentless efficiency.[29]

Resilient photosynthesizers were relied on and employed in formal, clandestine, and gradual efforts to change urban landscapes. In New York City, the Department of General Services deployed "Operation GreenThumb" workers to build berms around debris-strewn "city-owned wastelands" in Brooklyn and the Bronx and seed them with rye, fescue, clover, and fruit-bearing bushes. The *Times* applauded this "inexpensive way to heal a scarred urban landscape." Some New Yorkers acted as "Green Guerrillas" who both tended to community gardens and promoted wildflower growth. They harvested seeds from Queen Anne's lace, chicory, evening primrose, black-eyed Susan, butter-and-eggs, jewelweed, goldenrod, asters, moth mullein, cinquefoil, and feral petunias; poured them into glass Christmas balls; supplemented them with commercial seed mixes; added a little soil; and lobbed these "seed grenades" onto "derelict lots . . . to try to dress up ugly places that get dumped on." A Brooklynite deadheaded "inner city wild flower[s]" and pitched them in neglected front yards and used-car lots to seed the next summer's "streetside micromeadows" in her neighborhood. Because such sites were sometimes fenced off, dangerous, or in difficult to reach locations, the spaces could not be prepared or tended. The guerrillas relied on these plants' abilities to grow in "unpromising, even antagonistic circumstances." One guerrilla described the efforts as a "public gardening" program "ideally suited to this florally deprived urb." While these metropolitan photosynthesizers could bring more color into cities, encouraging their growth undermined the relationship delineated between people and plants in municipal weed ordinances and possibly created spaces that reinforced other city dwellers' perception of weeds as agents of dereliction. More traditional garden spaces, in contrast, sometimes attracted building rehabilitators. Chicago's "weed-whacking activist" Fredrika Lightfoot transformed six vacant lots by steadily pulling up a few "ugly weeds" each day but also letting "beautiful ones" like Queen Anne's lace, sunflowers, and wild petunias remain. She planted pink and purple asters around condemned buildings. Lightfoot believed these green spaces made her Englewood "neighborhood safer."[30]

Although weed-filled cities seemed to be biologically impoverished, they were not necessarily ecologically inferior to wealthier neighbors. Perceptions and measures of ecological health and of environmental quality in metropolitan areas were many and contradictory. What many Americans considered pleasant environments in which to live were not necessarily ecologically sound. To conservation biologists, metropolitan landscapes were vast urban spaces that contained few of the "vegetation communities" found in "natural areas[s]." For example, even though the city of Chicago's borders had been largely unchanged since the 1910s, scientists reported that urban land in the Chicago region had increased by almost 15 percent from 1972 to 1985. During this time, "unassociated vegetation"—plants that "do not occur together naturally, either historically or as associates in self-perpetuating communities"—increased. Happenstance plants on vacant lots were not any more "unassociated" than a property with buckthorn hedges, Japanese barberry along the side of the house, a white pine tree and a red maple tree on a large lawn, and beds of tulips and daylilies. Residential grounds could be ecologically incoherent. Home owners could plant sod and saplings the same day, while in landscapes where people exerted minimal influence, these plants could be separated by decades of ecological time. Grasses often predominated after severe disturbances, while trees took years to emerge. Indeed, manicured lawns repressed the processes by which grand trees emerged. Lawn grasses were perhaps the most popular and common species in urbanized environments. By the early twenty-first century, they covered an estimated 23 percent of the nation's urban space. While weed-free lawns were a source of accomplishment to some individuals, such spaces collectively reinforced disdain for wild plants. In the Akron, Ohio, metropolitan area, the most desirable residential areas were near "large extensively wooded lots," with an overstory canopy of black, red, and white oaks. The ailanthuses, silver maples, and white mulberries found all over the region—including in poor urban areas, if any trees grew there—were dismissed as "fast growing 'weed' trees." Taken together, the lawns and trees dominating valuable landscapes made lightly, slightly managed ground and urban herbs seem undesirable.[31]

Americans' conventional view that stable environments contained desirable vegetation and disturbed urban spaces were filled with weeds made it difficult for most city dwellers to notice when seemingly natural, pleasant open spaces were becoming ecological problems. Most of the animosity against happenstance plants arose from fears that weeds were evidence of the deterioration of urban life. However, in the last decades of the century, some of these plants were able not only to undermine social order but also to injure nature and undermine ecological order. Americans who envisioned purer future environments were alarmed by "unstable or disturbed" vacant tracts, urban parks, and metropolitan edges that allowed invasive plants to flourish and spread. They were concerned that as these plants moved along waterways, across ranges, onto farms, into forests, and alongside roads, they altered ecosystems and landscapes, killed native plants, and reduced biodiversity.[32] Some of these plants seemed capable of wrecking the infrastructure that facilitated urban life. Thick masses of *Arundo donax* became entangled, trapped debris flowing in streams and rivers, and put pressure on bridge supports. *Cyperus rotundus* became established in asphalt and concrete, accelerated road deterioration, and increased maintenance costs. A federal agency reported that Japanese knotweed rhizomes had "grown through 2 inches of asphalt!"[33]

Throughout the first six decades of the century, enthusiasm for exotic plants exceeded worries about them. For a decade or more after the end of World War II, newspaper garden advisors promoted such "tough plants" as capable of surviving the "smoke, fumes, and less natural light" common in "large populated areas." Chicagoans learned that purple loosestrife would "grow for anyone, virtually anywhere" and that buckthorn was an ideal bush to grow in smoky areas. The *Washington Post* informed readers that they could rely on tatarian honeysuckle and Chinese wisteria to beautify their residences. Exotic, novel, and beautiful ornamental plants were widely established in parks and gardens, providing them with future opportunities to spread across and beyond cities. The early complaints about these plants involved their ability to injure other cultivated plants and the labor necessary to control them. More strident denunciations of them came from landscape purists like Jens Jensen, who detested the spread of Japanese honeysuckle during the 1930s in the woodlands of Maryland. However, many enemies of exotic plants simply detested weeds in general. *Science News Letter* columnist Frank Thone advocated the destruction of any "such open and acknowledged pests as poison ivy, ragweed and Japanese honeysuckle." In 1939, Thone wrote that people's use and abuse of environments made room for "uninvited immigrants" and could encourage "native species . . . to become vegetable gangsters." Thone argued that "these aggressive foreigners, and natives gone to the bad" could occupy places once home to rare, beautiful plants and "plants of greatest interest and importance from the scientific point of view." Similar aspersions came from herbicide-spraying contractors like the McMahon Brothers. At a 1959 weed science conference, Thomas McMahon suggested that herbicide sprayers could justify their work with the claim that it protected native plants. In a presentation about ragweed—a native plant—he declared, "The disturbance is already here, the disturbance of foreign invaders. Our native plants are disappearing." McMahon attributed the "vigor of weeds from Europe, Asia and elsewhere" to their release of chemicals from roots that injured nearby plants. McMahon believed that "maybe the hope of the native flower, at least along roadsides, is, to some degree, a weed spray program." McMahon seemed to forget that conference participants had been working since at least 1950 to find ways to kill "perennial weeds" on roadsides, such as New England asters, violets, field horsetail, milkweed, woodland strawberries, ferns, goldenrods, and narrowleaf cattails—all of which were native plants.[34]

For much of the century, few scientists carefully studied exotic plants, especially in cities, where in the short term, the plants increased biodiversity, provided habitat for birds, attracted and fed pollinators, prevented erosion, and contributed other ecosystem services. Ecologist Frank Egler, by setting aside the "botanico-emotional extremism" of the "lovers of natives" and "those entranced by exotics," argued that with proper management, common neonatives—dandelions, lamb's quarter, Queen Anne's lace, red and white clovers, water hyacinth, ailanthus, and Japanese barberry; plants disparaged in the past as weeds or on their way to being condemned as invaders—could do wonders for the landscape. Egler thought it was unreasonable to "bar the daisy, simply because man, not God, put it in America." In St. Louis, the Missouri Botanical Garden's Viktor Muhlenbach found 393 synanthropes—"adventive plants not native to . . . Missouri"—in his fifteen years of studying the flora of freight yards, switching tracks, and trunk lines of railroads. Many of these plants were annuals that had arrived with railcars and remained rare. He considered about 22 percent of the species naturalized. Among the plants living in all or all but one section of the city were twenty-seven grasses, as well as white mulberry, burningbush, Japanese knotweed, multiflora rose, nodding plumeless thistle, and tall tumblemustard. Muhlenbach commented that Japanese honeysuckle, which he saw in every area except one section on the northern edge of downtown, "can hardly be described as a pernicious weed." He reported that three small groups of plumeless thistles had become twenty-four colonies, some "of them huge and having persisted for many years" and having spread to vacant lots. Muhlenbach remarked that multiflora rose was "quite widespread in St. Louis," which he attributed to the state conservation department's use of it "along rural fences and roadsides for . . . creating wildlife cover."[35]

Many Americans who prized native plants sought to protect them in bucolic and wilderness landscapes. When their zeal increased and spread into cities that were perceived as degraded places, plants that had been improving cities became plants marring urban environments. City dwellers committed to purifying and rehabilitating nature in cities began eliminating exotic plants suspected of damaging urban parks. In the late 1970s, a Central Park task force organized volunteer groups, often children from nearby schools, to remove exotic plants. Restorationists working in the District of Columbia's Dumbarton Oaks Park in the 1990s decided that although multiflora rose, Japanese honeysuckle, and oriental bittersweet had been part of the designers' original landscape, they had to be removed because of "their aggressive growth habits" and replaced with "similar" but "non-invasive plants." In Philadelphia, a private organization involved in managing Wissahickon Park financed herbicide spraying to kill Japanese knotweed, which a journalist described as "a super-hardy alien weed" that had seized "vast stretches" of parkland. One volunteer was delighted when the knotweed died and native plants such as skunk cabbage and poison ivy began growing in its place. Although the plant's enemies acknowledged that knotweed had been incorporated into Philadelphia's landscape in the 1880s, they argued that it did not get "loose" in Wissahickon Park until the 1980s. In the 1990s, city dwellers from wildly different social worlds—Seattle Earth Day observers and Florida Department of Corrections inmates—destroyed exotic plants in urban environments. Such efforts in San Francisco's Golden Gate National Park, however, were suspected of being undermined by a "Johnny Weedseed" who was spreading South African capeweed, gorse, water hyacinth, watsonia, artichoke, thistle, and narcissus. Although many plants condemned for "invading" natural spaces were, in a longer view of plant transfers, relatively recent introductions, well-established metropolitan photosynthesizers were accused of similar malfeasance. A federal government publication on "invasive plants" described the dandelion as "a serious weed of urban areas throughout the United States."[36]

Efforts to eliminate particular exotic plants established a new relationship with happenstance plants in cities. City dwellers who welcomed poison ivy as a native plant and cheered its resurgence were prepared to exchange human safety and comfort for ecosystem health. Unlike urban Americans who celebrated the tenacity of plants as inspiring nature and seemed happy to see any and all green things, these prowild, pronature, but antiweed environmentalists conceived of native nature as righteous nature. Chicago native plant landscaper Jack Schmidling declared, "It's important that we make little islands of wilderness in our urban sprawl." They thought that wildness was good as long as the wildness was native. Desirable wildness was a characteristic only of plants that lived in North America before Europeans arrived. Exotic plants introduced to create some wildness were scorned as pollution by the end of the century. Native landscapers and fierce urban weed-control advocates shared a desire for order. The former often attempted to return wild-looking plants to domestic landscapes in controlled fashions, while the latter rigidly controlled nature in refined symbolic forms of lawns and trees. This shift, especially in park and garden management, redefined out of place plants. Belonging was predetermined by the geographic origin of species rather than the usefulness of species in managing and shaping landscapes and environments.[37]

City dwellers fighting exotic weeds contributed to a larger native-nature protection movement. Although protecting wilderness had become compromised by the internal logical contradictions of preserving culturally created eternal spaces in the midst of an ever-changing natural world, some environmentalists decided that re-creating the natural world in terms of where organisms were found in geographical space and geological time preceding particular historical events was more sustainable. Reworking and managing environments inside cities according to this ecological ideal, however, was ambitious, given that so many environmentalists had concluded that cities were undesirable places to live. If urban environments were as impoverished as scientists believed, fights against exotic plants in them showed city dwellers' desires to help purify nature. Removing such plants from metropolitan landscapes helped prevent the dispersion of transformer species beyond the urban world. For example, in the Chicago metropolitan area, conservation biologists reported that buckthorn and purple loosestrife were "pervasive regional threats" to be resisted. When biologist Robert DeCandido surveyed the botanical stock of New York City's parkland at the end of the century, he estimated that 43 percent of the wildflowers once found in the city had disappeared. DeCandido believed that these trends were important because "every day the world becomes more and more like New York City, not the reverse." With trade and time, nonnative species' richness increased. These ecological dynamics suggested that protecting the nation's native nature depended in part on containing exotic nature in cities and reducing it. Whether distinguishing urban metropolitan environments, countrysides, and wilder places in these terms was useful was uncertain since past and ongoing hierarchies of purity, livability, and naturalness generated population shifts, landscape modification, and new environmental dangers.[38]

Despite the changes throughout metropolitan areas, laws regulating weeds in municipalities in the postwar United States remained in force. Weed nuisance laws across the nation indicated that happenstance plants remained urban environmental problems. For example, in the St. Louis region, weed laws inside and beyond St. Louis demonstrated a shared animosity and imposed similar restrictions. In 1960, the City of St. Louis replaced its periodically amended turn-of-the-century law with an ordinance that banned Russian and Canadian thistles, wild lettuce, wild mustard, wild parsley, ragweed, milkweed, and ironweed. This specificity was unnecessary since the law also prohibited all "poisonous plants or shrubs, and all other noxious weeds and vegetation." On "any lot or lands within the City" where these plants topped five inches, properties violated the law. Surrounding St. Louis County's weed ordinance resembled the city's law but permitted plants up to twelve inches tall. The county's law governed unincorporated county land for which there were legally recorded subdivision plats and the property was within one hundred feet of such subdivisions' outer boundaries. Such laws indicated that St. Louisans living in many different types of environments held a similar contempt for weeds. However, the consensus that weeds were undesirable at best and irritating and dangerous most of the time became the basis for conflicts with city dwellers who recognized the utility of urban herbs and tolerated them on or welcomed them to their properties. Although there were far more spaces and parcels than municipal and county officials could regulate, they protected their power to abate nuisance plants and regulate weeds.[39]

The legal status of happenstance plants in postwar United States was determined and reasserted in urbanizing Lincoln, Nebraska. Settled by pioneers in the mid-1850s and incorporated in 1859 as Lancaster, the place was renamed Lincoln in 1869. When the young botanists Frederic Clements and Roscoe Pound surveyed Lincoln's vegetation, they found vigorous growers such as sunflowers, ragweed, and butterweed. In other places, pigweed, sorrels, and dock bunched together. They spotted masses of jimson weed and black mustard on large dumps. Fields of panicgrass bordered the city's paved streets. Some of these plants were targets of Lincoln women's "incessant war . . . against weeds," which included ascertaining which weeds were the worst nuisances and pressuring the city to perform the "otherwise neglected duty" of removing them. With modest populations of 40,169 in 1900 and 43,973 in 1910, Lincoln officials, unlike their counterparts who governed denser, older cities, faced few challenges in applying City Beautiful design ideals. The Commercial Club boasted of Lincoln's broad streets, ample parking, and "shade trees everywhere in abundance." Lincolnites buried prairie under miles of streets, and they planted elms, maples, pin oaks, pines, and spruces to beautify these roadways. Yet happenstance plants remained abundant. Anthropologist Loren Eiseley grew up in Lincoln, and during his boyhood he romped around with friends in a sunflower thicket southwest of downtown. Leaders continued to refine the city's landscape by creating public parks. In 1950, 98,884 Lincolnites enjoyed 1,361 acres of parks.[40]

Producing and maintaining a tidy city included clearing weeds, although the costs of this work sometimes exceeded the costs of just cutting the plants. In 1943, city workers mowed a lot and left mangled vegetation on a sidewalk. The next morning, a cafe worker tripped over the plants, cut her face, bruised a knee, broke a finger, and sued the city for negligence. Although such hazards could appear across the city, these mishaps did not impede weed destruction. In a twelve-month period spanning 1947 and 1948, vegetation was removed from 1,230 properties. In the early 1950s, this work returned city officials to the courts. In September 1950, city officials investigated a property "overgrown with various tall weeds, six to eight feet in height." Ben Greenwood, a truck driver, and Ruby, his wife, received a notice about this violation. They cleared some of their vegetation, but in October, city laborers returned with a tractor to decimate the offending plants. This nuisance-abatement method destroyed raspberry bushes with which the Greenwoods supplemented their income. When the Greenwoods asked the city to compensate them for $2,500 in damages, the city ignored their claim. In litigation, a district court jury awarded the couple $1,250 in damages.[41]

Court records do not document how the condition of the Greenwoods' property became a concern for Lincoln officials. Perhaps they were responding to a complaint made by a Veterans Hospital gardener who lived a few doors down from the Greenwoods. In 1948 and 1949, the city only occasionally dispatched workers to clear lots on blocks in the Greenwoods' vicinity. In those years, the Greenwoods' property apparently did not violate the ordinance. However, from the fall of 1950 to the summer of 1951, the city cleared twelve lots in the area, including the Greenwoods' property. Their raspberry bushes and other vegetation may have seemed out of place to neighborhood boosters proud of their "nice homes, attractively kept yards, strong churches, [and] good schools."[42]

Just as, perhaps, the Greenwoods' neighbors could not stand the vegetation, the city could not let the court decision stand. The Nebraska Supreme Court heard the case, and Justice Paul Boslaugh reversed the lower court's decision and cited *St. Louis v. Galt* to uphold the city's power. He wrote, "The power and duty of the city to require the destruction and removal of weeds . . . could not be" questioned. The court also ruled that the city did not have to compensate the Greenwoods for its zealous weed control. *Greenwood v. Lincoln* reaffirmed the power of cities to regulate vegetation. The validation of Lincoln's law thrust turn-of-the-century reasoning into postwar life. While the court's decision reinforced concerns with weeds as public nuisances, in Lincoln, the Greenwoods' plants constituted a nuisance unrelated to public health. Lincoln's 1940 ordinance required landowners to cut and to clear "all weeds and worthless vegetation" four times throughout the summer and fall to prevent plants from reaching six inches in height; the law referenced neither allergies nor infectious diseases. A city official who inspected the

Greenwoods' property had found grasses, sunflowers, ragweed, and "all kinds of wild vegetation that is not included in natural growth." The appraisal of "natural growth" and "natural vegetation" as "worthless" indicated that the only plants that belonged in the city were those that had some value, whether in the fruits they produced or as objects managed to cultivate conformity or beauty. Plants such as "natural weeds" that people had not added to the land added no value to property. In seeking damages rather than challenging the premises of the city's ordinance, the Greenwoods seemed to share city officials' attitudes toward nature.[43]

In subsequent years, governments small and large relied on *Galt* and *Greenwood* to defend their attacks on weeds. In 1954, officials in Beaver Falls, Pennsylvania, a city of about 17,000 people some thirty miles northwest of Pittsburgh, prevailed in a county court against property owners who had not complied with a 1948 ordinance making "weeds and similar vegetation not edible or planted for some useful or ornamental purpose" illegal. Although the property owners claimed that the ordinance was "wholly illegal, unconstitutional and void, and that its enforcement would cause irreparable loss to their property," a Beaver County judge decided that the state legislature had conferred the powers on all municipalities to combat public nuisances such as the "weed evil" in Beaver Falls. The judge determined that the deficiencies of the ordinance were not indicating at what height the plants violated the law and not requiring the city to notify offenders. The judge discussed *Galt* and *Greenwood* to demonstrate acceptable municipal weed legislation. In 1970, a Florida court upheld Metropolitan Dade County's 1962 "lot clearing" ordinance, which was used to regulate unincorporated land outside of Miami's city limits. The ordinance prohibited "untended growth of weeds, undergrowth or other dead or living plant life upon any lot, tract or parcel of land, improved or unimproved, within one hundred feet (100') of any improved property." Such vegetation was a public nuisance not only because it threatened public health, safety, and welfare but also because it "adversely affect[ed] and impair[ed] the economic welfare of adjacent property." Judges maintained that such a law regulated property "for the benefit of the community as a whole" and was "readily ascertainable by the average person as to its meaning." They also cited *Galt* and *Greenwood* to assert the validity of the ordinance. Although grounded in turn-of-the-century environmental perceptions and public nuisance principles, local weed laws covered the diverse urban spaces in metropolitan America.[44]

The winnowing of metropolitan nature to particular aesthetic forms also generated conflict when city dwellers welcomed plants thought to belong in disordered environments and permitted them to traverse natural and social boundaries. Property owners who attempted to integrate unconventional, ordinary plants into residential landscapes unnerved vigilant, and sometimes vengeful, neighbors. Disputes among home owners, neighbors, and city officials in the Buffalo area and inside Chicago demonstrated how entrenched concerns about weeds as public health and safety nuisances continued to define legal and appropriate vegetation. Court cases pitting cities against their citizens not only involved issues that had arisen in *Galt*, such as constitutional rights and the human control of and relationship with nature, but also examined whether weeds were liabilities that reduced the financial value of abutting property. Even though fortuitous flora existed throughout metropolitan areas, the insistence that weeds were associated primarily with the most degraded areas convinced many city dwellers and officials that these plants did not belong in comfortable residential areas, even when property cultivators desired their growth.

A volatile mix of Henry David Thoreau's writings and contemporary environmental awareness inspired Stephen Kenney's cultivation of wildflowers to protest prosaic lawns. At the University at Buffalo (SUNY), Kenney immersed himself in Thoreau's writing as dissertation material. Thoreau's vow to "improve every opportunity to wonder and worship as a sunflower welcomes the light" served as the epigraph to Kenney's dissertation. A spirituality engaged with the natural world, Kenney wrote, was one that might offer "a viable alternative to . . . the multitude of horrors now threatening this wondrous and teaming sapphire and emerald world," such as the polluted Love Canal, about ten miles north of Kenmore. Kenney expressed these ideas not just on paper but in his wildflower-rich front yard and backyard vegetable garden. He planted asters, bachelor's buttons, black-eyed Susans, clover, coneflowers, cornflowers, dandelions, gentians, goldenrod, ox-eye daisies, and poppies, to list a few, hoping to produce a less toxic residential environment and to introduce agents of ecological change.[45]

Kenney's neighbors in Kenmore, a municipality just a couple of miles beyond Buffalo's northern border, were less interested in his intellectual endeavors and more concerned about the condition of his property. They feared the yard might result in rodents and insect infestations. The wildflowers appeared to make a built environment full of automobiles even more dangerous. Kenney's plants seemed to be a safety hazard because older drivers exiting their driveway could not see children playing on the sidewalk. Neighbors who believed Kenney's front yard approximated urban dereliction worried about their property values. One complained that the space looked "like an overgrown vacant lot in a blighted part of town." Village inspectors warned David Tritchler, Kenney's landlord, in June 1983 that the property had "undergrowths and accumulations of plant growth which are noxious and detrimental to health" and told him in May 1984 that "grass and plant growth must be controlled—yard must be kept broom clean." The village fined Kenney fifty dollars each day that the wildflowers remained in place, a fine that Kenney, and his landlord, refused to pay. Kenney's wildflowers may not have been artfully arranged, but officials' reasoning regarding why the plants did not belong

Figure 4.5. Steven Kenney's yard, 1984. Printed with permission of *Ken-Ton Bee*.

in Kenmore was even less artful. In court, one village inspector called the vegetation "undergrowth" and defined "undergrowth" as "weeds that grow underneath your grass." The inspector also stated that Kenney's plants were "noxious to look at." Another Kenmore resident called Kenney's property "depressing." After a two-day trial in September 1984, Judge H. Walker Hawthorne decided that the yard was not a "front lawn" but a "weed patch" and ruled that the wildflowers had to be mowed.[46]

The patch of plants provoked diverse expressions of competing environmental values. Kenney thought that his plants were beautiful and that his yard was "ecologically superior." According to Kenney, the toxic hazards of chemically treated lawns—not his pesticide-free meadow—were what needed regulation. Kenney believed that he had the right not only to plant what he wished but to say what he wished with the plants; yard-making was a First Amendment right. The plants, however, did not necessarily speak for themselves. To assert the appropriateness of the plants in Kenmore and beyond, Kenney planted a sign that described the vegetation as "a natural yard, growing in the way God intended." The village ordered its removal. Kenmore's prosecutor, Thomas Viksjo, claimed that a front yard was an inappropriate space for such expression because there was "virtually no likelihood . . . the average individual" would "understand Mr. Kenney's message." He argued that Kenney's message was one of incoherent nihilism. Viksjo said that the "lawn says nothing. It represents nothing and it symbolizes nothing." Such a denunciation, if accurate, suggested that the village was indeed making something out of nothing. Although Viksjo's contortions indicated that Kenmore's residents, many of whom were wage earners who worked at nearby airplane and automobile factories, could not comprehend symbolic nature, they did see the plants as symbols of urban disarray and lawlessness. James Kiouses, a self-labeled environmentalist who owned "land out in the country," stated that environmental regulations promoted social peace. He claimed that environments—urban and otherwise—would be a "total disaster" if people acted without restraint on their desires. Kiouses hoped that Kenney would not be sent to jail, lest he become a "martyr" for whom environmentalists would "erect a weed in his honor." To residents like Kiouses, an everyday environment was supposed to be one of urban order; if city dwellers wanted to experience nature, they could drive to it. In contrast, Kenney thought a touch of wildness was useful.[47]

Kenney and his just plants challenged Kenmore's established ecological order. Kenney's belief that such plants belonged in ordinary spaces and his attempt to use them to improve the urban environment threatened to change nature in and the nature of Kenmore. Residents and village officials were adamant that Kenney's land stewardship did not fit the community's standards. While both sides presented evidence as to whether the vegetation was "noxious," in the trial, Justice Hawthorne stated that his decision would be based on whether the property fit "into the nature of the community." As the protector of a residential community in both its actual and its ideal conditions—"what it is and what we expect it to be"—Hawthorne decided he would not permit Kenney's ecological experiment to degrade community norms or prosperity. Hawthorne believed that if good-natured neighbors adhered to fair laws, the community could remain healthy and happy. The judge called Kenney's desire to emulate mid-nineteenth-century Walden a "selective ecology" that injured his neighbors in the present and endangered Kenmore's future. Another defender of Kenmore denounced Kenney's property as a "menace to health, safety, and stability of the community." Environmental change, however well intentioned, was a threat. Community order determined the nature of the landscape in Kenmore and the inappropriateness of romantic visions of the past, even if such a sense of community was anachronistic. Although Hawthorne accused Kenney of desiring a mythic individualism, Hawthorne seemed just as smitten with a distant past of fissureless communities that prevented his recognizing that urban landscapes and environmental values could—and perhaps needed—to change.[48]

The appropriateness of Kenney's plants seemed to be evaluated less on the plants' capabilities than based on Kenmore residents' contempt for Kenney's personal style. The people who disliked Kenney's plants also seemed to dislike Kenney's long hair and aviator sunglasses. Anonymous notes in his mailbox taunted him as a homosexual with girl's hair, despite the fact that Kenney was engaged. That Kenney was unwelcome and out of place in Kenmore was evident when one woman shouted "Go to Hippietown" outside the courtroom. Hawthorne considered Kenney an intellectual transient, an individual whose "tenancy," in Hawthorne's words, "could very well be short-lived." Even though Kenney saw his yard as a protest against toxic environments, Hawthorne saw Kenney as a toxic agent whose rebelliousness was turning friends and neighbors into enemies and whose persistent toxicity might "torment future communities." Handling Kenney as an outsider simplified Hawthorne's evaluation of whether the plants were weeds. Appeals and references to spiritual authority seemed to intensify personal animosities. Kenney's environmental spirituality, which credited a creator for inspiring his yard, may have offended other Kenmore residents as blasphemous. When his neighbors shot birds in his yard and sprayed his rose bush with herbicides, Kenney quipped that such destruction drew "the wrath of God." *Buffalo News* columnist Margaret Sullivan assessed who was wrong and who was righteous when she stated that Kenney had violated the "quintessential suburban commandment: Mow Thy Lawn."[49]

Assertions of the supremacy of community order stimulated Kenney's defiance and his neighbors' lawlessness. Thoreau's spirit of civil disobedience animated Kenney, who pledged to serve jail time before he would destroy the plants. He declared, "I ain't paying and I ain't cutting." Instead, he appealed. When his case reached Erie County's Judge Joseph Forma in June 1985, Kenney had incurred more than twelve thousand dollars in fines. The encouragement of botanists, environmentalists, libertarians, and even a neighbor or two helped Kenney sustain his resistance. Forma heard arguments but delayed issuing a ruling. When Kenney was on vacation in July, two neighbors attacked the wildflowers one afternoon. In a more legendary version of the story, a "posse of neighbors armed with lawn equipment converged on the home, launching a midnight mowing mission." When Kenney filed vandalism charges, Hawthorne terminated Kenney's complaint by deciding the plants were not property. A neighbor commented, "It could have been worse. It could have been burned." In November 1985, Forma upheld Hawthorne's decision. He decided that a municipality "can regulate its citizens on basis of aesthetics" when "regulations bear a reasonable relationship to preserving the character of the community as well as the general health and welfare." By the time Kenney's appeal to the New York State Court of Appeals was denied in April 1986, Kenney and his wife had moved away to eastern New York. In the summer of 1986 and in the following summers, local lawn "vigilantes" organized an annual block party called "Kenmore Clipping Day" and offered to mow neighbors' lawns for free.[50]

That Kenney's wildflowers were weeds—unacceptable, unhealthy, unsafe, and despised—emphasized that weeds were figments of American culture. The outcome of Kenney's trial may not have surprised the attendees of the 1961 Northeastern Weed Control Conference, who heard Howard Campbell describe the scourge of weeds on Long Island. Campbell expanded the conventional definition of a weed to include "any plant growing where it is not wanted" by adding "any plant which may endanger the social status of a homeowner." Although Campbell likely meant the weeds on a property owner's grounds, *People v. Kenney* showed that this definition encompassed his neighbors' grounds as well. In proper urban environments, "high vegetation" seemed to ruin adjacent "trimmed lawns." The democratization of the lawn made disdain for plants a widespread environmental value and made weeds widespread dangers to a neighborhood's socioeconomic status. The *Ken-Ton Bee* opined that Kenmore's urban environment where houses were "quite close" to each other was not "the proper place" to use yard space as "a wildflower preserve." This reasoning appeared in *Goldbecker v. Fairfax County Board of Supervisors,* when a Virginia court ruled that an ordinance prohibiting owners of residential parcels of "less than one-half acre" from letting the vegetation "of any grass or lawn area" top twelve inches was a "valid use" of police power because it protected "property values and marketability" and prevented "harmful conditions and the attraction of undesirable pests." In both cases, jurists protected the rights of those who defined middle-class status in terms of a trim lawn by declaring unconventional yard vegetation out of place on smaller lots. In denying people of modest means wildflower yards, opinion shapers and courts limited the right to cultivate wilder nature to those who could afford large properties. Only in parts of metropolitan areas where lot sizes were so large that property owners' socioeconomic status was not in doubt could happenstance plants escape capricious weed definitions. Some of the distance between cities and nature arose from legal-spatial demarcations and class-defined limits on liberties.[51]

While Kenney aroused his neighbors' ire for challenging lawn norms, Marie Wojciechowski upset Chicago officials with her adventuresome wildflower cultivation and prairie restoration. From 1989 to 1994, Chicago officials tangled with Wojciechowski over the plants around her home in the northeastern corner of gentrifying Bucktown. Her cultivation of native vegetation seemed to re-create environments that officials and neighbors associated with dangerous neglected areas. When summoned to the circuit court of Cook County in 1989, Wojciechowski, as a would-be latter-day Smith P. Galt without wealth or legal acumen, made a pro se appearance. After Wojciechowski violated the law again in 1990, she secured pro bono representation from a local attorney who was a Sierra Club member and a gardening enthusiast.

Wojciechowski's desire to welcome prairie plants to her property conflicted with her neighbors' and city officials' perceptions of an orderly urban landscape. A July 1990 inspection of Wojciechowski's two North Honore Street properties revealed "weeds in excess of 10 inches in height growing randomly throughout the entire lot." Neighbors worried that her "natural lawn" was a fire hazard. A renter who lived next to Wojciechowski's lot complained that the weeds obstructed a walkway he used and exacerbated his allergies. What irritated neighbors was future wild beauty to Wojciechowski. Although she originally planned to turn the vacant lot next to her home into an organic vegetable garden, after Wojciechowski noticed prairie plants growing, she decided to turn the land into bird habitat. She was also inspired by plants growing in a yard on the western edge of Lincoln Park. She worked to earn a certificate from the federal government's Backyard National Wildlife Habitat Program.[52]

Chicago's weed ordinance was designed to identify vegetation to be controlled on vacant land, but not to avoid conflict with property owners eager to grow something other than lawns. The law required people to "cut or otherwise control all weeds . . . so that the average height of such weeds does not exceed 10 inches." Once plants surpassed ten inches they became potential public nuisances. However, the inadequate botanical and ecological knowledge of city officials and property owners alike confused which plants were weeds and why. Hugh Ziomek, an inspector who conceded he was a sanitation laborer and not a scientist, gathered samples to be identified by a botanist of unknown abilities: "the City of Chicago." Fred Karlinsky, another inspector, admitted he could not tell a chrysanthemum from a snapdragon. Wojciechowski's botanical knowledge was also imperfect. When a judge hypothesized that even if cannabis were a Chicago prairie plant it would be illegal to grow, Wojciechowski responded that "marijuana is ragweed" and that it did not occupy her land. Wojciechowski confused ecology and history when she stated that "pioneer plants" were those that arrived with European immigrants in centuries past. In fact, "pioneer plants" were ones that grew first on bare ground; they could be plants that had existed in the Chicago region far before Europeans' arrival in the New World, as well as plants from anywhere in the world that had arrived since then.[53]

Although many plants on Wojciechowski's property exceeded ten inches in height, whether they were weeds depended partly on evidence gathered from elsewhere in the city, her garden, and books about the region. Wojciechowski claimed to have obtained seeds for her property from grasses and wildflowers cultivated at a nearby South Side animal control center. Her pro bono attorney, Bret Rappaport, argued that there was no basis for a citizen to assume that plants grown on city, county, and state grounds would be illegal for them to grow. Patricia Armstrong, a consulting ecologist who had earned her M.S. at the University of Chicago, stated that Wojciechowski was actively increasing the prairie species and decreasing "alien non-prairie species." Armstrong detected neither neglect nor a violation of the ordinance. The city employed the *Plants of the Chicago Region* to argue that a number of plants on the property were "common weedy plants growing in disturbed or abandoned habitats." The city used the book *Natural History of Vacant Lots* to show that "shepherd's purse was one of the most widespread weeds" and that "prickly lettuce is also a common weed." Wojciechowski never denied that her property harbored weeds. She believed that the ones growing there were inconsequential. When asked in her October 1989 trial whether dandelions, knotweed, horseweed, wild lettuce, crabgrass, bristlegrass, goosefoot, shepherd's purse, and wild carrots grew on her property in large numbers, Wojciechowski answered they grew there and grew in large numbers because she had a large amount of land.[54]

Wojciechowski defended her vegetation with statements about the significance of native plants. If the plants in her yard had always grown in Chicago, they could not be out of place—only not yet displaced, a change that would perhaps diminish environmental quality. Wojciechowski claimed that a portion of her land had never been excavated or built on. When she found goldenrod, milkweed, coneflower, and snow on the mountain "growing spontaneously," she was amazed that they had survived Chicago's history. Armstrong's property evaluation recorded twenty-seven plants indigenous to regional ecosystems; the yard scored an 18.86 on the Natural Area Rating Index, a figure that approached preservation quality. When Rappaport challenged the city to specify exactly where on the property each weed grew and to provide both the common name and the scientific name for these weeds, the city objected but eventually listed dropseed, common milkweed, hairy aster, annual fleabane, garden sunflower, common evening primrose, garden phlox, tall goldenrod, lamb's quarter, common dandelion, St. John's wort, brown-eyed Susan, and woodland knotweed. The city's list made no distinction between plants with long roots in Chicago and plants more recently arrived from abroad, which all mingled together next to and behind the house. Since the law defined weeds based on height and patterns of growth, all the plants on Wojciechowski's property were weeds, whether they were originally from Chicago or had immigrated from afar.[55]

Since the city did not and could not attempt to control what plants property owners planted, the city tried to control the owners' control of their plants. Since many of Wojciechowski's plants were in the right place geographically and ecologically, the city's effort to persuade her to conform to others' expectations emphasized how the plants appeared and how their growth in space violated the law. In court, Wojciechowski was asked if she had placed the plants in rows. When a city lawyer stated that they looked "randomly ranched," Wojciechowski replied, "That's the way I want them." Judge Michael Bolan told Wojciechowski that she could have "a prairie garden," but not "a prairie garden in a random fashion." The disorder of the space seemed to be more of a nuisance than any particular plant. After Judge Bolan explained that one job of his court was to determine when Chicagoans' property uses infringed on the rights of the general public, he informed Wojciechowski that what was tolerable in rural environments or nature preserves might not be tolerable in urban neighborhoods. Bolan likened prairie cultivation to construction of a tannery to make leather jackets. Nevertheless, he decided she could have "special plants that are described as prairie plants" if she eliminated "common weeds like dandelions." Bolan's zoning guidelines for nature dictated how close city dwellers could come to nature in Chicago.[56]

Legal arguments over whether certain plants were weeds were based on the assumption that all weeds were dangerous and prevented carefully examining whether weeds belonged in Chicago or could benefit urban environments. The defense argued that a law declaring that all "weeds" over a certain height harmful unnecessarily prohibited plants that were actually useful. Wojciechowski's attorneys argued that, compared to lawns, "natural landscapes" were "generally more environmentally beneficial to Chicago and its residents." Rather than harming her neighbors, she was protecting them from toxic herbicides, pesticides, fertilizers, and petroleum products. She did not operate a lawnmower that generated noise pollution. Her land required little water to maintain and saved the city money in hauling "grass clippings and other yard waste." As long as Chicago was invested in perpetuating antiquated attitudes, the city could not realize the savings of a revegetated urban environment.[57]

The solutions that judges devised for Wojciechowski's disordered land attempted to make space for prairie plants but to prevent sanctioning neglect of land that could allow prairies to reappear throughout Chicago. In 1989, Judge Bolan encouraged Wojciechowski to become a more discriminating gardener and to recognize the distinction between gardening with prairie plants and re-creating a prairie or relying on nature to create a prairie for her. Bolan recognized that Illinois was a prairie state and that "unique" prairie plants could be grown if Wojciechowski avoided experiments that violated the law. He recommended using only "prairie plants that are not also noxious weeds." When Wojciechowski violated the weed ordinance in 1994, the court listed which plants were weeds and had to be removed, including awl aster, Canadian goldenrod, common evening primrose, common milkweed, common sunflowers, jumpseed, seaside goldenrod, and three-lobed coneflower. The list outlawed particular native plants that might regularly be found on untended vacant lots, as well as dandelions and lamb's quarter. Prohibited species were limited to "the growth of no more than one member or individual plant . . . per three square yards of property." The court also attempted to restrict the visibility of her property—perhaps to appease neighbors but also perhaps to prevent mainstreaming such a landscape—by ordering construction of a six-foot-high privacy fence. Strangely, the court bounded her prairie by requiring the perimeters of property to be "maintained free of plants greater than 10 inches in height," although these perimeters could have trees and shrubs. While the city forced Wojciechowski to remove some native herbaceous plants as weeds, the city's refusal to regulate shrubs and trees abetted the spread of purple loosestrife, buckthorn, and Norway maple—plants that were becoming the weeds of the future.[58]

Chicago officials' crackdown on Wojciechowski's wild-looking urban prairie may also have been a strategy to reassert municipal authority to regulate unconventional vegetation. While Chicago's second case against Wojciechowski was proceeding, five other natural landscapers sued Chicago over the "threats" that they received from officials regarding their plants. Ultimately, federal judges decided that the city's law did not harm these gardeners' rights. With Chicago's authority to police urban vegetation reaffirmed, officials claimed they would search for ways to allow Chicagoans to cultivate prairie grasses and "native lawns" without violating the city's ten-inch regulation. Possibilities included issuing permits to such property owners or devising an ordinance regarding "the planting and maintenance of natural landscapes" to avert such conflict in the future. However, such legal means for producing appropriate vegetation for cities were as imperfect as both sides' grasp of the past. The city was committed to imposing past people's ideas about ordered urban environments on the present, while native plant enthusiasts attempted to cultivate landscapes that rejected the ecological dynamics that had emerged through centuries of history and decades of urbanization. In the drive to win acceptance for native plants and "natural gardens" on their premises, environmentally sensitive gardeners contributed to the hostility toward other longtime city-dwelling happenstance plants.[59]

Since Kenney's and Wojciechowski's cases did not reach state supreme courts or federal courts, their cases did not become widely acknowledged precedents like *Galt* or *Greenwood*. However, another Lincoln, Nebraska, case renewed the validity of ingrained animosities against weeds and denied the validity of ecologically informed vegetation-management practices. Lincoln officials continued to prosecute property owners who failed to control "weeds or worthless vegetation." When officials interviewed Robert Howard in 1988 about his violation of the law, Howard stated his plants' role as "ground cover" constituted their "worth" when asked if they "had value." When Howard failed to remove the vegetation, the city destroyed it and assessed him for the work. To prove Howard's vegetation lacked value, city officials hired a Ph.D.-holding agronomist who verified that seventeen plants were textbook weeds and "grew unchecked and randomly." Nebraska Supreme Court judges noted that Howard's vegetation "more closely resembled a briar patch than a bed of prize roses." Acknowledging the similarity of Howard's case to *Greenwood,* the court quoted a passage from that decision that originated in *Galt:* "The word 'weed' has a common, everyday, meaning to the mind of every man." *Howard v. Lincoln* used *Greenwood* as a bridge to transport the logic of *Galt* across nine decades. This distorted *Galt* because that case was not about worthless plants, but about whether weeds were "indisputably and universally . . . unobnoxious and harmless." By the late twentieth century, when many Americans had developed some ecological awareness, the question might have been whether ordinary plants were indisputably and universally worthless and useless. Such a conclusion could only be reached by denying that the ecological work done by these plants lacked any value.[60]

The more deeply rooted in history judges imagined weeds to be, the more obvious it seemed that the timeless plants did not belong in cities. In 1986, U.S. District Court judge Myron Gordon ruled that ordinances ordering the destruction of weeds were not "impermissibly vague" and that "the word 'weeds' . . . is sufficiently well-rooted in the English language" that law-abiding citizens did not need elaborate explanation to comply with such laws. In 1993, a U.S. Appeals Court judge decided that Chicago's weed law was not "unconstitutionally vague nor irrational." In *Howard,* the Nebraska Supreme Court came to the same conclusion about Lincoln's law and added that listing every prohibited plant by its scientific name "would be extremely confusing" to ordinary people. An unstated but not unreal benefit of preserving established reasoning was rescuing city officials from having to evaluate the implications of new environmental values for municipal regulations. Similarly, courts employed outmoded language to reject evolving environmental values. When courts occasionally sided with property owners, they identified problems with the wording of the laws rather than their intent. In Licking County, Ohio, where city of Newark officials stated that "noxious weeds" were "unintentional plants whose bad qualities outweigh their good qualities," a judge decided that Newark's ordinance was out of line with state law in not specifying plants or height. By the end of the twentieth century, several courts had protected the prerogative of municipalities to define weeds with extreme, if not infinite, vagueness.[61]

Although Kenney and Wojciechowski resembled Galt, Green, and Greenwood in asserting their property rights, their unconventional garden-yards not only delighted their creators but also functioned to benefit other city dwellers and urban organisms, even if other people did not comprehend or appreciate their efforts. Their plants were longtime if increasingly rare city inhabitants, ecologically appropriate, and environmentally useful. To authorities, however, the plants remained inappropriate in and antithetical to built environments. Defeating their cultivators preserved municipal power for the sake of preserving municipal power.

The municipalities and courts allied against Kenney and Wojciechowski simply assumed the validity of the public nuisance purpose of the laws and avoided scientific evaluation of whether the vegetation presented a danger. To the limited extent that authorities acknowledged the significance of unique regional plants in Wojciechowski's case, this begrudging tolerance was for plants on well-tended plots—not a recognition of their desirability throughout a city changing with ecological time. Municipal efforts to conserve power and preserve peace were missed opportunities to reevaluate the relationships of people and plants and to use ecological time to their benefit.

Dealing with wildflower-eyed city dwellers was a nuisance to cities struggling to deal with violent drug crime. In addition, some cities losing battles with weeds in vacant lots were intensely fighting human weeds creating dangerous cities. In 1991, metaphor-minded law-enforcement professionals endowed their human-weed-control efforts with a name that expressed its transformative ends. The U.S. Department of Justice's creation of Operation Weed and Seed (OWS) targeted criminals responsible for urban disorder. Modeled on Philadelphia's efforts to reduce drug trafficking, OWS was a plan to "weed out" criminals through law enforcement and prosecution, as well as to "seed" neighborhoods with prevention and rehabilitation services. OWS officials abridged human weeds as "weeds." In 1992, OWS director Deborah Daniels noted the problem of bringing "a neighborhood back to life if it's overrun with crime and criminals." In OWS parlance, drug-dealing criminals were "weeds" that undermined community health. They thwarted a community's ability to grow into a safe place to live. Daniels described OWS as an attempt to "pull out the weeds that choke off opportunity, and begin to sow the seeds of community revitalization." In 2002, Robert Samuels, the acting director of OWS, claimed that OWS inspired the everyday courage necessary to make neighborhoods safer, "to grow gardens from weeds overtaking cracked asphalt," and "to sustain a faith that people can make a better future for their children." OWS officials believed that the "root cause of criminal behavior" was an absence of "core values such as self-restraint and respect for the rights of others." Yet OWS did not specify whether this "root cause" was an inherited characteristic or itself evidence of the environmental problems that produced such weeds.[62]

OWS grants allowed cities to intensify police presence in targeted neighborhoods and to support social programs to reverse "crime, poverty, social disorganization, and urban decay." OWS administrators preferred allocating resources to eliminating weeds. A 1996 review of OWS programs indicated that three-quarters of grant monies went to "weeding"—community policing, police-overtime funds, and prosecution work. Police worked to arrest street dealers of small amounts of crack cocaine or heroin. Expedited prosecution of "the worst offenders" as federal criminals was an attempt to remove them from cities for as long as possible. Swift investigation and strict punishment of criminals was also considered a deterrence measure that reinforced the moral transformation of the community. Community involvement was limited to "seeding activities" such as neighborhood beautification projects and identification of problems for code enforcement. In Springfield, Illinois, OWS assigned prison laborers to support neighborhoods "clearing overgrown shrubbery, mowing, planting and caring for flowers."[63]

The program concentrated on "weeding" because establishing and cultivating new communities was more difficult than imprisoning criminals. "Seeding" results depended on a community's leadership and institutional resources, crime level, economic development prospects, and demographic factors. For example, the target population in Salt Lake City consisted primarily of immigrants who were "transient, poorer, less well-organized." OWS considered such people easy to remove and their communities in need of less "seeding." In contrast, Eastern industrial cities that had struggled with crime demanded intensive "weeding" to produce security. Drug problems, economic stresses, and deteriorated built environments were so severe that "seeding activities alone could not have overcome them." Even though OWS analysts reported that "weeding" had to be sustained over time and that "seeding" had to be evident early on, problems developed where it was assumed that "weeding" had to occur before strengthening the community, which Justice Department analysts noted "may be encouraged, unfortunately by the OWS name itself." In Seattle, protesters compelled city officials to acknowledge that the program name was "offensive." Activists in the Los Angeles Urban Strategies Group stated that "designating many youths in communities of color as weeds" resembled "the racist ideology that demonized and dehumanized Asian populations." However terribly individuals behaved, the group doubted the efficacy of a program based on the idea that low-income youth were "simply 'weeds' to be rooted out and burned." Critics pointed out that OWS reinforced the inequalities that contributed to problems it purported to solve. Indiscriminate "weeding" left juveniles with lengthy criminal records. While middle-class people were governed by local and state laws, OWS placed minority neighborhoods under harsher federal laws.[64]

However well-intentioned OWS participants were in restoring order in violent communities, the OWS analogy fused the depravity of weeds with hopeless urban environments. OWS efforts to transform communities may have been more successful in dispersing the seeds of urban ills throughout cities than in imprisoning weeds. One of the largest OWS operations in Chicago targeted the Housing Authority's Ida B. Wells Homes in April 1992. Three hundred law-enforcement officers made sixty-five arrests. This "weeding," however, did not save the community. The OWS war on human weeds did not make weeds disappear. Demolition of the homes began in the early 1990s and continued into 2008. Meanwhile, fears of weeds, drug dealing, and violence persisted in the vicinity. In 1995, a resident pointed out the weeds into which "a lady 5 foot tall could disappear," a space where a neighborhood advocate said criminals could "stash drugs, or hide, or toss guns." Across from the Darrow Homes, where there had once been an A&P, a McDonald's, and a drug store, a *Sun-Times* reporter saw only "vacant lots overgrown by towering weeds as far as a child's eye can see." The Fourth Ward, in which Wells, Darrow, and other projects were sited, had 1,644 vacant lots. The city employee assigned to control such plants in the ward was also responsible for an adjacent ward that contained 2,816 vacant lots. To emphasize the affinity between weeds and criminals, a *Sun-Times* reporter wrote, "Not everyone wants the weeds cut down," when informed by a city official that "alleged gang-bangers" harassed children in a summer jobs program that attempted to clean up one vacant lot. It appeared that where weeds remained, human weeds would use them to foment disorder. Implying that human weeds and weeds conspired with each other indicated that while metropolitan America had changed much over the century, Americans themselves may not have changed as much as they imagined.[65]

CONCLUSION

A Brief History of Weeds in the Twenty-First Century

Throughout the twentieth century, many city dwellers insisted that happenstance plants were weeds and worked to eliminate them. However, the plants survived these attacks and ongoing urbanization. Their resilience indicated that cities could not negate ecological time and that metropolitan landscapes could not easily be perfected. Regardless of how much distance Americans wished there was, imagined there to be, or actually put between their cities and nature, fortuitous flora sustained ecological time in cities. In the first half of the century, ever-greater numbers of Americans resided in cities, and in the second half, many Americans relocated to smaller cities, newer cities, or sunnier cities of lower densities. Yet wherever Americans were shaping cities, just plants remained city dwellers because property owners and municipal officials had not been able to develop, regulate, and manage all the land inside their borders. When Americans used *weeds* to describe deteriorating cities, they suggested that these plants were suddenly growing where they had not been before or had not been for a long time. Without recognizing the place of these plants in cities in the decades beforehand and the persistence of ecological time across time, they reinforced perceptions of urban decline and possibly accelerated disinvestment. Throughout the century, the attitude

Figure C.1. Lower Ninth Ward, New Orleans, August 2008. Courtesy of "Infrogmation of New Orleans."

that weeds did not belong in cities and that their presence in them sullied urban environments was an antiurban one that may have accomplished nothing more than reinforcing the redistribution of city dwellers over more space and producing more spaces for happenstance plants.

The advent of a new millennium and a new century did not suddenly transform the ecological conditions in metropolitan America or metropolitan Americans' experiences of the natural world. Many twenty-first-century Americans' attitudes and behaviors toward weeds had been bequeathed to them from past city dwellers. Urban Americans continued to attack weeds but also still marveled at fortuitous flora. These plants settled in environments transformed by tremendous storms and inhabited landscapes stunted by financial turbulence, especially in Arizona, California, Florida, and Nevada. Expectations of and desires for weed-free cities provoked outbursts of frustration with metropolitan photosynthesizers from the Atlantic coast to the Pacific coast. Increasing attention to global environmental problems made urban herbs indicators of worrisome global trends that seemed to make weeds ever more harmful.

In New Orleans, Hurricane Katrina's floodwaters destroyed homes and created more space for happenstance plants to grow. When a retired postal worker returned to the devastated Ninth Ward to live in a Federal Emergency Management Agency trailer, she had a "broad view of nothing but weeds and a few far-

Figure C.2. Stratland Estates in Gilbert, Arizona, April 2008 (*above*) and September 2008 (*opposite*). Courtesy of Housingdoom.com.

away rooftops." She was disconcerted by how "high weeds" made the area "just a jungle." When city officials decided to permit residents to return to their old neighborhoods without a phased redevelopment, some people returned to places where they had no neighbors—places where, according to a *Times-Picayune* columnist, "weeds . . . stretch for miles, stretch above your head, stretch to nowhere, everywhere." When two *Newsweek* writers observed in 2008 that the Lower Ninth Ward had become "cleaned-up vacant lots of weeds and surprising bursts of pink wildflowers," they sustained the frivolous distinction between useless and beautiful plants.[1]

Metropolitan photosynthesizers proliferated throughout California's Central Valley when urban development slowed as demand for and means to purchase new

homes rapidly declined. In 2007, the *Sacramento Bee* announced that "blighted scenes" were "growing like weeds." In sections of booming Elk Grove, rapid growth yielded rapid neighborhood blight. One realtor claimed that irresponsible renters, broken windows, and weeds created a "feeling of chaos." More than thirty homes that had never been occupied or had been abandoned became "indoor marijuana farms." Members of a local church spent some of their Sundays battling "weeds that were almost eye-level high." In Merced, a city invigorated by a new University of California campus, there were more houses than people to buy them, as well as too many homes priced above what the local population could afford. Land set aside for a firehouse and commerce became "an immense scrubby field" where adolescents and "vagrants" started fires. Where homes might have been surrounded by "suburban putting greens," instead there was a "dried-out meadow" of "crackly tan . . . knee-high grass" and "prickly" weeds "springing up like, well, weeds." Merced's fortuitous flora grew larger and more dangerous as the months went by. A reporter warned that "dense weeds reminiscent of the lush Amazon jungle" would become "a dangerous fire hazard" if they went uncut. The fire chief blamed out-of-town speculators for the "verdant expanses of waist-high weeds" that violated the two-inch height permitted in the weed ordinance. A tax-phobic resident upset by "tumble weeds . . . as big as cars" along a main thoroughfare proclaimed the solution to this public nuisance was to "stop paying property taxes" but did not explain how such disobedience would eliminate the weeds. At the southern end of the valley, in the Bakersfield area, a reporter who visited a Greg Norman–designed golf course saw "weeds, some taller than your wife . . . in the fairways." Real estate market turmoil undermined the war against weeds on a terrain where people ordinarily fought the plants tenaciously.[2]

As home sales plummeted and foreclosures proliferated throughout the Phoenix metropolitan area, "thousands of tumbleweeds" returned to the Valley of the Sun. Happenstance plants such as Russian thistles grew "waist-high or taller" and defied local laws limiting weeds to six inches in height. In March 2008, municipal officials were inundated with complaints about "out-of-control weeds" taking over yards, while home owners' associations (HOAs) struggled like city bureaucracies in decades past had to eliminate weeds. A Surprise resident living in an "otherwise pristine . . . subdivision" complained that "bulldozers" would be necessary to rip out the "tap roots" of weeds that an ineffective HOA had failed to eradicate. When an HOA in Gilbert neglected to remove eight-foot-tall weeds on a foreclosed property and instead hassled a fee-paying home owner about the weeds in his yard, the disrespected resident exposed the HOA's impotence by massacring the weeds on the adjacent property with a chainsaw. Happenstance plants proliferated most wildly where there were too few solvent home owners to support HOA operations or where HOAs collapsed when builders went bankrupt. In 2010, Phoenix and Tempe lent equipment to "do-it-yourself weed whackers." Chandler's code-enforcement manager insisted that neighbors had to help neighbors and implied that they had to tend to absentee owners' yards because the city lacked the funding to "become the lawn service" and using the law to compel abatement was "time-consuming and expensive."[3]

While troubled speculative developments in the West became filled with fortuitous flora, weeds in and around metropolitan environments all across the country irritated twenty-first-century Americans. A blogger who had left Buffalo and its "urban prairies" for Clinton Hill, Brooklyn, was dismayed to find not only vacant lots with motley vegetation but also "an insane amount of grasses and weeds growing up out of the sidewalk areas that are paved with old bricks." In

Niskayuna, New York, a town about twelve miles northwest of Albany, a rebuilt section of highway landscaped with new trees and perennial flowers became "prime real estate for weeds." In the District of Columbia, volunteers from universities and corporations removed weeds from the grounds of the Woodrow Wilson High School in Tenleytown and the inner-city Perry School Community Center. On the southeastern edge of Jacksonville, Florida, residents complained about the "mountains of weeds" occupying power-line easements behind their properties, which "caused a snake hazard." In Chicago, a neighborhood leader complained about weeds "as big as trees" hiding a stashhouse for criminals' weapons, while a councilwoman argued that criminals hid "drugs in the weeds" and left murder victims in "weeded" areas. A Denver official attempted to deflect criticism of the condition of city parks by claiming that heavy rainfall and snowmelt had "made weeds a lot more aggressive." Pocatello, Idaho's mayor told a local television station that "weeds are not created equal with beautiful landscapes. . . . Weeds destroy the beauty of our community, and we will leave no weed standing." In West Jordan, Utah, a city southwest of Salt Lake City, a woman in a wheelchair fought weeds along a construction-area fence so she could pass by. In San Jose, California, drivers complained about "millions of tall weeds" in highway medians and along roadsides that were "as high as" car windows.[4]

While most urban Americans continued to detest weeds in general, some metropolitan Americans reserved their contempt for transformer species. In Baltimore, "urban weed warriors" spent nearly sixty-seven hundred hours in 2007 removing plants such as stilt grass, wineberry, kudzu, and porcelainberry from 142 sites in city parks and watersheds. Near the District of Columbia, Cabin John Regional Park officials trained volunteers to search for *Artemisia vulgaris* and destroy the plant and warned other park users who harvested it that "removing things from county property is illegal." The mugwort that weed warriors were eradicating was the souk that recent Korean American immigrants were harvesting to flavor seafood soups and to dye rice cakes. The possibility that a plant might variously be considered a nuisance to uproot, appear on a dinner plate, or be protected as property demonstrated the complex nature of so-called weeds. Residents of Milwaukee's Story Hill neighborhood spent a Saturday morning "beat[ing] back the evil weed" garlic mustard by pulling it out and stuffing it into garbage bags. They also removed "invasive, hazardous buckthorn," which interfered with their plans to eradicate the garlic mustard. Transformer species were environmental problems that Seattle mayor Greg Nickels thought threatened to transform Seattle into "the city 'formerly known as Emerald.'" The Green Seattle Partnership warned that Seattle was becoming "a city of weeds." In response, EarthCorps organized volunteers on Martin Luther King Jr. Day to clear "invasive" plants from a park that had been home to "hobos" during the Great Depression and was being frequented by drug-selling gangs and prostitutes. EarthCorps also restored twenty-five hundred acres of "the weediest city forests" by removing English ivy and Himalayan blackberry. In this nationwide destruction of exotic plants, however, "native" plants such as ragweed were sometimes regarded as "invasive" plants.[5]

Happenstance plants were most often perceived as localized environmental threats to people's health, safety, property, and prized public spaces. However, as global warming proceeded, the problems created by metropolitan photosynthesizers threatened to grow worse. Big-carbon-footed Americans were advancing weed evolution by producing "supercharged weeds." Government scientists informed Congress that carbon dioxide emissions benefited "'invasive' weeds" more than "most cash crops." Lewis Ziska, a U.S. Department of Agriculture plant physiologist, found that the heat-trapping effects of cities made the plants in them more vigorous than vegetation in other environments. Lamb's quarter grew to eight feet on a farm but reached twelve feet in a city. The conditions risked making plants harmful to people more dangerous. Ziska warned that the oil in poison ivy tissue would become "more toxic." *Newsweek*'s Sharon Begley reported that higher greenhouse gas levels not only made "urban ragweed . . . three to five times bigger than rural ragweed" but also allowed the plants to spew "allergenic pollen weeks earlier each spring" and produced plants that released "10 times more pollen." Despite the work cities had done to clean up their environments in the twentieth century, greenhouse gas–induced climate shifts seemed to indicate that the centers of metropolitan areas remained less pure than their more pleasant edges. However, some scientists believed heavily exploited sites could be examined to provide critical information about how climate change and pollution shaped "future plant communities." Rutgers University–Newark ecologists studied a contaminated, abandoned rail yard inside Jersey City's Liberty State Park where over four decades a unique and highly fortuitous flora had emerged from heavy-metal-laden and compacted soil.[6]

While scientists researched how people were driving the evolution of plants, writers and filmmakers imagined how the plants would change the world if people died off. Alan Weisman feared that the planet could "degenerate into something resembling a vacant lot, where crows and rats scuttle among weeds." Weisman described how apocalyptic nature would disassemble Manhattan. Mustard, shamrock, and goosegrass would move beyond Central Park, grow in cracked pavement, expand the fissures, and disintegrate sidewalks. After lightning started fires that burned structures to the ground, nutrient-enriched dirt would feed more plants. Reduced air pollution would accelerate the growth of Virginia creeper and poison ivy, plants that would scale, weaken, and topple structures. "Within two centuries," oak and maple forests would return. In *I Am Legend*, rampant photosynthesizers replace idling taxis and limousines on Fifth Avenue. Yet the destructive capabilities of happenstance plants also impressed those who observed their handiwork in the here and now. A Catholic woman confessed her fascination with urban herbs "upspringing, invading, taking over, breaking down the grim order we have created." She noted that although some people saw such city ruins as "discouraging, because it says something about the transience of human works," she found it "greatly cheering, because it speaks to me of life's abundance. It also appeals to my wicked side." Her admission might not have surprised a psychologist, who, when pondering why weeds were so "unwanted," concluded it was because their tendency to "spread wildly and crowd out everything else" indicated they were "basically selfish." Although this observation was offered to explain how people should avoid letting their best character traits crowd "out the beauty" of other people's strengths, in failing to consider environmental conditions, it misstated the nature of weeds. The unruliness believed to be stored in weeds seemed to be more powerful than two decades earlier, when one biologist thought the prospects of "nature's living demolition squad" were bleak because Americans' technology had "finally overwhelmed nature's tremendous regenerative forces."[7]

Which plants would raze cities after human extinction depended in part on which plants survived the extinction episode that people were presiding over. The dynamics of a posthuman ecological free-for-all were of less concern to scientists who studied metropolitan America to assess biological diversity. A pair of botanists who measured the biodiversity levels of Ellis Island found that nonnative species outnumbered native species. Their selection of a place involved in establishing the nation's cultural diversity and symbolic of that diversity—coupled with their statement that "the precise contribution of Ellis Island's immigrants to present day non-native flora . . . never can be determined accurately"—hinted that their concern was with how human demographic change influenced the fate of plant populations. A less ethnoconscious assessment of urban environments found that cities shared a common biological poverty, regardless of "age, size, location, and surrounding habitats." As American cities and cities around the globe became more intensely developed and as each generation lost appreciation of the natural world that had preceded them, people's standards and sense of "ecological health" declined. Over time, this dynamic made future environmental degradation probable because the largest population increases came from the poorest people in the environments with the least localized diversity.[8]

Persistent concerns about weeds did not inhibit other twenty-first-century Americans' admiration for and appreciation of such plants. They used them as symbols of past cultural shifts, present dissatisfactions, and future transformations. Upon her 2007 induction into the Rock 'n' Roll Hall of Fame, Patti Smith described the tumult of the late 1960s in terms of vigorous plants busting through engineered constraints. She had left southern New Jersey for New York City in 1967, a time when "rock 'n' roll . . . was a fusion of intimacies. Repression bloomed into rapture like raging weeds shooting through cracks in the cement." A blogger who got frustrated with stopped highway traffic and sun-baked shopping center parking lots looked for cracks in the pavement to find "weeds pushing through the asphalt, squeezing their roots in and stretching and breaking the concrete." She explained, "Part of me is rooting for the weeds. I want them to win, I guess." One of her readers concurred that it was "comforting to know that whatever we do, ultimately, we are no match for the earth." Another reader added, "Nature is fighting back against us[,] and someday it will win. Go weeds!" A western Pennsylvania anarcho-primitivist who was preparing for the demise of civilization wore a tattoo of skyscrapers with the words, "We are the weeds in the sidewalk." A business writer encouraged managers to value their "high-performing weeds" who did not need instructions on how "to get from one side of a cement sidewalk to the other." Such "low-maintenance employees" possessed self-reliance and the sense of accountability necessary to accomplish their goals. Indeed, she concluded that a success-seeking manager could also "be a weed" by claiming responsibility for at least 85 percent of successes and failures alike. While weedroots activists may have seemed morbidly destructive, law-abiding Americans were possibly more destructive, just less aware of their role in ravaging the Earth. Historian of science Robert Kohler described humans as "weeds . . . who build and inhabit airports and autobahns, chain stores and malls, suburbs, and cyberspace." He thought that the "weedy culture" of our "weedy species" could do little more than construct "placeless places." Kohler's outlook echoed Ralph Audy's remarks at the 1963 Sierra Club Wilderness Conference, physicist Chet Raymo's claim that "we are weeds par excellence" as evidenced by our intrusion "into every nook and corner of the planet," and David Quammen's simple and inclusive "Weeds are Us." Twenty-first-century metropolitan Americans could embrace, be unaware of, or deny the notion that they were as opportunistic as kudzu—that they were "consummate" weeds.[9]

Yet twenty-first-century self-botanomorphization was not inherently subversive or self-deprecating. After Hurricane Katrina, an engineer dedicated to rebuilding New Orleans extracted lessons about the past and the future from sunflowers and thistles—"the wildest of weeds"—transforming brown wastelands with splotches of color. He considered the failed levees to be evidence of the hubris of trying to dominate nature with human-made contrivances. He noted the renewal of this attitude when the city passed a law prohibiting "unkempt lawns" to compel property owners to reassert control of their land. Yet the blogger's reluctance to dominate nature was not surrendering to nature. He believed that "we ARE nature," and he considered building and rebuilding to be natural behaviors.

Protecting New Orleans from future hurricanes, he speculated, could be accomplished with more intelligent consideration of and less belligerent actions toward nature, such as building "houses up off the ground and strong enough to withstand the wind." Whereas others saw the city's thistles as pests or pollution, the engineer saw determination. He suggested that whether New Orleanians would "be viewed as volunteers or as weeds" depended in part on whether they could respect the thistles rising "defiantly on the vacant land here, growing where some would not approve, reaching where some would not dare." A tuba-playing California transplant to New Orleans dubbed it the "sunflower city" because the plants marked some of the first life to emerge after the storm and because people, like sunflowers, brought color with them as they returned. By using sunflowers to express their hopes for a vibrant reurbanization, New Orleanians interpreted the plants much differently than Edgar Anderson had some six decades earlier, when he saw sunflowers as evidence of the excesses of urbanization. St. Louis was at that time the nation's "sunflower city" because it had been "boomed by war industries," settled by "country-bred war workers," and filled with sunflowers beautifying "areas made ugly by civilization."[10]

The twenty-first century will have to drift far away from the twentieth before it will be possible to discern whether drawing inspiration from "the wildest of weeds" changed the nature of cities for the better. However, if most Americans continue to handle weeds in ways initiated by their nineteenth-century predecessors—incessantly waging erratic wars that disrupt ecological time—city dwellers who admire happenstance plants might only be able to establish enough space and enough time to create "strange new niches" for ordinary vegetation to thrive here and there. Although Harvard University entomologist and conservation advocate E. O. Wilson has recommended that Americans "get rid of" the term *weeds*, expunging the word from dictionaries may not change how Americans perceive and treat "any plant that interferes with some human purpose." The perspectives on the evolution of environments offered by environmental historians, themselves occupants of "strange new niches," are necessary to help people reevaluate their ideas about the natural world. More important than the word *weeds* is what people do with the plants. The supposedly useless plants can be used to reduce or accelerate the destruction of cities and other wonderful places. The years to come may reveal whether it is more useful to understand why weeds are just plants or to present weed-ridden and weed-producing cities as threats to millennia of human evolution.[11]

NOTES

ARCHIVAL MATERIALS

American Museum of Natural History, New York

Central Files

Aton Forest, Norfolk, CT

Frank E. Egler Papers

Beinecke Rare Book and Manuscript Library, Yale University

Rachel Carson Papers
Julia Ellsworth Ford Papers

City of St. Louis Microfilm Archival Library, St. Louis

Cook County, IL, Clerk of Court Records, Chicago

Chicago v. Wojciechowski Case Files

Library of Congress, Washington, DC

Luther Burbank Papers

Missouri Botanical Garden, St. Louis

Edgar Anderson Papers

Missouri Historical Society, St. Louis

George Edward Kessler Papers

Missouri State Archives, Jefferson City

Missouri Supreme Court Judicial Case Files, Case no. 10769

National Archives at College Park, MD

Records of the Office of the Secretary of Agriculture, Record Group 16
Records of the Bureau of Plant Industry, Record Group 54

Records of the Works Progress Administration, Record Group 69
Records of the United States Public Health Service, Record Group 90
Records of the Drug Enforcement Agency, Record Group 170
Records of the District of Columbia, Record Group 351

National Museum of Natural History, Department of Botany Library, Washington, DC

Lyster Dewey Field Book

National Park Service National Capital Region Office, Washington, DC

Palisades Park Reservation No. 404, District of Columbia, Land Record No. 250

Nebraska State Historical Society, Lincoln

Nebraska Supreme Court Judicial Case Files, Case nos. 32673 and 33175

New York City Municipal Archives, New York

Mayor Laguardia Papers
Mayor O'Dwyer Papers
Records of the New York City Department of Health
Records of the New York City Department of Parks

New York State Library, Albany

Special Collections Research Center, University of Chicago

Ernest W. Burgess Papers
George Damon Fuller Papers
Robert Ezra Park Papers

Village of Kenmore, NY

People v. Kenney Case Files

Government Documents and Publications

Annual Proceedings, Board of Supervisors of the County of Sullivan, New York
Annual Report of the Health Department, City of Chicago
Proceedings, Lincoln City Council
Annual Report of the Health Department, City of New York
Annual Report of the Health Department, City of St. Louis
Annual Report of the Police Department, City of New York
Annual Report of the Public Works Department, City of Chicago

ABBREVIATIONS

AMCAEP	American Memory Collection, American Environmental Photographs
AMNH	American Museum of Natural History
ARSLHD	Annual Report of the St. Louis Health Department
AZREP	*Arizona Republic*
BDE	*Brooklyn Daily Eagle*

BON	Bureau of Narcotics, Subject Files
CCMW-322079	*City of Chicago v. Marie Wojciechowski*, Case File 90 MC1 322079,
CDFD	*Chicago Defender*
CT	*Chicago Daily Tribune*
EA-MBG	Edgar Anderson Papers
EWBP	Ernest W. Burgess Papers
FEEP	Frank E. Egler Papers
GC-SOA	General Correspondence of the Secretary of Agriculture
GDFP	George Damon Fuller Papers
JEFP	Julia Ellsworth Ford Papers
JLPUE	*Journal of Land and Public Utility Economics*
KTB	*Ken-Ton Bee*
LAT	*Los Angeles Times*
LOC	Library of Congress
MHS	Missouri Historical Society
MM	*Meehan's Monthly*
MODP	Mayor O'Dwyer Papers
MSC-10769	*City of St. Louis v. Smith P. Galt*, October term, 1903, Case No. 10769
NEWCC	*Proceedings of the Northeastern Weed Control Conference*
NSC-NSHS	Nebraska Supreme Court, Judicial Case Files
NYDOH	Records of the New York City Department of Health
NYDOP	Records of the New York City Department of Parks
NYT	*New York Times*
PCALMD-1923	*Placerville (CA) Mountain Democrat*, 14 July 1923
RC-YC	Rachel Carson Papers
REPP	Robert Ezra Park Papers
RHO-RDCD	Report of the Health Officer of the District of Columbia
RMPD-RDCD	Report of the Metropolitan Police Department of the District of Columbia
SANE	*San Antonio Express*
SANLI	*San Antonio Light*
SFNM	*Santa Fe New Mexican*
SLPD	*Saint Louis Post-Dispatch*
VK-SK	*Village of Kenmore v. Stephen A. Kenney, David Tritchler*
WES	*Washington Evening Star*
WP	*Washington Post*
WPA	Records of the Works Progress Administration

INTRODUCTION

1. Alan Havig, "Presidential Images, History, and Homage: Memorializing Theodore Roosevelt, 1919–1967," *American Quarterly* 30 (1978): 514.

2. Barry Mackintosh, *C&O Canal: The Making of a Park* (Washington, DC: Department of the Interior, 1991), 18–20, 35–97; *The Douglas Letters: Selections from the Private Papers of Justice William O. Douglas,* ed. Melvin I. Urofsky (Bethesda: Adler and Adler, 1987), 236–40; 64 Stat. at L. 903, chap. 984; *The Future of Georgetown* (n.p., 1924), 3–7, 11–14. A recent account of the hike is included in Adam M. Sowards, *The Environmental Justice: William O. Douglas and American Conservation* (Corvallis: Oregon State University Press, 2009). On preservation efforts in Georgetown, see David Allan Hamer, *History in Urban Places: The Historic Districts of the United States* (Columbus: Ohio State University Press, 1998), 93–99; Dennis Earl Gale, "Restoration in Georgetown, Washington, D.C., 1915–1965" (Ph.D. diss., George Washington University, 1982), 71–88, 136–44, 216, 231–32; Eileen Zeitz, *Private Urban Renewal: A Different Residential Trend* (Lexington: Lexington Books, 1979), 63–66. Historians have written more about how, before World War II, Americans tried to protect the natural environment from automobiles rather than the urban environment from automobiles. Cf. Paul S. Sutter, *Driven Wild: How the Fight against Automobiles Launched the Modern Wilderness Movement* (Seattle: University of Washington Press, 2002); Hamer, *History in Urban Places,* 5; Charles B. Hosmer, *Preservation Comes of Age: From Williamsburg to the National Trust, 1926–1949* (Charlottesville: University Press of Virginia, 1981), 1–2. Hosmer notes that "the revolutionary element in the preservation picture was the arrival of the automobile."

3. Palisades Park Reservation No. 404, District of Columbia, Land Record No. 250, 30 Sept. 1957, National Park Service National Capital Region Office; Kathleen M. Lesko, Valerie Melissa Babb, and Carroll R. Gibbs, *Black Georgetown Remembered* (Washington, DC: Georgetown University Press, 1991), 91–93; Charles Weller, *Neglected Neighbors* (Philadelphia: Winston, 1909), 242–46; Gale, "Restoration in Georgetown," 58, 88, 240; "K St. Viaduct in Georgetown to Be Rushed," *Washington Post* (*WP* hereafter), 14 Feb. 1937; Mackintosh, *C&O Canal,* 3; Mary Mitchell, *Chronicles of Georgetown Life* (Cabin John, MD: Seven Locks, 1986), 109–10. Damage from an 1889 flood and competition from railroads had undermined the canal's economic utility before it ceased operating.

4. *District of Columbia v. Green,* 29 App. D.C., 296 (1907); Joseph R. Passonneau, *Washington through Two Centuries: A History in Maps and Images* (New York: Monacelli Press, 2004), 164; *Future of Georgetown,* 6; Lyster Dewey Field Book, entries 300–303, Department of Botany Library, National Museum of Natural History, Washington, DC; U.S. Congress, Senate Committee on the District of Columbia, "Removal of Weeds from Lands in the District of Columbia," 55th Cong., 3rd sess., 1899, Senate Report 1550, 3; *Washington Evening Star* (*WES* hereafter), 4 Mar. 1899.

5. 30 Stat. at L. 959, chap. 326; Steven Radosevich and Jodie Holt, *Weed Ecology,* 2nd ed. (New York: Wiley, 1997), 4–6.

6. Historians who have studied weeds have focused on regional landscapes and discussed weeds as pests interfering with crop and livestock production. See Mark Fiege, "The Weedy West: Mobile Nature, Boundaries, and Common Space in the Montana Landscape," *Western Historical Quarterly* 36 (2005): 22–47; Clinton L. Evans, *The War*

on *Weeds in the Prairie West: An Environmental History* (Calgary: University of Calgary Press, 2002); L. Van Sittert, "'The Seed Blows About in Every Breeze': Noxious Weed Eradication in the Cape Colony, 1860–1909," *Journal of Southern African Studies* 26 (2000): 655–74; Freda Knoblock, *The Culture of Wilderness* (Chapel Hill: University of North Carolina Press, 1996), 113–45; F. Timmons, "A History of Weed Control in the United States," *Weed Science* 18 (1970): 294–307.

7. An argument for considering land "beyond the political boundaries of the city" as urban land was made as early as 1928, the same year that planner Benton Mackaye defined the metropolitan environment as a "wilderness—the wilderness not of an integrated, ordered nature, but of a standardized, unordered civilization." See Herbert Dorau and Albert Hinman, *Urban Land Economics* (New York: Macmillan, 1928), 151–52; Benton MacKaye, *The New Exploration: A Philosophy of Regional Planning* (New York: Harcourt, Brace and Company, 1928), 161. The blurry borders and edges of cities in the United States already existed then and are still frequently discussed. See, e.g., Grady Clay, *Real Places* (Chicago: University of Chicago Press, 1994), 154–59; John Stilgoe, *Outside Lies Magic: Regaining History and Awareness in Everyday Life* (New York: Walker and Company, 1998), 21–58; N. McIntyre, K. Knowles-Yanez, and D. Hope, "Urban Ecology as an Interdisciplinary Field: Differences in the Use of 'Urban' between the Social and Natural Sciences," *Urban Ecosystems* 4 (2000): 9–15; Jon C. Teaford, *The Metropolitan Revolution: The Rise of Post-Urban America* (New York: Columbia University Press, 2006), 1–7; Steven Clemants and Gerry Moore, "Patterns of Species Richness in Eight Northeastern United States Cities," *Urban Habitats* 1 (2003): 5, http://www.urbanhabitats.org. Clemants and Moore define urban areas as "all the contiguous counties in and around the city with more than 86% of their populations living in urban areas, as measured by the United States Geological Survey (2000)." On "ecology in" versus "ecology of," see Nancy B. Grimm et al., "Integrated Approaches to Long-Term Studies of Urban Ecological Systems," *Bioscience* 50 (2000): 573–75.

8. On the ecological relationships of people and plants, see Clint Evans, "The 1865 Thistle Act of Upper Canada as an Expression of a Common Culture of Weeds in Canada and the Northern United States," *Canadian Papers in Rural History* 10 (1996): 127–48. Many environmental historians emphasize the relationship of human bodies to their environments, rather than the relationships of people with other organisms. Examples include Linda Nash, *Inescapable Ecologies: A History of Environment, Disease, and Knowledge* (Berkeley: University of California Press, 2006); Ellen Stroud, "Dead Bodies in Harlem: Environmental History and the Geography of Death," in *The Nature of Cities*, ed. Andrew C. Isenberg (Rochester: University of Rochester Press, 2006), 62–76; Sylvia Hood Washington, *Packing Them In: An Archaeology of Environmental Racism in Chicago, 1865–1954* (Lanham: Lexington Books, 2004); Conevery Bolton Valencius, *The Health of the Country: How American Settlers Understood Themselves and Their Land* (New York: Basic Books, 2002).

9. Two recent overviews of urban environmental history are Dorothee Brantz, "The Natural Space of Modernity: A Transatlantic Perspective on (Urban) Environmental History," in *Historians and Nature: Comparative Approaches to Environmental History*, ed. Ursula Lehmkuhl and Hermann Wellenreuther (Oxford: Berg, 2007), 195–225; and Bernd Herrman, "City and Nature and Nature in the City," in Lehmkuhl and Wellenreuther,

Historians and Nature, 226–56. In contrast to the many works of urban environmental history that focus on an array of environmental changes in one city, this book focuses on an ecological dynamic common across cities. Studies of particular American cities and regions include Matthew W. Klingle, *Emerald City: An Environmental History of Seattle* (New Haven: Yale University Press, 2007); Martin V. Melosi and Joseph A. Pratt, eds., *Energy Metropolis: An Environmental History of Houston and the Gulf Coast* (Pittsburgh: University of Pittsburgh Press, 2007); Michael F. Logan, *Desert Cities: The Environmental History of Phoenix and Tucson* (Pittsburgh: University of Pittsburgh Press, 2006); William Deverell and Greg Hise, eds., *Land of Sunshine: An Environmental History of Metropolitan Los Angeles* (Pittsburgh: University of Pittsburgh Press, 2005); Joel A. Tarr, ed., *Devastation and Renewal: An Environmental History of Pittsburgh and Its Region* (Pittsburgh: University of Pittsburgh Press, 2004); Ari Kelman, *A River and Its City: The Nature of Landscape in New Orleans* (Berkeley: University of California Press, 2003); Matthew Gandy, *Concrete and Clay: Reworking Nature in New York City* (Boston: MIT Press, 2002); Char Miller, ed., *On the Border: An Environmental History of San Antonio* (Pittsburgh: University of Pittsburgh Press, 2001); Craig E. Colten, ed., *Transforming New Orleans and Its Environs: Centuries of Change* (Pittsburgh: University of Pittsburgh Press, 2000); Andrew Hurley, ed., *Common Fields: An Environmental History of St. Louis* (St. Louis: Missouri Historical Society Press, 1997); William Cronon, *Nature's Metropolis: Chicago and the Great West* (New York: Norton, 1991). Scientific statements about the significance of urban ecosystems include S. T. A. Pickett et al., "Urban Ecological Systems: Linking Terrestrial Ecological, Physical, and Socioeconomic Components of Metropolitan Areas," *Annual Review of Ecology and Systematics* 32 (2001): 127–57; Grimm et al., "Integrated Approaches," 571–84. Studies of natural life and environmental science by environmental historians and historians of science suggest the care needed to use ecological ideas and findings to explain the past. See Joel B. Hagen, "Teaching Ecology during the Environmental Age, 1965–1980," *Environmental History* 13 (2008): 704–23; Sharon E. Kingsland, *The Evolution of American Ecology, 1890–2000* (Baltimore: Johns Hopkins University Press, 2005); Joel B. Hagen, *An Entangled Bank: The Origins of Ecosystem Ecology* (New Brunswick: Rutgers University Press, 1992); Gregg Mitman, *The State of Nature: Ecology, Community, and American Social Thought, 1900–1950* (Chicago: University of Chicago Press, 1992); Ronald Tobey, *Saving the Prairies: The Life Cycle of the Founding School of American Plant Ecology, 1895–1955* (Berkeley: University of California Press, 1981).

10. Will Turner, Toshihiko Nakamura, and Marco Dinetti, "Global Urbanization and the Separation of Humans from Nature," *BioScience* 54 (2004): 585–90; Yi-Fu Tuan, "The City: Its Distance from Nature," *Geographical Review* 68 (1978): 1–12; Thomas B. Simpson, "Ecological Restoration and Re-Understanding Ecological Time," *Ecological Restoration* 23 (2005): 46–51; Martin Zobel, "Plant Species Coexistence: The Role of Historical, Evolutionary and Ecological Factors," *Oikos* 65 (1992): 314–20; Gary McDonough, "The Geography of Emptiness," in *The Cultural Meaning of Urban Space*, ed. Robert Rotenberg and Gary McDonough (Westport, CT: Bergin and Garvey, 1993), 3–15. On the perception of time in urban environments, see Edward Relph, *The Modern Urban Landscape* (Baltimore: Johns Hopkins University Press, 1987), 1–11; Kevin Lynch, *What Time Is This Place?* (Boston: MIT Press, 1973), 29–37, 66–72, 119–28.

11. Edgar Anderson, *Plants, Man, and Life* (Boston: Little, Brown and Company, 1952), 16, 150; Alfred W. Crosby, *Ecological Imperialism: The Biological Expansion of Europe, 900–1900* (New York: Cambridge University Press, 1986), 170. On ruderal plants, see J. P. Grime, *Plant Strategies, Vegetation Processes, and Ecosystem Properties*, 2nd ed. (Chichester: Wiley, 2001), 132.

12. "A True and Sincere Declaration," reprinted in Alexander Brown, *The Genesis of the United States: A Narrative of the Movement in England, 1605–1616* (Boston: Houghton-Mifflin, 1890), 1: 352; George Chalmers, *Political Annals of the Present United Colonies, from Their Settlement to the Peace of 1763* (London, 1780; New York: Ayer Publishing Company, 1968), 36, 65, 73, 111 (citations to the 1968 edition); Charles Skinner, *Nature in a City Yard* (New York: Century, 1897), 90; Thomas Hooker, *The Application of Redemption by the Effectual Work of the Word, and Spirit of Christ, for the Bringing Home of Lost Sinners to God* (London: Peter Cole, 1656), 380, accessed in Early English Books Online; William Cronon, *Changes in the Land: Indians, Colonists, and the Ecology of New England* (New York: Hill and Wang, 1983), 56–57; Pastorius quoted in E. Gordon Alderfer, "Pastorius and the Origins of Pennsylvania German Culture," *American-German Review* 17 (Feb. 1951): 9; Tom Wessels, *Reading the Forested Landscape: A Natural History of New England* (Woodstock, NY: Countryman Press, 1997), 34–37; Carolyn Merchant, *Ecological Revolutions: Nature, Gender, and Science in New England* (Chapel Hill: University of North Carolina Press, 1989), 79–86; Carolyn Merchant, *Reinventing Eden: The Fate of Nature in Western Culture* (New York: Routledge, 2004), 99–100. Chalmers perpetuated the "vile weed" and "contemptible weed" designations.

13. Benjamin Rush to Walter Jones, 30 July 1776, in *Letters of Delegates to Congress, 1774–1789*, ed. Paul Smith (Washington, DC: Library of Congress, 1979), 4: 582; *Gazette of the United States*, 10, 21 June 1800; Harry Warfel, *Noah Webster, Schoolmaster to America* (New York: Macmillan, 1936), 292; Noah Webster, *An American Dictionary of the English Language* (New York: Converse, 1828), s.v. "weed"; Richard Rollins, "Words as Social Control: Noah Webster and the Creation of the American Dictionary," *American Quarterly* 28 (1976): 415–30; *Hudson (NY) Balance*, 28 Aug. 1804; Linda K. Kerber, *Federalists in Dissent: Imagery and Ideology in Jeffersonian America* (Ithaca: Cornell University Press, 1970), 20; Richard Wade, *The Urban Frontier: The Rise of Western Cities* (Cambridge: Harvard University Press, 1959), 136; Thomas Benton, "The Political Career of Andrew Jackson (United States Senate, January 12, 1837)," in *The World's Best Orations: From the Earliest Period to the Present Time*, ed. David Josiah Brewer (St. Louis: Kaiser, 1899), 2: 415, 424.

14. Richard W. Judd, *Common Lands, Common People: The Origins of Conservation in Northern New England* (Cambridge: Harvard University Press, 1997), 94; "Destroy Your Weeds," *Farmer's Cabinet* 4 (Sept. 1839): 34; William Darlington, *American Weeds and Useful Plants*, rev. ed., with additions by George Thurber (New York: Orange Judd and Company, 1859), xiii; Knobloch, *Culture of Wilderness*, 122–34.

15. *The Great Battle between Slavery and Freedom* (Boston: Greene, 1856), 21–22; Henry Steele Commager, *Theodore Parker, Yankee Crusader* (Boston: Little, Brown, and Company, 1936), 199–200; George Cheever, *God against Slavery: and the Freedom and Duty of the Pulpit to Rebuke against It, as a Sin against God* (New York: Joseph Ladd,

1857), 171; Martha Griffith Browne, *Autobiography of a Female Slave* (New York: Redfield, 1857), 288; *Henry Ward Beecher, Patriotic Addresses in America and England, 1850–1885*, ed. John Howard (New York: Fords, Howard, and Hulbert, 1888), 334; Charles Fessenden Morse, *Letters Written during the Civil War* (n.p., 1898), 85; *The Reminiscences of Carl Schurz*, ed. Frederic Bancroft and William Dunning (New York: McClure Company, 1908), 3: 167, 198.

16. Thomas Bender, *A Nation among Nations: America's Place in World History* (New York: Hill and Wang, 2006), 156–57, 182–88; Asa Gray, "Pertinacity and Predominance of Weeds," *American Journal of Science and Arts* 18 (1879): 161–62; John Burroughs, *Pepacton* (Boston: Houghton, Mifflin, 1909), 212–31; Grant Allen, "American Jottings," *Fortnightly Review*, new series, 240 (1886): 731–36; Henry Mann, *The Land We Live In* (New York: Christian Herald, 1896), 327; Alice Eastwood, "The Plant Inhabitants of Nob Hill," *Erythea* 6 (1898): 61–63; Willis Blatchley, *Indiana Weed Book* (Indianapolis: Nature Publishing, 1912), 13. On American perceptions of exotic plants in the eighteenth and early nineteenth centuries, see Philip J. Pauly, *Fruits and Plains: The Horticultural Transformation of America* (Cambridge: Harvard University Press, 2007), 10–32.

17. Ralph Waldo Emerson, *Fortune of the Republic: Lecture Delivered at the Old South Church, March 30, 1878* (Boston: Houghton, 1879), 3–4; Byron Halsted, "Migration of Weeds," *Proceedings of the American Association for the Advancement of Science* 39 (1890): 304–12; Lyster Dewey, "Legislation against Weeds" in *Yearbook of Agriculture* (Washington, DC: 1896), 3–12; Gerald McCarthy, "American Weeds," *Science* 20 (1892): 38; "Nature Study Round Table," *School Journal* 60 (1912): 381; Elmer Campbell, "What Is a Weed?" *Science* 58 (1923): 50; O. A. Stevens, "What Is a Weed?" *Science* 59 (1924): 360–61; "Why Is a Weed a Weed?" *Science* 83, supp. (24 Apr. 1936): 10; Edgar Anderson, "A Classification of Weeds and Weed-Like Plants," *Science* 89 (1939): 364–65; J. de Wet, "The Origin of Weediness in Plants," *Proceedings of the Oklahoma Academy of Sciences* 47 (1966): 14.

18. J. Horace McFarland, "Gassing the Garden," *Independent* 94 (1918): 210; Chase Osborn, "Shall We Deport Our Undesirables?" *Outlook* 124 (1920): 77; William Wood, *Captains of the Civil War* (New Haven: Yale University Press, 1921), 57–58; Theodore R. Bassett, "The Third National Negro Congress," *Communist* 19 (1940): 549; Arthur Wickwire, *The Weeds of Wall Street* (New York: Newcastle Press, 1933), frontispiece; "January 3, 1934 Annual Message to Congress," in *The Public Papers and Addresses of Franklin D. Roosevelt*, vol. 3, *The Advance of Recovery and Reform* (New York : Random House, 1938), 129; Herbert Hoover, *Memoirs of Herbert Hoover, 1920–1933* (New York: Macmillan, 1952), 223; Herbert Hoover, *Addresses upon the American Road* (New York: Scribner's Sons, 1938), 53.

19. Gerald Phelan, "How Can Scholarship Contribute to the Relief of International Tensions," in *Learning and World Peace*, ed. Lyman Bryson, Louis Finkelstein, and R. MacIver (New York: Conference, 1948), 117–18; Philip Murray, "American Labor and the Threat of Communism," *Annals of the American Academy of Political and Social Science* 274 (1951): 125; Morrie Ryskind, "Attacks Pollsters," *Waterloo Daily Courier*, 25 May 1964.

20. "A Week for Earth," *Pasadena Star-News*, 15 Apr. 1974; Adam Rome, "The Genius of Earth Day," *Environmental History* 15 (2010): 194–205.

CHAPTER ONE

1. *Overland Monthly* 29 (1897): 658; Charles Skinner, *Flowers in the Pave* (Philadelphia: Lippincott, 1900), 8–9, 30, 64–71, 191; Charles Skinner, *Nature in a City Yard* (New York: Century, 1897), 138–39; "Now, Look at the Fields," *Brooklyn Daily Eagle* (hereafter *BDE*), 24 Mar. 1895. On the emergence of Brooklyn as an "urban residential community," see Marc Linder and Lawrence Zacharias, *Of Cabbages and Kings County: Agriculture and the Formation of Modern Brooklyn* (Iowa City: University of Iowa Press, 1999), 4, 112–34.

2. Michael Kammen, *A Time to Every Purpose: The Four Seasons in American Culture* (Chapel Hill: University of North Carolina Press, 2004), 26; Thomas R. Dunlap, *Faith in Nature: Environmentalism as Religious Quest* (Seattle: University of Washington Press, 2004); Skinner, *Nature in a City Yard*, 58; Skinner, *Flowers in the Pave*, 202–11.

3. John Bealle, "Another Look at Charles M. Skinner," *Western Folklore* 53 (1994): 112; Skinner, *Flowers in the Pave*, 13, 37; "Sanitary Sketches: A Summer Ramble," *BDE*, 18 June 1873; Andrew Hurley, "Creating Ecological Wastelands: Oil Pollution in New York City, 1870–1900," *Journal of Urban History* 20 (1994): 340–64; Matthew Gandy, *Concrete and Clay: Reworking Nature in New York City* (Cambridge: MIT Press, 2003), 19–113; "Kissena Grove," http://www.nycgovparks.org/sub_your_park/historical_signs/hs_historical_sign.php?id_12242; Elizabeth Britton, "Vanishing Wild Flowers," *Contributions from the New York Botanical Garden* 16 (1899–1902) 1: 85–94; Minnie Reynolds, "Wild Flowers of New York," *New York Times Magazine*, 2 Feb. 1902; "Wild Flowers Becoming Extinct," *Chicago Daily Tribune* (hereafter *CT*), 26 May 1912; William Rich, "City Botanizing," *Rhodora* 10 (Sept. 1908): 150–55; Edward Steele, "Sixth List of Additions to the Flora of Washington," *Proceedings of the Biological Society Washington* 14 (1901): 52, 58. On the scientific significance of this botanizing, see Robert E. Kohler, *All Creatures: Naturalists, Collectors, and Biodiversity, 1850–1950* (Princeton: Princeton University Press, 2006), 30–40. Skinner grew up in Cambridge, MA, and Hartford, CT, cities home to fewer than forty thousand people, before joining Brooklyn's nearly four hundred thousand people in 1881.

4. W. W. Bailey, "About Weeds," *American Naturalist* 12 (1878): 742; Steele, "Sixth List," 75; Rich, "City Botanizing," 150–55; Lester Ward, *Flora of Washington and Vicinity* (Washington, DC: Government Printing Office, 1881), 90; Theodore Dreiser, "New Knowledge of Weeds," *Ainslee's* 8 (Jan. 1902): 533–38; Alice Eastwood, "Plant Inhabitants of Nob Hill," *Erythea* 6 (1898): 61–63. The concern with weeds in cities was rarely expressed in terms of the "invasion" of weeds from beyond the United States and their role in the destruction of "native" plants. Botanists who collected plants, dried them, and sold them may have been most sensitive to and observant of the disappearance of "native species," in part because such medicinal herbs were sources of income. See M. Bebb, "Recently Introduced Plants in and about Rockford, Ill.," *Botanical Gazette* 7 (June 1882): 69. This history is based on textual records and generally accepts past people's data as botanically correct, whether they were amateurs or academically trained botanists. This study has not cross-checked their writings with specimens in local herbaria and occasionally guesses at what plants they encountered when their reported names do not match present-day nomenclature. Studies of cities and their environments by scientifically trained botanists are probably the most reliable sources. Examples include Henri Hus, "An

Ecological Cross Section of the Mississippi River in the Region of St. Louis, Missouri," *Annual Report of the Missouri Botanical Garden* 19 (1908): 214–50; John Harshberger, "A Phyto-Geographic Sketch of Extreme Southeastern Pennsylvania," *Bulletin of the Torrey Botanical Club* 31 (1904): 125–59. On the development of botanical study, see Elizabeth Keeney, *The Botanizers: Amateur Scientists in Nineteenth-Century America* (Chapel Hill: University of North Carolina Press, 1992); Sharon E. Kingsland, *The Evolution of American Ecology, 1890–2000* (Baltimore: Johns Hopkins University Press, 2005); Sally Gregory Kohlstedt, "Nature, Not Books: Scientists and the Origins of the Nature-Study Movement in the 1890s," *Isis* 96 (2005): 324–52.

5. Britton, "Vanishing Wild Flowers," 90–91; Skinner, *Flowers in the Pave*, 185; Skinner, *Nature in a City Yard*, 63–68, 88–89, 128.

6. Lyster Dewey, "Weeds in Cities and Towns," in *Yearbook of Agriculture* (Washington, DC, 1898), 197–99. Another version of this essay appeared in the Oct. 1899 *Sanitarian*, and it was summarized in the *St. Louis Medical Gazette* 4 (1900): 341. Newspapers also disseminated Dewey's findings. See, e.g., "Weeds of Many Cities," *WP*, 7 July 1899; "Nurseries of Imported Weeds," *Los Angeles Times* (hereafter *LAT*), 31 Mar. 1901 (initially published in *CT*); entry 305, Lyster Dewey Field Book, Department of Botany Library, National Museum of Natural History, Washington, DC.

7. John Henry Hepp IV, *The Middle-Class City: Transforming Space and Time in Philadelphia, 1876–1926* (Philadelphia: University of Pennsylvania Press, 2003); Eric Sandweiss, "Mind Reading the Urban Landscape: An Approach to the History of American Cities," in *Historical Archaeology and the Study of American Culture*, ed. Bernard Herman (Winterthur: Winterthur Museum, 1996), 323; Roger D. Simon, *The City-Building Process: Housing and Services in New Milwaukee Neighborhoods 1880–1910*, rev. ed. (Philadelphia: American Philosophical Society, 1996), 1–7; David R. Contosta, *Suburb in the City: Chestnut Hill, Philadelphia, 1850–1990* (Columbus: Ohio State University Press, 1992), 33–34; Edward Relph, *The Modern Urban Landscape* (Baltimore: Johns Hopkins University Press, 1987), 3, 260; Richard A. Walker, "A Theory of Suburbanization: Capitalism and the Construction of Urban Space in the United States," in *Urbanization and Urban Planning in Capitalist Society* (London: Methuen, 1981), 384–89.

8. Thomas J. Campanella, *Republic of Shade: New England and the American Elm* (New Haven: Yale University Press, 2003), 69–121; Fred E. H. Schroeder, *Front Yard America: The Evolution and Meanings of a Vernacular Domestic Landscape* (Bowling Green, OH: Bowling Green State University Popular Press, 1993), 33–39.

9. Carole Shammas, "The Space Problem in Early United States Cities," *William and Mary Quarterly* 57 (2000): 537–42; Anne Krulikowski, "'Farms Don't Pay': The Transformation of the Philadelphia Metropolitan Landscape, 1880–1930," *Pennsylvania History* 72 (2005): 204–5; John W. Reps, *Saint Louis Illustrated* (Columbia: University of Missouri Press, 1989); John W. Reps, *Washington on View* (Chapel Hill: University of North Carolina Press, 1991); Joseph R. Passonneau, *Washington through Two Centuries: A History in Maps and Images* (New York: Monacelli Press, 2004). Camille Dry and Richard Compton's *Pictorial St. Louis* (St. Louis: Compton and Co., 1876) was exceptional in revealing to viewers St. Louis's unbuilt pockets and expanses. The volume aspired to render "accurately" both buildings and the "topography of unoccupied property." Illustrations show sinkholes, trees between homes, fences, and even leaping dogs. The authors' pens

manicured ground as smooth, not unlike airbrushes applied to photographed faces. This outer city was space inside St. Louis's municipal limits. Sometimes past people called such space "suburban" to convey its spatial distance from cities' dense centers. In this book, "outer city" is used to distinguish land inside of cities from suburbs that were autonomous entities outside of major cities. The adjective *suburban* conveys little significant about space, including its ecological time. Important efforts to conceptualize this space include Kohler, *All Creatures*, 30–46; Melanie L. Simo, *Forest and Garden: Traces of Wildness in a Modernizing Land, 1897–1949* (Charlottesville: University of Virginia Press, 2003); John R. Stilgoe, *Borderland: Origins of the American Suburb, 1820–1939* (New Haven: Yale University Press, 1988); Kenneth T. Jackson, *Crabgrass Frontier: The Suburbanization of the United States* (New York: Oxford University Press, 1985), 11.

10. Eric Sandweiss, *St. Louis: The Evolution of an American Urban Landscape* (Philadelphia: Temple University Press, 2001), 29–36, 67–69, 147; Hus, "Ecological Cross Section," 214–50; *St. Louis Post-Dispatch* (hereafter *SLPD*), 10 Aug., 2 Sept. 1905.

11. Kenneth R. Bowling, *Creating the Federal City, 1774–1800* (Washington, DC: American Institute of Architects Press, 1988), 7, 23–60, 80–98; Sarah Luria, *Capital Speculations: Writing and Building Washington, D.C.* (Durham: University of New Hampshire Press, 2006); Passonneau, *Washington through Two Centuries*, 30, 38, 78, 90, 106, 125, 164; Constance McLaughlin Green, *Washington: Capital City, 1879–1950* (Princeton: Princeton University Press, 1963), 11–12; Timothy Davis, "Inventing Nature in Washington, D.C.," in *Inventing for the Environment*, ed. Arthur P. Molella and Joyce Bedi (Boston: MIT Press, 2003), 34–51; Dewey, "Weeds in Cities and Towns," 198; *Flora Columbiana, or Catalogue of Plants Growing without Cultivation* (Washington, DC: Columbia Press, 1876); Ward, *Flora of Washington*, 64–111. Washington City and the District of Columbia were initially not the same geographic entities and did not become so until 1904.

12. M. Kent, R. Stevens, and L. Zhang, "Urban Plant Ecology Patterns and Processes: A Case Study of the Flora of the City of Plymouth, Devon, U.K.," *Journal of Biogeography* 26 (1999): 1282–86, 1295; J. Bastow Wilson and Warren McG. King, "Human-Mediated Vegetation Switches as Processes in Landscape Ecology," *Landscape Ecology* 10 (1995): 191–96; Matthew Vessel and Herbert Wong, *Natural History of Vacant Lots* (Berkeley: University of California Press, 1987), 13–21; Theodore Sudia, *The Vegetation of the City* (Washington, DC: Department of the Interior, 1974), 6; Herbert Sukopp, Hans-Peter Blume, and Wolfram Kunick, "The Soil, Flora, and Vegetation of Berlin's Waste Lands," in *Nature in Cities: The Natural Environment in the Design and Development of Urban Green Space*, ed. Ian Laurie (Chichester: Wiley, 1979), 116; Wolfgang Aey, "Historical Approaches to Urban Ecology: Methods and First Results from a Case Study," in *Urban Ecology: Plants and Plant Communities in Urban Environments*, ed. Herbert Sukopp and Hejny Slavomil (The Hague: SPB Academic Publishing, 1990), 119–21; Thomas B. Simpson, "Ecological Restoration and Re-Understanding Ecological Time," *Ecological Restoration* 23 (2005): 47; J. P. Grime, *Plant Strategies, Vegetation Processes, and Ecosystem Properties*, 2nd ed. (Chichester: Wiley, 2001), 63–64, 84, 87–98, 117, 127–28, 135, 245–46; Franz Rebele, "Colonization and Early Succession on Anthropogenic Soils," *Journal of Vegetation Science* 3 (1992): 204–6.

13. Booth Courtenay and James Zimmerman, *Wildflowers and Weeds* (New York: Van Nostrand Reinhold, 1972), xvi.

14. F. A. Bazzaz, *Plants in Changing Environments: Linking Physiological, Population, and Community Ecology* (New York: Cambridge University Press, 1996), 38–45; Sudia, *Vegetation of the City*, 6; Aey, "Historical Approaches," 119–21; Ingo Kowarik, "Some Responses of Flora and Vegetation to Urbanization in Central Europe," in Sukopp and Slavomil, *Urban Ecology*, 51–53; Vessel and Wong, *Natural History of Vacant Lots*, 1–29; Matthew T. M. Crowe, "Lots of Weeds: Insular Phytogeography of Vacant Urban Lots," *Journal of Biogeography* 6 (1979): 169–81; Grime, *Plant Strategies*, 85–87, 118–35, 185; S. T. A. Pickett, M. Cadenasso, and S. Bartha, "Implications from the Buell-Small Succession Study for Vegetation Restoration," *Applied Vegetation Science* 4 (2001): 43–45; Anthony Krzysik et al., "A Primer of Successional Ecology," *Landscape Architecture* 71 (1981): 482–86; Rebele, "Colonization and Early Succession," 204–6.

15. Crowe, "Lots of Weeds," 175–76; Vessel and Wong, *Natural History of Vacant Lots*, 19; Grime, *Plant Strategies*, 87–124, 179–256, 337; Pickett et al., "Implications," 47–49; Harshberger, "Phyto-Geographic Sketch," 158.

16. Linder and Zacharias, *Of Cabbages and Kings County*, 220–44; Contosta, *Suburb in the City*, 41; Thomas Meehan, "Distribution of Weeds," *Bulletin of the Torrey Botanical Club* 10 (1883): 24; Grime, *Plant Strategies*, 83–84. The formation of "seed banks" in the ground and the theory of initial floristic development suggest that the plants that flourished when crop production ceased were deposited during periods when the land had been farmed.

17. Zachary J. S. Falck, "Controlling the Weed Nuisance in Turn-of-the-Century American Cities," *Environmental History* 7 (2002): 626 n. 6; Robert D. Lewis and Richard Walker, "Beyond the Crabgrass Frontier: Industry and the Spread of North American Cities, 1850–1950," *Journal of Historical Geography* 27 (2001): 3–19; Ann Durkin Keating, *Building Chicago: Suburban Developers and the Creation of a Divided Metropolis* (Columbus: Ohio State University Press, 1988), 68–70; Michael J. Doucet, "Urban Land Development in Nineteenth-Century North America," *Journal of Urban History* 8 (1982): 306–15, 329; Contosta, *Suburb in the City*, 33–41; "Sweet Clover," *WP*, 7 Aug. 1896.

18. Simon, *City-Building Process*, 60, 79; William S. Worley, *J. C. Nichols and the Shaping of Kansas City: Innovation in Planned Residential Communities* (Columbia: University of Missouri Press, 1990), 1–5; Sandweiss, *St. Louis*, 144–65. For one estimate of ragweed seed production, see France Royer and Richard Dickinson, *Weeds of the Northern U.S. and Canada* (Edmonton: University of Alberta Press, 1999).

19. Falck, "Controlling the Weed Nuisance," 626 n. 6; John R. Stilgoe, *Metropolitan Corridor: Railroads and the American Scene* (New Haven: Yale University Press, 1983), 3; Clay McShane, *Down the Asphalt Path: The Automobile and the American City* (New York: Columbia University Press, 1994), 14–27; Stilgoe, *Borderland*, 70–80; Victor Mühlenbach, "Contributions to the Synanthropic (Adventive) Flora of the Railroads in St. Louis, Missouri," *Annals of the Missouri Botanical Garden* 66 (1979): 6, 92–102; J. Bastow Wilson and King, "Vegetation Switches," 194; Dix Noel, "Nuisances from Land in Its Natural Condition," *Harvard Law Review* 56 (1943): 778–81; *State v. Boehm*, 100 N.W. 95 (1904); *Doeppenschmidt v. International & G. N. R. Co.*, 101 S.W. 1080 (1907); E. Hill, "Notes on the Flora of Chicago and Vicinity," *Botanical Gazette* 17 (1892): 246–52; Hus, "Ecological Cross Section," 180–82, 199–200; E. Hill, "Notes on Migratory Plants," *Bulletin of the*

Torrey Botanical Club 29 (1902): 569; Dewey, "Weeds in Cities and Towns," 193; Harshberger, "Phyto-Geographic Sketch," 155–57.

20. Martin V. Melosi, *The Sanitary City* (Baltimore: Johns Hopkins University Press, 2000), 194–200; Martin V. Melosi, *Garbage in Cities,* rev. ed. (Pittsburgh: University of Pittsburgh Press, 2005), 23–34, 146–53; Brian D. Crane, "Filth, Garbage, and Rubbish: Refuse Disposal, Sanitary Reform and Nineteenth-Century Yard Deposits in Washington, D.C.," *Historical Archaeology* 34 (2000): 20–38; Craig E. Colten, "Chicago's Waste Lands: Refuse Disposal and Urban Growth, 1840–1990," *Journal of Historical Geography* 20 (1994): 124–33; Philadelphia Department of Public Health and Charities, *Digest of the Statutes and Ordinances Relating to Public Health* (Philadelphia, 1911), 164; Dewey, "Weeds in Cities and Towns," 197–99; Hill, "Notes on the Flora of Chicago," 247; Addison Brown, "Plants Introduced with Ballast and on Made Land," *Bulletin of the Torrey Botanical Club* 6 (1878): 255–58; *BDE,* 8 Sept. 1895; "Ragged Edges of the City," *Craftsman* 24 (1913): 635; Clay McShane and Joel Tarr, *The Horse in the City: Living Machines in the Nineteenth Century* (Baltimore: Johns Hopkins University Press, 2007), 122–27; H. Skeels, "A Weed Study," *American Botanist* 8 (1905): 26–28.

21. J. C. Nichols, *Real Estate Subdivision* (Washington, DC: American Civic Association, 1912), 10–11; Linder and Zacharias, *Of Cabbages and Kings County,* 25; Olmsted's 1884 remark to B. Ramsey quoted in Robert Fogelson, *Bourgeois Nightmares: Suburbia, 1870–1930* (New Haven: Yale University Press, 2005), 28; *BDE,* 8 Sept. 1895; Richard Bushman, *The Refinement of America: Persons, Houses, Cities* (New York: Knopf, 1992), 353–68. Dismay over this landscape began as early as 1855, when the *Crayon* observed that in the United States the "city grows into country; we never know when we leave one or enter the other." See Jackson, *Crabgrass Frontier,* 71.

22. "Definition of a Weed," *American Botanist* 5 (1903): 39; Gerald McCarthy, "American Weeds," *Science* 20 (15 July 1892): 38; Skinner, *Nature in a City Yard,* 91–92; "Weeds," *Meehan's Monthly* (hereafter *MM*) 4 (1894): 190; "Weeds," *MM* 9 (1899): 126. The artificiality of places without weeds was evident in horticulturist Liberty Hyde Bailey's remark that outside of cultivated areas even "species that are habitual weeds . . . can scarcely be called weeds." See Liberty Hyde Bailey, "Weeds," in *Cyclopedia of American Horticulture* (New York: Macmillan, 1900), 4: 1972.

23. This value was recognized, although not priced, in Joseph Cocannouer, *Weeds, Guardians of the Soil* (New York: Devin-Adair, 1950). Today, weeds might be seen as providing disturbance regulation. See Robert Costanza et al., "The Value of the World's Ecosystem Services and Natural Capital," *Nature* 387 (1997): 254.

24. Campanella, *Republic of Shade,* 69–121; H. Cleveland, *Landscape Architecture* (Chicago: Jansen, McClurg and Co., 1873), 12–13; Frederick Law Olmsted, "Village Improvement," *Atlantic* 95 (June 1905): 799–803; Charles Mulford Robinson, *The Improvement of Towns and Cities,* 4th ed. (New York: Putnam's Sons, 1913), 133–41; Mrs. Louis Marion McCall, "Making St. Louis a Better Place to Live in," *Chautauquan* 36 (1903): 405; William Wilson, *The City Beautiful Movement* (Baltimore: Johns Hopkins University Press, 1989); Bonj Szczygiel, "'City Beautiful' Revisited: An Analysis of Nineteenth-Century Civic Improvement Efforts," *Journal of Urban History* 29 (2003): 107–32. On the City Beautiful in St. Louis, see Sandweiss, *St. Louis,* 183–230.

25. *WES,* 29 Sept. 1900; *Tree Planting in St. Louis* (St. Louis: Nixon-Jones, 1902), 5–8, 34; *LAT,* 14 Oct. 1900; "Improve the School Grounds," *Santa Fe New Mexican* (*SFNM* hereafter), 18 May 1905; *SFNM,* 11 July 1907; "St. Louis Realty Market," *St. Louis Builder* 8 (May 1901): 17; "Objects of West End Residents' Protective Association," Missouri Historical Society (hereafter MHS), St. Louis; *Civic Improvement Bulletin* 1 (Apr. 1902): 1; "Victoria 50 and 25 Years Ago," *Victoria Advocate,* 1 Mar. 1954; "For Cleaning Up City Vacant Lots," *LAT,* 14 Aug. 1903; Mary Gray, "Putting Your Civic House in Order," *Craftsman* 30 (June 1916): 283; Nichols, *Real Estate Subdivision,* 13.

26; Phyllis Andersen, "The City and the Garden," in *Keeping Eden: A History of Gardening in America,* ed. Walter T. Punch (Boston: Little, Brown, 1992), 162–65; Virginia Tuttle Clayton, *The Once and Future Gardener: Garden Writing from the Golden Age of Magazines, 1900–1940* (Boston: David Godine, 2000), xiv–xxii; Ann Leighton, *American Gardens of the Nineteenth Century* (Amherst: University of Massachusetts Press, 1987), 228–60; Schroeder, *Front Yard America,* 17–40; I. Tabor, "Garden Wonders in a 25 x 35 Backyard," *Garden Magazine* 2 (1906): 271; *Trillia* 4 (1914): 61; Ida Bennett, *The Making of a Flower Garden* (New York: Stokes Company, 1919), 114–15.

27. Elizabeth Strang, "Developing a City Garden," *House and Garden* 29 (May 1916): 42; Charles Skinner, *Little Gardens: How to Beautify City Yards and Small Country Spaces* (New York: Appleton, 1904), 24–25, 80; "Wild Garden," in *Standard Cyclopedia of Horticulture* (New York: Macmillan, 1900), 1976–78; George Klingle, "A Weedy Garden," *House and Garden* 24 (July 1913): 6–7, 57; Grace Tabor, "Wild Flowers in the Garden," *House and Garden* 24 (July 1913): 29–31.

28. Bushman, *Refinement of America,* 137–38; Alexander Jackson Downing, *Treatise on the Theory and Practice of Landscape Gardening,* 6th ed. (New York: Moore, 1859), 286, 422–24; "A Good Lawn," advertisement, *Garden and Forest* 5 (11 May 1892): iii; Robert Sterling, "Care of the Lawn," *House and Garden* 12 (Nov. 1907): 184; Virginia Scott Jenkins, *The Lawn: A History of an American Obsession* (Washington, DC: Smithsonian Institution Press, 1994), 37–85, 103; Ted Steinberg, *American Green: The Obsessive Quest for the Perfect Lawn* (New York: Norton, 2007), 11–13; Schroeder, *Front Yard America,* 101–5. Jenkins notes that even some lawn proponents accepted dandelions as part of the home grounds and that the 1921 pamphlet "A Lawn without Dandelions" marked the "new standard for weed-free lawns."

29. Robinson, *Improvement of Towns and Cities,* 133; Henry Sargent, "Supplement to the Sixth Edition," in Downing, *Treatise,* 429; Olmsted, "Village Improvement," 803; *House Beautiful* 18 (June 1905), reprinted in Clayton, *Once and Future Gardener,* 131; *Scribner's Magazine* 56 (July 1914), reprinted in Clayton, *Once and Future Gardener,* 128; Thorstein Veblen, *The Theory of the Leisure Class* (New York: Macmillan, 1899), 132; "Now, Look at the Fields," 6.

30. Jacob Riis, *How the Other Half Lives* (New York: Scribner's Sons, 1890), 163; Charles Frederick Weller, *Neglected Neighbors* (Philadelphia: Winston, 1909), 215–16; Benjamin Rush, *An Inquiry into the Various Sources of the Usual Forms of Summer and Autumnal Disease* (Philadelphia: Conrad and Co., 1805), 5–10; Annual Report of the St. Louis Health Department (hereafter ARSLHD), in *Mayor's Message* (1880–81), 354; St. Louis Ordinance 18,415 (approved 9 Apr. 1896), ARSLHD (1895–96), 129; "Sweet Clover"; "Weeds," *WP,* 8 July 1898; Dewey, "Weeds in Cities and Towns," 197–98;

"Our Superiority in Weeds," *WP*, 23 Aug. 1898. Kirkwood, a suburb of St. Louis, passed a law closely resembling St. Louis's. See "Revised Ordinances of the City of Kirkwood," chapter 7, section 146 (Kirkwood Tablet, 1909).

31. ARSLHD (1880–81), 354; *SLPD*, 30 Aug. 1900, 7 Aug. 1905; *WES*, 4 Aug. 1906; "New Argument against Weeds," *American Journal of Public Health* 3 (1913): 714–15; Falck, "Controlling the Weed Nuisance," 626–27 n. 7.

32. *Washington Times*, 21 Aug. 1904; Dewey, "Weeds in Cities and Towns," 197–98; Charles Pollard, *The Families of Flowering Plants* (Washington, DC: Plant World Co., 1902), 250–53; "Hay Fever," *St. Louis Medical and Surgical Journal* 71 (1896): 59; "Report of the Health Officer of the District of Columbia," in *Report of the Commissioners of the District of Columbia* (hereafter RHO-RDCD) (Washington, DC, 1905–6), 41; "Weeds of Many Cities"; *SLPD*, 11 Aug. 1905; "Pollen Theory of Hay Fever Untenable," *St. Louis Courier of Medicine* 26 (1902): 136; "Great Necessity of Cleaning This City," *SFNM*, 14 Sept. 1907; "Minor City Topics," *SFNM*, 14 Aug. 1907; "Killed by Poisonous Weeds," *BDE*, 5 Sept. 1900; *SLPD*, 6 Aug. 1912; Charles Saunders, "Poison Ivy and Its Extermination," *American Botanist* 5 (1903): 53–54.

33. Eugene McQuillin, *Municipal Code of St. Louis* (St. Louis, 1901), 552–53; Senate Committee on the District of Columbia, "Removal of Weeds from Lands in the District," 55th Cong., 3rd sess., 1899, S. Report 1550, 3; *SLPD*, 10 Aug. 1905; *SLPD*, 28 July 1905. "Moral environmentalism" is examined in Paul Boyer, *Urban Masses and Moral Order in America, 1820–1920* (Cambridge: Harvard University Press, 1978), and Stanley Schultz, *Constructing Urban Culture* (Philadelphia: Temple University Press, 1989). Ecologist Frederic Clements labeled plants such as sunflowers, switchgrass, wild rye, goldenrod, evening primrose, beggartick, boneset, compass plant, and jimson weed "vigorous growers." See Frederic Clements, *Research Methods in Ecology* (Lincoln: University Publishing Company, 1905), 312–13. In dry Western cities, weeds were considered "highly combustible covering on the ground," which threatened property. See "Clean the Vacant Lots," *LAT*, 15 Aug. 1903.

34. Charles Towne, *Manhattan* (New York: Kennerley, 1909), 44; *State v. Flutcher*, 166 Mo. 582; "Annual Report of the St. Louis Park Department," in *Mayor's Message* (1892–93), 311; Senate Committee, "Removal of Weeds from Lands in the District," 3; "Report of the Major and Superintendant of the Metropolitan Police Department," in *Report of the Commissioners of the District of Columbia* (hereafter RMPD-RDCD) (Washington, DC, 1896–97), 152; RMPD-RCDC (1902–3), 221; "Same Here, Precisely," *WP*, 27 July 1897; *CT*, 6 Oct. 1914, 15 July 1922; Chicago Commission on Race Relations, *The Negro in Chicago* (Chicago: University of Chicago Press, 1922), 481–84. The way a New York prison official described the dangers posed by tramps—"they disseminate disease [and] perpetuate and encourage crime"—closely resembled the way urban Americans described the dangers posed by weeds See O. Lewis, "The Tramp Problem," *American Academy of Political and Social Science Annals* 40 (1912): 218.

35. *San Antonio Daily Light*, 18 May 1899; *SLPD*, 10, 12 Aug. 1905, 17 July 1914; Stilgoe, *Metropolitan Corridor*, 150; Eric H. Monkkonen, *The Dangerous Class: Crime and Poverty in Columbus, Ohio, 1860–1885* (Cambridge: Harvard University Press, 1975), 159; Linder and Zacharias, *Of Cabbages and Kings County*, 245.

36. Lee Meriwether, *The Tramp at Home* (New York: Harper and Brothers, 1889), 51; *BDE*, 21 June 1895; *SLPD*, 7 Aug. 1897, 3 Aug. 1906; Roger Lane, *Murder in America: A History* (Columbus: Ohio State University Press, 1997), 119, 190, 234–35; Roger Lane, *Violent Death in the City: Suicide, Accident and Murder in Nineteenth-Century Philadelphia*, 2nd ed. (Columbus: Ohio State University Press, 1999), 96–100; "Only Skeleton Left," *WP*, 7 Aug. 1904; "Found Dead in Weeds," *SLPD*, 15 Sept. 1905. Lane attributes declines in the number of infants abandoned and found dead in cities to the increasing availability of pasteurized cow's milk, more births taking place in hospitals, and adoptions by celebrities.

37. Falck, "Controlling the Weed Nuisance," 627 n. 8; "Farming of City Lots," *BDE*, 19 Mar. 1896; Frances Smith, "Continued Care of Families," in *Proceedings of the National Conference of Charities and Corrections at the Twenty-Second Annual Session* (1895): 87; "From Marie Etienne Burns," in *Memories of Jane Cunningham Croly*, ed. Caroline Morse (New York: Putnam's Sons, 1904), 190; Arthur von Briesen, "Respect for Law in the United States," *Annals of the American Academy of Political and Social Science* 36 (1910): 211. The farming of vacant lots was not always characterized as a charitable relief measure; it was also imagined as an antiurban policy. One Brooklyn commentator hoped that tenement dwellers impressed with the bounty of vacant-lot garden patches might leave the city for "the health and freedom of the country." See "Vacant Lot Farms," *BDE*, 30 Apr. 1899.

38. *SLPD*, 30 Aug. 1900, 4 Aug. 1897, 9 Aug. 1905; "St. Louis Has Many Weeds to Cut," *SLPD*, 11 Aug. 1905; "Weeds," *WP*, 8 July 1898; "Our Superiority in Weeds," *WP*, 23 Aug. 1898; "Cut the Weeds!" *WES*, 4 Aug. 1906; "How to Grow Weeds," *SLPD*, 9 Sept. 1913; *Chicago Defender* (hereafter *CDFD*), 9 July 1911; Kevin K. Gaines, *Uplifting the Race: Black Leadership, Politics, and Culture in the Twentieth Century* (Chapel Hill: University of North Carolina Press, 1996), 11–20; Kimberly Smith, *African American Environmental Thought: Foundations* (Lawrence: University of Kansas Press, 2007), 5; Daniel Burnstein, *Next to Godliness: Confronting Dirt and Despair in Progressive Era New York City* (Urbana: University of Illinois Press, 2006), 32–53.

39. Controlling weeds could be listed alongside improvement work ranging from cleaning candy factories to removing snow from sidewalks. See Louisa Spalding Millspaugh, "Women as a Factor in Civic Improvement," *Chautauquan* 43 (1906): 312–19; Burton Hendrick, "Taking the American City Out of Politics," *Harper's* 137 (1918): 106–13.

40. F. Timmons, "A History of Weed Control in the United States," *Weed Science* 18 (1970): 300–301; Lyster Dewey, "Legislation against Weeds," in *Yearbook of Agriculture* (Washington, DC, 1896), 3–22, 37–53; *Charter and Ordinances of the City of Battle Creek* (Review and Herald Office, 1861); "Laws against Weeds," *MM* 8 (1898): 32; "Injurious Weeds," *MM* 8 (1898): 104. A retrospective analysis found that in instances when farmers or other private landowners tried to recover damages from weeds that had allegedly spread to their property from another property or railroad, courts found that the allegedly offending landowners were not liable, although states could pass laws against permitting certain plants from spreading that also allowed landowners to recover damages from violation of this law. See Noel, "Nuisances from Land," 772–79, 791–92.

41. *State of Ohio, General and Local Laws* (Willis, 1884), 17; *Ordinances of the City of Columbus, Ohio, Revised, Codified, and Consolidated*, section 149 (Westbote, 1896);

Municipal Code of Lincoln, 1889, sec. 975 (State Journal, 1889); *General Ordinances of the City of Indianapolis,* sec. 4060 (Burford, 1904); Willis Blatchley, *Indiana Weed Book* (Indianapolis: Nature Publishing, 1912), 16.

42. Horace Wood, *Practical Treatise on the Law of Nuisances in Their Various Forms,* 3rd ed. (San Francisco: Bancroft-Whitney, 1893), 11–24, 44–46, 104; John Dillon, *Commentaries on the Law of Municipal Corporations,* 5th ed. (Boston: Little, Brown, and Company, 1911), 2: 1045; Eugene McQuillin, "Abatement of Smoke Nuisance in Large Cities by Legislative Declaration," *Central Law Journal* 60 (1905): 344. See also Louise Halper, "Untangling the Nuisance Knot," *Environmental Affairs* 26 (1998): 91; *SLPD,* 10 Aug. 1905, 11.

43. Wood, *Practical Treatise,* 148–49; *Giles v. Walker,* 24 Q.B. Div 656 (1890); Adam W. Rome, "Coming to Terms with Pollution: The Language of Environmental Reform, 1865–1915," *Environmental History* 1 (1996): 6–28; David Christian, *Maps of Time: An Introduction to Big History* (Berkeley: University of California Press, 2004), 106–38; Theodore Steinberg, *Slide Mountain; or, The Folly of Owning Nature* (Berkeley: University of California Press, 1995), 14–19.

44. Dewey, *Legislation against Weeds,* 17; Falck, "Controlling the Weed Nuisance," 625 n. 2; *Municipal Ordinances, Rules, and Regulations Pertaining to Public Health* (Washington, DC: Government Printing Office, 1917), 473; "Cleaning Up and Staying Clean," *American Journal of Public Health* 4 (Aug. 1914): 708; 30 Stat. at L. 959, chap. 326; *Ordinances of the City of Columbus;* Ernst Freund, *Police Power: Public Policy and Constitutional Rights* (Chicago: Callaghan, 1904), iii; Dillon, *Commentaries,* 994; Christopher Tiedeman, *Treatise on The Limitations of Police Power in the United States* (1886; reprint, New York: Da Capo Press, 1971), 426–27. After Los Angeles passed a 1912 ordinance requiring weed removal in April and August, the assessment levied when the city did the work was called the "weed tax." See "Taxing the Weeds in Los Angeles," *National Municipal Review* 2 (1913): 154.

45. *City of St. Louis v. Galt,* 179 Mo. 8 (1903); "Appellant's Abstract," 7, 11, and "Respondent's Statement," 3, both in *City of St. Louis v. Smith P. Galt,* Oct. term, 1903, Case No. 10769 (hereafter MSC-10769), Judicial Case Files, Missouri Supreme Court, RG 600, Missouri State Archives, Jefferson City; *SLPD,* 30 Aug. 1900; ARSLHD (1900–1901); William Hyde and Howard Conrad, *Encyclopedia of the History of St. Louis,* 4 vols. (New York: Southern History Company, 1899); 1: 191–92, 2: 823–26, 860–61, 1030–33, 4: 1943–44, 2517–18; *SLPD,* 30 Aug. 1900, 11; ARSLHD (1900–1901); *ES,* 19 Aug. 1901. Violators of Washington's law included financier Jay Cooke and John R. Dos Passos, the father of writer John Dos Passos.

46. *City of St. Louis v. Galt,* 179 Mo. 11; "Respondent's Statement," 5, and "Motion for Rehearing," 2, 5, both in MSC-10769; Missouri Botanical Garden, Trustees Records, collection 1, RG 1, ser. 11, box 7, folder 4. Galt's argument regarding the economy of nature hinted at the ecological consequences of destroying weeds. Galt was not alone in this sentiment. When Denver's health officer began destroying weeds on city streets and vacant lots, beehive owners showed "strong opposition" to the destruction of the blossoms on which their bees fed. See "Crusade against Weeds," *BDE,* 15 Sept. 1899.

47. "Appellant's Abstract," 9, in MSC-10769; Hus, "Ecological Cross Section," 230–33;

Grime, *Plant Strategies*, 121–22, 135; Edgar Anderson, *Plants, Man, and Life* (Boston: Little, Brown and Company, 1952), 186–201; "Weeds Are Being Cut," *SLPD*, 10 Aug. 1905; "Weedless St. Louis," *SLPD*, 27 Nov. 1903; "Goldenrod Not a Weed," *SLPD*, 12 Aug. 1905; "Vandeventer Place," box 6, George Edward Kessler Papers, MHS; "Catlin Residence," box 4, folder 4, Kessler Papers; James Neale Primm, *Lion of the Valley, St. Louis, Missouri*, 2nd ed. (Boulder: Pruett Publishing, 1990), 365–66. By the end of twentieth century, Missouri officials considered Japanese barberry to be one of the state's "noxious invasive weeds." See "*Berberis thunbergii* (Japanese barberry)," http://www.nps.gov/plants/alien/list/b.htm.

48. "Weeds," *MM;* "Weedless St. Louis"; S. Sherer, "The 'Places' of St. Louis," *House and Garden* 4 (Apr. 1904): 189; *Tree Planting in St. Louis*, 6, 34; Hus, "Ecological Cross Section," 232; Campanella, *Republic of Shade*, 55, 99–102; Relph, *Modern Urban Landscape*, 261. For a description of a wildflower as a rarity rather than a plant of uncertain beauty, see "Wild Flowers Becoming Extinct." Animals common in the country, such as chickens and cows, were also banned in cities, sometimes even earlier than the turn of the twentieth century. The "pig nuisance" highlights the uniqueness of the weed nuisance. The "pig nuisance" was created by active use of land to produce an economic resource. Weeds arose after farmers stopped producing crops or after speculators owned the land. While the "weed nuisance" was a problem associated with the production of wealth—profits from real estate sales—it arose from not using the land or directly manipulating nature. See Linder and Zacharias, *Of Cabbages and Kings County*, 251–56; John Duffy, "Hogs, Dogs, and Dirt: Public Health in Early Pittsburgh," *Pittsburgh Magazine of History and Biography* 87 (1963): 294–305; Joanna Dyl, "The War on Rats versus the Right to Keep Chickens," in *The Nature of Cities: Culture, Landscape, and Urban Space*, ed. Andrew C. Isenberg (Rochester: University of Rochester Press, 2006), 38–62; Fogelson, *Bourgeois Nightmares*, 180.

49. Davis, "Inventing Nature in Washington," 34–51; *ES*, 4 Mar. 1899; Passonneau, *Washington through Two Centuries*, 110–30; Brief for Plaintiff in Error, *District of Columbia v. Green*, D.C. Appellate Court Case no. 1738 (1907); John Wallace, "Growth and Development of a Late Nineteenth Century High Status Suburb: The Case of Kalorama" (master's thesis, University of Maryland, 1991), 123–31; Senate Committee, "Removal of Weeds from Lands in the District." Green and Ridout's argument regarding uniformity among residents and nonresidents may have been inspired by the invocation of the uniformity clause by Eastern nonresident speculators in Midwestern land who hoped to protect their investments from "discriminatory" taxation. See Robert Swierenga, "Land Speculation and Frontier Tax Assessments," *Agricultural History* 44 (1970): 253–66. A real inequality was that while a resident often spent time or money to comply with the law or risked being summoned to court, nonresident owners rarely made such expenditures. In St. Louis, one landowner who spent "three hours destroying weeds" on Sunday mornings was demoralized by weeds flourishing on adjacent land, on which the city had posted (disregarded) notices to clear. See *SLPD*, 28 July 1905.

50. "War of Extermination Waged against Thistles," *CT*, 23 Oct. 1901; "City Threatened by Thistles," *CT*, 18 July 1902; "Waging a War on Weeds," *CT*, 17 July 1903; *New York Times* (*NYT* hereafter), 7 Oct. 1903; "Tell of Year's Reforms," *CT*, 21 Oct. 1903;

People Ex Rel. NC Van Slooten v. Board of Commissioners of Cook County et al., 77 N.E. 915. The Circuit Court of Cook County and the Illinois Supreme Court found the law invalid because the per-lot and per-acre fines were not uniform and not in proportion to taxable values.

51. *CT*, 7 June 1901; Craig Turnbull, *An American Urban Residential Landscape, 1890–1920: Chicago in the Progressive Era* (Amherst, NY: Cambria Press, 2009), 155; "Plan Anti-Weed Movement," *WP*, 16 Oct. 1905.

52. *In re Opinion of Justices*, 69 A. 627 (Maine); *State v. Morse*, 80 A. 189 (Vermont); 195 N.W. 544 (Wisconsin); *St. Louis Gunning Advertising Company v. St. Louis*, 137 S.W. 929; James Ely Jr., *The Guardian of Every Right: A Constitutional History of Property Rights* (New York: Oxford University Press, 1992), 85–90; William Novak, *The People's Welfare: Law and Regulation in Nineteenth-Century America* (Chapel Hill: University of North Carolina Press, 1996), 228–33.

53. St. Louis City ordinances 18,415 and 19,418. St. Louis excluded "any fields used for farming or gardening purpose." This loophole indicated that landowners who cultivated certain plants at certain times of the year ignored happenstance plants at other times and other certain areas. See Senate Committee, "Removal of Weeds from Lands in the District," 3; *SLPD*, 10, 11 Aug. 1905; RHO-RCDC (1903), 22; RHO-RCDC (1910), 24.

54. Falck, "Controlling the Weed Nuisance," 619–23; RHO-RCDC (1913), 105. Critics of agricultural weed laws like Meehan and Bailey insisted that regular tilling was the only weed control necessary as it prevented weed reproduction. See Liberty Hyde Bailey, "Coxey's Army and the Russian Thistle: A Sketch of the Philosophy of Weediness," in *The Survival of the Unlike* (New York: Macmillan, 1896), 199; "Laws against Weeds"; "Injurious Weeds." However simple the task, when the number of properties in cities with uncut plants were aggregated, the costs were not insignificant.

55. *SLPD*, 9, 10 Aug. 1905; *WES*, 2 July 1906, 24 Aug. 1910; Mary Wilkins, "Protests at Weed Crop," *WP*, 18 July 1910; Subcommittee of House Committee on Appropriations, District of Columbia Appropriation Bill for 1919, 65th Cong., 2nd sess., 1917, 273; "Weed War Opens Soon," *WES*, 28 May 1924. From the time the District's weed law went into effect, Woodward was skeptical that the department would be able to do much more than respond to particular complaints. Woodward had initially been skeptical of Providence's superintendent of health Charles Chapin's belief that filth nuisances were neither pleasant nor "wholesome" but also were not significant disease vectors. However, Woodward eventually accepted this viewpoint and used it to justify limited enforcement by contending that weeds had "practically nothing to do with sanitary matters" and denying the very per se "universally . . . harmless" reasoning of *Galt*. The health officer of Richmond, VA, changed his mind in the opposite manner. While he once thought the department's job of issuing weed violations was "absurd," he concluded that "long weeds" were not "a silly matter" after inspecting a patch and finding "piles of human feces . . . a swarm of flies . . . [and] hundreds of mosquito larvae." See James Cassedy, *Charles V. Chapin and the Public Health Movement* (Cambridge: Harvard University Press, 1962), 98; RHO-RCDC (1903), 22; Senate Committee on the District of Columbia, "Removal of Weeds from Lands in the City," 57th Cong., 1st sess., 1902, S. Report 749, 2; Ernest Levy, "Present Views of the Importance of Municipal Sanitary Inspection for the Abatement of Nuisances," *American Journal of Public Health* 2 (1912): 10–11.

56. John Stevenson, "Organization of Health Department, Sanitary Legislation, and the Abatement of Nuisances," *Public Health Papers and Reports* 10 (1884): 307; ARSLHD (1906–7), 30; RHO-RCDC (1910), 24; RHO-RCDC (1916), 30; *St. Louis Gunning Advertising Company v. St. Louis,* 235 Mo. 135, 140–43. In the courts, city officials used the same nuisance language to describe billboards that they had used to describe weeds; they were "constant menaces to the public safety and welfare of the city; . . . endanger[ed] the public health, [and] promote[d] immorality, constitute[d] hiding places and retreats for criminals and all classes of miscreants. They [were] also inartistic and unsightly."

57. W. Loucks, "Increments in Land Values in Philadelphia," *Journal of Land and Public Utility Economics* (*JLPUE* hereafter) 1 (1925): 469–77; Howard Shannon and H. Bodfish, "Increments in Subdivided Land Values in Twenty Chicago Properties," *JLPUE* 5 (1929): 29–47.

58. Grime, *Plant Strategies,* 80–88, 249; Bazzaz, *Plants in Changing Environments,* 108–46.

59. "What Is a Weed?" *NYT,* 7 Nov. 1915; Skinner, *Nature in a City Yard,* 62–63, 128; Dreiser, "New Knowledge of Weeds," 533–58; "Now, Look at the Fields," 6; W. W. Bailey, "About Weeds," 743.

60. Fogelson, *Bourgeois Nightmares,* 181, which follows turn-of-the-century land economist Richard Hurd's 1903 *Principles of City Land Values* (New York: Record and Guide, 1903), 117; Alexander von Hoffman, "Weaving the Urban Fabric: Nineteenth-Century Patterns of Residential Real Estate Development in Outer Boston," *Journal of Urban History* 22 (1996): 224; Sears Roebuck, *Honor Bilt Modern Homes* (Chicago, 1927), 137; Jackson, *Crabgrass Frontier,* 174.

CHAPTER TWO

1. Karl Koessler and O. Durham, "A System for an Intensive Pollen Survey," *Journal of the American Medical Association* 86 (1926): 1204–5; Paul Standley, *Common Weeds* (Chicago: Field Museum, 1934), 2–31; Edith Gladfelter, *Physiographic Ecology of the Drainage Canal Dumps* (master's thesis, University of Chicago, 1903), 1–9; A. Getman to Wiley Mills, 24 June 1925, North Austin Social and Betterment Club Records, Chicago Historical Society; Henry Cowles, "The Plant Societies of Chicago and Vicinity," *Bulletin of the Geographic Society of Chicago* 2 (1901): 41–43; *International Phytogeographic Excursion in America: The Vegetation of the Chicago Region* (1913): 9–10; Herman Pepoon, *Annotated Flora of the Chicago Area* (Chicago: Chicago Academy of Sciences, 1927), 292, 523.

2. John Dewey, "Evolution and Ethics," in *The Essential Dewey,* vol. 2, *Ethics, Logic, Psychology,* ed. Larry Hickman and Thomas Alexander (Bloomington: Indiana University Press, 1998), 225–35; *Report on the Proposed Sand Dunes National Park* (Washington, DC: Government Printing Office, 1917), 44; George Fuller, "Some Undesireable Plant Immigrants," *Montreal Family Herald and Weekly Star,* 10 Dec. 1902, George Damon Fuller Papers (hereafter GDFP), University of Chicago; Victor Cassidy, *Henry Chandler Cowles: Pioneer Ecologist* (Chicago: Kedzie Sigel Press, 2007), 74–75; "The City," n.d., folder 3, box 18, Robert Ezra Park Papers (hereafter REPP), University of Chicago; W. McDougall, *Plant Ecology* (Philadelphia: Lea and Febiger, 1927), 206, reprinted in R. McKenzie, *Readings in Human Ecology* (Ann Arbor: Wahr, 1934), 5. These scholars'

studies have resulted in intensive study of their work. The relationship of ecological and sociological knowledge emerged partly from their not unrelated intellectual origins. Auguste Comte, Herbert Spencer, and Lester Ward, intellectuals who held interest in both nascent disciplines, were among the nineteenth century's protosociological and protoecological thinkers who examined ideas about organisms and equilibrium. Useful and insightful introductions include Andrew Delano Abbott, *Department and Discipline: Chicago Sociology at One Hundred* (Chicago: University of Chicago Press, 1999); Matthias Gross, "Early Environmental Sociology: American Classics and Their Reflections on Nature," *Humboldt Journal of Social Relations* 25 (1999): 1–29; Emanuel Gaziano, "Ecological Metaphors as Scientific Boundary Work: Innovation and Authority in Interwar Sociology and Biology," *American Journal of Sociology* 101 (1996): 874–907; Dorothy Ross, *The Origins of American Social Science* (New York: Cambridge University Press, 1991); Robert Bannister, *Sociology and Scientism: The American Quest for Objectivity, 1880–1940* (Chapel Hill: University of North Carolina Press, 1987); Marlene Shore, *The Science of Social Redemption: McGill, the Chicago School, and the Origins of Social Research in Canada* (Toronto: University of Toronto Press, 1987); Fred Matthews, *Quest for an American Sociology: Robert E. Park and the Chicago School* (Montreal: McGill-Queen's University Press, 1977), 37–39, 134; Cynthia Russett, *The Concept of Equilibrium in American Social Thought* (New Haven: Yale University Press, 1966). Among historians' criticisms of Chicago sociologists' writings are the inaccuracy, simplicity, and inflexibility of their models; their naturalization of hierarchy, capitalist competition, and racial stratification; and their lack of utility for urban environmental history. See Andrew C. Isenberg, ed., *The Nature of Cities: Culture, Landscape, and Urban Space* (Rochester: University of Rochester Press, 2006), xii–xiv; Mary Beth Prudup, "Model City? Industry and Urban Structure in Chicago," in *Manufacturing Suburbs: Building Work and Home on the Metropolitan Fringe*, ed. Robert D. Lewis (Philadelphia: Temple University Press, 2004), 74–75; Martin V. Melosi, "The Place of the City in Environmental History," *Environmental History Review* 17 (1993): 8; Thomas Bender, *Community and Social Change in America* (New Brunswick: Rutgers University Press, 1978), 7–11. The most significant works on the university's ecologists include Gregg Mitman, *The State of Nature: Ecology, Community, and American Social Thought, 1900–1950* (Chicago: University of Chicago Press, 1992), and J. Ronald Engel, *Sacred Sands: The Struggle for Community in the Indiana Dunes* (Middletown, CT.: Wesleyan University Press, 1983).

3. Mitman, *State of Nature*, 2, 46; Cowles, "Plant Societies," 9–10; Robert Park, "Succession, an Ecological Concept," in *Human Communities: The City and Human Ecology* (Glencoe: Free Press, 1952), 224–27 (originally published in 1936). For a more recent analysis of the agency and interactions in plant neighborhoods, see F. A. Bazzaz, *Plants in Changing Environments: Linking Physiological, Population, and Community Ecology* (New York: Cambridge University Press, 1996), 128–46, 192–93.

4. Cowles, "Plant Societies," 7; Helen MacGill, "Land Values as an Ecological Factor in the Community of South Chicago" (master's thesis, University of Chicago, 1927), 64; "The Community," n.d., folder 3, box 18, REPP.

5. Charles Adams and George Fuller, "Henry Chandler Cowles, Physiographic Plant Ecologist," *Annals of the Association of American Geographers* 30 (1940): 41; Cowles, "Plant Societies," 9; Ernest W. Burgess, "Can Neighborhood Work Have a Scientific

Basis?" in *The City*, ed. Robert Park (Chicago: University of Chicago Press, 1967), 148; Rolf Lindner, *The Reportage of Urban Culture: Robert Park and the Chicago School*, trans. Adrian Morris (New York: Cambridge University Press, 1996), 73–74; R. D. McKenzie, "Ecological Approach to the Study of the Human Community," in Park, *City*, 76–79; "Study of an Area in Transition," folder 2, box 155; Ernest W. Burgess Papers (hereafter EWBP), University of Chicago Library; "Effect of Group Displacement on Property Values," 5, folder 3, box 155, EWBP.

6. George Fuller, *The Vegetation of the Chicago Region* (Chicago: University of Chicago, 1917), 3. Although Fuller was born and educated partially in Canada, at the turn of the century he espoused a thoroughly Emersonian outlook, writing, "Every plant . . . is beautiful and good if we can but discover the proper standpoint from which to view them." See George Fuller, "Plant Life," *Montreal Family Herald and Weekly Star*, 26 Nov. 1902, GDFP; Pepoon, *Annotated Flora of the Chicago Region*, 215. By the second half of the 1920s, Chicago botanists and plant ecologists had become cautious in their analogies and metaphors because of the embrace of "phytosociology" by Eastern Europeans. One reason they searched for new metaphors for their science may have been changing geopolitical contexts. Chicago-trained ecologist William Cooper introduced a hydrological analogy: "The vegetation of the earth is presented as a flowing braided stream. Its constituent elements branch and interweave, disappear, and reappear." See G. Fuller, "Origin and Development of Plant Sociology," *Botanical Gazette* 85 (1928): 229–32. On the relationship of ecological work to social thought, politics, and cultural change, see Mitman, *State of Nature;* Gregg Mitman, "Defining the Organism in the Welfare State: The Politics of Individuality in American Culture, 1890–1950," in *Biology as Society, Society as Biology: Metaphors*, ed. Sabine Maasen, Everett Mendelsohn, and Peter Weingart (Dordrecht: Kluwer Academic, 1995), 249–78; Malcolm Nicolson, "The Development of Plant Ecology, 1790–1960" (Ph.D. diss., University of Edinburgh, 1983); Ronald Tobey, *Saving the Prairies: The Life Cycle of the Founding School of American Plant Ecology, 1895–1955* (Berkeley: University of California Press, 1981). A gap between Chicago's ecologists and sociologists also emerged, partly because the former were skeptical about plant ecologist Frederic Clements's superorganism concept, while the latter contemplated Clements's *Plant Succession* and *Plant Indicators* as possible "model[s] for similar studies in human ecology." Clements's intellectual associations informed his interest in ecology's similarities with sociology. He wrote, "Sociology is the ecology of a particular species of animal, and has in consequence, a similar close connection with plant ecology. The widespread migration of man and his social nature have resulted in the production of groups or communities which have much more in common with plant formations than do formations of other animals. . . . The laws of succession are essentially the same for both plants and man." See Frederic Clements, *Research Methods in Ecology* (Lincoln: University Publishing Company, 1905), 16; Robert Park and Ernest Burgess, *Introduction to the Science of Sociology* (Chicago: University of Chicago Press, 1924), 555; Tobey, *Saving the Prairies*, 77–136. Although Park arrived at Chicago after Dewey had left for Columbia, Park had studied with Dewey at Michigan before the latter departed for Chicago.

7. Peter J. Bowler, *Evolution: The History of an Idea*, 3rd ed. (Berkeley: University of California Press, 2003), 318–20; Thomas Carlyle Dalton, *Becoming John Dewey: Dilemmas*

of a Philosopher and Naturalist (Bloomington: Indiana University Press, 2002), 29–36; T. H. Huxley, *Collected Essays*, vol. 9, *Evolution and Ethics* (New York: Appleton, 1911), 9–11; Donald Worster, *Nature's Economy: A History of Ecological Ideas* (Cambridge: Cambridge University Press, 1977), 176–79.

8. Andrew Hurley, *Environmental Inequalities: Class, Race, and Industrial Pollution in Gary, Indiana, 1945–1980* (Chapel Hill: University of North Carolina Press, 1995), 16–22; *Report on the Proposed Sand Dunes*, 43–46; Engel, *Sacred Sands*, 107–72, 234–36; "Effect of Cement Dust on Pines, Buffington, Indiana," University of Chicago Department of Botany Records; American Memory Collection, American Environmental Photographs (hereafter AMCAEP), 1891–1936, Library of Congress (hereafter LOC), http://hdl.loc.gov/loc.award/icuaep.inn2; "Pinus Banksiana Killed by Cement Dust, Buffington, Indiana," AMCAEP, http://hdl.loc.gov/loc.award/icuaep.inn3; "Effect of Tramping, Succession with City Growth, Gary, Indiana," AMCAEP, http://hdl.loc.gov/loc.award/icuaep.inn119; "Stand of Tramped Trees, Succession with City Growth, Gary, Indiana," AMCAEP, http://hdl.loc.gov/loc.award/icuaep.inn118; "Succession with City Growth, Gary, Indiana," AMCAEP, http://hdl.loc.gov/loc.award/icuaep.inn120; "Main Street, Gary, Indiana, Succession with City Growth," AMCAEP, http://hdl.loc.gov/loc.award/icuaep.inn121; Robert Croker, *Pioneer Ecologist: The Life and Work of Victor Ernest Shelford, 1877–1968* (Washington, DC: Smithsonian Institution Press, 1991), 8–31; Arthur Itterman, "Southwest Section of the East Side," Mar. 1933, 3–4, folder 2, box 160, EWBP.

9. Charles Saunders, "Botanizing on City Vacant Lots," *MM* 12 (1902): 184–85.

10. Christopher Tiedeman, *Treatise on the Limitations of Police Power in the United States* (1886; reprint, New York: Da Capo Press, 1971), 122; Maasen, Mendelsohn, and Weingart, *Biology as Society, Society as Biology*, 2. On the spatial dimensions of metaphorical language, see Tim Cresswell, "Weeds, Plagues, and Bodily Secretions: A Geographical Interpretation of Metaphors of Displacement," *Annals of the American Association of Geographers* 87 (1997): 330–45.

11. *Marlborough v. Sisson*, 31 Conn. 332 (1863); *Saturday Evening Post*, 27 Oct. 1860; John Burroughs, "Notes of a Walker," *Scribner's* 19 (1880): 689; Joel Chandler Harris, *Life of Henry W. Grady* (New York: Cassell Publishing, 1890), 184; "Rank Growth of Cads," *Sacramento Daily Record-Union*, 22 June 1893; "Breadth and Dignity of Character," *Fitchberg Daily Sentinel*, 2 July 1895.

12. Washington Irving, *Knickerbocker's History of New York* (New York: Putnam, 1860), 230 (the misspelling of *dandelions* persisted until Anne Moore corrected it in 1928); John Burroughs, "A Bunch of Herbs," in *Pepacton* (Boston: Houghton, Mifflin, 1895), 196–97; Ledyard Bill, *Minnesota: Its Character and Climate* (New York: Wood and Holbrook, 1871), 12, 177; Charles Skinner, *Nature in a City Yard* (New York: Century, 1897), 61–63; Charlie Samolar, "Argot of the Vagabond," *American Speech* 2 (1927): 386–92; *Industrial Worker*, 29 Apr. 1909; David Grayson, *The Friendly Road* (New York: Doubleday, 1913), 321–34; Edmond Kelly, *The Elimination of the Tramp* (New York: Putnam, 1908), 103–7; Charles Canaday, "Weeding Out the Tramps," *American City* 14 (1916): 270. Burrough's speculations were the basis for children's nature study lessons as well. Of autumn "tall weeds," children learned, "They love the road-side. Here they are safe. They are ragged

and dusty tramps." Skinner contemptuously compared tramps to other nonhuman creatures: "Squirrels, rats, snakes, spiders, elephants, and tramps have been partly domesticated." Rural Americans compared migrant workers to birds such as swallows. See John Burroughs, Charles Dudley Warner, and Mary Burt, *Little Nature Studies for Little People* (Boston: Ginn, 1895), 56; Charles Skinner, *Flowers in the Pave* (Philadelphia: Lippincott, 1900), 103–4; Frank Tobias Higbie, *Indispensable Outcasts: Hobo Workers and Community in the American Midwest, 1880–1930* (Urbana: University of Illinois Press, 2003), 25. On the relationship of transportation corridors and migrants, see Tim Cresswell, *The Tramp in America* (London: Reaction Books, 2001), 171–93; Roger Bruns, *Knights of the Road: A Hobo History* (New York: Methuen, 1980), 204 and photograph "Jungle, Downer's Grove, Illinois, 1924"; John R. Stilgoe, *Metropolitan Corridor: Railroads and the American Scene* (New Haven: Yale University Press, 1983), 140–51. Some migrating Americans also used *weed* as a verb roughly meaning "to give" as well as "to weed out," or to take more than one's fair share; see Samolar, "Argot of the Vagabond," 386–92; David Maurer, "Argot of the Underworld," *American Speech* 7 (1931): 117. On the use of environmental degradation to alienate poor Americans from the landscape, see Karl Jacoby, *Crimes against Nature: Squatters, Poachers, Thieves, and the Hidden History of American Conservation* (Berkeley: University of California Press, 2001). For the economic conditions producing rootless Americans, the social consequences of their proliferation, and cultural reaction to their plights, see Todd DePastino, *Citizen Hobo: How a Century of Homelessness Shaped America* (Chicago: University of Chicago Press, 2003), 3–58; Higbie, *Indispensable Outcasts*, 25–133; Cresswell, *Tramp in America;* Kenneth Kusmer, *Down and Out, on the Road: The Homeless in American History* (New York: Oxford University Press, 2003), 35–116; Eric Monkkonen, *Walking to Work: Tramps in America: 1790–1935* (Lincoln: University of Nebraska Press, 1984), 1–10.

13. *SLPD*, 10 Aug. 1905; Willis Blatchley, *Indiana Weed Book* (Indianapolis: Nature Publishing, 1912), 8–10; Melvin Gilmore, "Plant Vagrants in America," *Papers of the Michigan Academy of Science, Arts, and Letters* 15 (1931): 65–67; *Botanical Gazette* 34 (1902): 319; Albert Hansen, "Danger Lurks in City Weeds," *American City* 31 (1924): 16; Frank Thone, "Pokeberry," *Science News Letter* 14 (1928): 149; Oren Durham, *Your Hay Fever* (Indianapolis: Bobbs-Merrill, 1936), 145; *CT*, 14 Aug. 1935.

14. "Indiana Sand Dunes," *Logansport Pharos*, 4 Aug. 1903; Robert Hessler, "Plants and Man: Weeds and Diseases," *Proceedings of the Indiana Academy of Science* 20 (1910): 49, 53–55, 62–63. "Plants and Man" was published in a slightly different form as "Weeds and Diseases" in *Survey* 26 (1911): 51–63.

15. J. P. Grime, *Plant Strategies, Vegetation Processes, and Ecosystem Properties*, 2nd ed. (Chichester: Wiley, 2001), 61–64, 245–46; Clifton Phillips, *Indiana in Transition: The Emergence of an Industrial Commonwealth* (Indianapolis: Indiana Historical Bureau, 1968), 184–85, 541–42; William Watt, *The Pennsylvania Railroad in Indiana* (Bloomington: Indiana University Press, 1999), 30–105; *Fifth Annual Report of the State Board of Health of Indiana* (Indianapolis: Burford, 1887), 22–23. The Logansport ordinance gave each health officer the duties of seeing "that all streets, gutters, alleys and public places within his precinct are put in a clean and healthy condition at the very earliest possible time, by the removal of all rubbish, stagnant water, weeds, or anything else that in the opinion of the Board of Health may in any way be detrimental to health." Logansport's population nearly

doubled as urbanization intensified across the nation; its population was 11,198 in 1880 and 21,626 by 1920. See Hessler, "Plants and Man," 49–55.

16. Hessler, "Plants and Man," 49, 54–62. On the "balance of nature," see Donald Worster, "Transformations of the Earth: Toward an Agroecological Perspective in History," *Journal of American History* 76 (1990): 1092–93; William Cronon, "Modes of Prophecy and Production: Placing Nature in History," *Journal of American History* 76 (1990): 1127.

17. *Logansport Pharos*, 22 Dec. 1911; James Inciardi and Karen McElrath, eds., *American Drug Scene*, 2nd ed. (Los Angeles: Roxbury Publishing, 1998), xi. Just one of the scores of the newspapers in which the ad appeared was the *Waukesha (WI) Freeman*, 3 Dec. 1903. See Hessler, "Plants and Man," 64.

18. Hessler, "Plants and Man," 61–68; Adam Rome, "Nature Wars, Culture Wars: Immigration and Environmental Reform in the Progressive Era," *Environmental History* 13 (2008): 432–53. In the emerging terminology, Hessler was dedicated to euthenics, "race improvement through environment," and minimally if at all committed to eugenics, "race improvement through heredity." Hessler remained a strong advocate of sanitary reform. He continued to use the weed metaphor in euthenic advocacy in *Dusty Air and Ill Health* (Indianapolis: Burford, 1912). The least practical of Hessler's ideas was going "back to nature"; he believed that returning to the land from cities was a likely way to recover "strong and healthy and prolific" citizens because rural life was more simple and sanitary. In 1934, he reasserted that individuals had no control over conditions that they had inherited, but that they could control their environments. See *Journal of Home Economics* 2 (1910): 697; Annie Dewey, "Standards of Living in the Home," *Outlook* 101 (1912): 491; Robert Hessler, "Hypertension," *Proceedings of the Indiana Academy of Science* 44 (1934): 220–22; Peter Schmitt, *Back to Nature: The Arcadian Myth in Urban America* (Baltimore: Johns Hopkins University Press, 1969), 185–88; Alexandra Minna Stern, "'We Cannot Make a Silk Purse Out of a Sow's Ear': Eugenics in the Hoosier Heartland," *Indiana Magazine of History* 103 (2007): 3–9.

19. Robert Johnson, "Fountains of Ashokan," *Natural History* 20 (1920): 91; *Waterloo (IA) Evening Courier*, 6 Oct. 1919; Katherine Pandora, "Knowledge Held in Common: Tales of Luther Burbank and Science in the American Vernacular," *Isis* 92 (2001): 502–16; Elizabeth Urquhart, "In the Garden with Burbank: Human Weeds," *Oakland Tribune*, 14 June 1926, 7; Luther Burbank, *Training of the Human Plant* (New York: Century, 1907), 48–50; Elizabeth Urquhart, "In the Garden with Burbank: Needs of the Garden," *Logansport Morning Press*, 6 Aug. 1926; Charles Ribton-Turner, *History of Vagrants and Vagrancy* (London: Chapman and Hall, 1887), 176; Epes Sargent, *Testimony of the Poets* (Boston: Mussey, 1854), 317. Although Urquhart's interviews with Burbank referenced in this text appeared after Burbank died in Apr. 1926, at least one "In the Garden" column appeared in the *Atlanta Constitution* in 1925. The interview material may have been procured even earlier, when Urquhart wrote about gardening for the *San Francisco Chronicle* in 1922. Burbank was known for freely talking with journalists. See Peter Dreyer, *A Gardener Touched with Genius: The Life of Luther Burbank* (New York: Coward, McCann and Geoghegan, 1975), 270–71.

20. *Placerville (CA) Mountain Democrat*, 14 July 1923 (hereafter *PCALMD*-1923); "Human Weeds Is Burbank Rating," *Sarasota Herald*, 9 Mar. 1926; "We're Human

Weeds, Declares Burbank," *Palm Beach Daily News,* 11 Mar. 1926; "We're Human Weeds, Declares Burbank," *LAT,* 12 Mar. 1926; *Philistine,* 29 Nov. 1909, 174; Burbank, *Training of the Human Plant,* 75; Luther Burbank, "Cultivate Children like Flowers," *Elementary School Teacher* 6 (1906): 458. The "field of wild human weeds" appeared in shorter and different form as "Luther Burbank Warns of Deterioration of Race" in the *Hammond (IN) Times,* 27 Dec. 1922, and the *Mexia (TX) Evening News,* 28 Dec. 1922. Many articles attributed to Burbank may only have been dictated to ghostwriters or drafted from interviews; some may have been entirely fabricated. According to Dreyer, Wilbur Hall was a "careless" amanuensis. How accurately Burbank's writings reflect Burbank's actual thinking is therefore uncertain. Such a distinction, however, would not have been evident to Americans who encountered Burbank's statements in newspapers or were not aware of this writing method. See Pandora, "Knowledge Held in Common," 485; Dreyer, *A Gardener Touched with Genius,* 90, 265.

21. "Luther Burbank Warns of Deterioration of Race"; Urquhart, "Needs of the Garden"; Urquhart, "Human Weeds"; *PCALMD*-1923; Pandora, "Knowledge Held in Common," 509–10; Hessler, "Plants and Man," 68; W. Howell, "Eugenics as Viewed by the Physiologist," in *Eugenics: Twelve University Lectures* (New York: Dodd, Mead, 1914), 94. Given the uncertainties about the production of Burbank's writings discussed supra note 20, the implications of Burbank's statement that "in my gardens, dangerous weeds are uprooted and destroyed, weak plants are removed, and no tainted individual is permitted to bear seed, or even to develop pollen, lest an entire field be affected and the strong, the good, and the beautiful be cursed with the evil influence of one depraved individual" are unclear, when following a statement that care could not transform all individuals with some inherited characteristics. See Luther Burbank with Wilbur Hall, *The Harvest of the Years* (Boston: Houghton Mifflin, 1927), 156–57.

22. Pandora, "Knowledge Held in Common," 502–8; Alexandra Minna Stern, *Eugenic Nation: Faults and Frontiers of Better Breeding in Modern America* (Berkeley: University of California Press, 2005), 22, 50–54, 120–21; J. Kellogg, "Needed—A New Human Race," in *Proceedings of the First National Conference on Race Betterment* (Battle Creek: Race Betterment Foundation, 1914), 431; Luther Burbank, *How Plants Are Trained to Work for Man: Plant Breeding* (New York: Collier, 1921), 32; Henry Williams, "Can the Criminal Be Reclaimed?" *North American Review* 163 (1896): 217; Charles Davenport, "The Eugenics Programme and Progress in Its Achievement," in *Eugenics: Twelve University Lectures,* 1; Jan Anthony Witkowski and John R. Inglis, eds., *Davenport's Dream: 21st Century Reflections on Heredity and Eugenics* (Cold Spring Harbor: Cold Spring Harbor Laboratory Press, 2008); Linda Gordon, *The Moral Property of Women: A History of Birth Control Politics in America* (Urbana: University of Illinois Press, 2002), 86–104; "No Human Weeds," *New York Sun,* 28 Apr. 1902; *Sermons of M. J. Savage* 6, no. 31 (2 May 1902): 9–10. On modern humans as a single species, see David Christian, *Maps of Time: An Introduction to Big History* (Berkeley: University of California Press, 2004), 120–33.

23. Ellsworth Huntington and Leon Whitney, *The Builders of America* (New York: William Morrow, 1927), 70–72; Christine Rosen, *Preaching Eugenics: Religious Leaders and the American Eugenics Movement* (New York: Oxford University Press, 2005), 12–23; Daniel J. Kevles, *In the Name of Eugenics: Genetics and the Uses of Human Heredity* (New York: Knopf, 1985), 64–89; Diane Paul, "Eugenics and the Left," *Journal of the History*

of Ideas 45 (1984): 585–90; Michele Mitchell, *Righteous Propagation: African Americans and the Politics of Racial Destiny after Reconstruction* (Chapel Hill: University of North Carolina Press, 2004), 81–88. Although Stern writes that "all eugenics was local," the "human weed" rhetoric and metaphor were transatlantic and ubiquitous. Prison managers calculated the costs of imprisoning "human weeds" in Ireland. Karl Marx wrote, "The cleanly weeded land, and the uncleanly human weeds, of Lincolnshire, are pole and counterpole of capitalist production." In German, the word is *menschenunkraut*. In his essay "Weeds, Human and Otherwise," James Friswell compared the way manure heaps generate and disseminate weeds to how "human weeds are drawn together in great cities, are there brought to live, warmed into action, and scattered over the country." English feminist Josephine Butler described children of intemperate and criminal parents as "useless . . . fibreless human weeds." In a short story entitled "Weeds," poorly managed institutions permit impoverished, enfeebled, diseased individuals to continue to produce children who become expensive wards of the state. Edith Kelley's 1923 novel *Weeds* depicts a community of tenant Kentucky tobacco farmers in which women are as worn out by rural motherhood as the soil is by tobacco. In 1931, British pragmatist philosopher Ferdinand Schiller, who considered positive eugenics the effort to render the human race "intrinsically better, higher, stronger, healthier, more capable," noted, "Negative eugenics aims at checking the deterioration to which the human stock is exposed, owing to the rapid proliferation of what may be called human weeds." See Stern, "'We Cannot Make,'" 6–7; *Irish Quarterly Review* 16 (1854): 1194; J. Friswell, *The Gentle Life* (London: Sampson Low, Son, and Marston, 1870), 346–47; Karl Marx, *Capital: A Critique of Political Economy*, ed. Frederick Engels (New York: Modern Library, 1906), 766; Josephine Butler, *An Autobiographical Memoir* (Bristol: Arrowsmith, 1909), 61; Richard Connell, "Weeds," *Birth Control Review* 6 (1922): 38–39, 61–62; Allison Berg, *Mothering the Race: Women's Narratives on Reproduction, 1890–1930* (Urbana: University of Illinois Press, 2002), 78–102; Schiller quoted in Donald Childs, *Modernism and Eugenics: Woolf, Eliot, Yeats, and the Culture of Degeneration* (Cambridge: Cambridge University Press, 2001), 3.

24. Huntington and Whitney, *Builders of America*, 70–71, 78–82; *PCALMD*-1923; Howell, "Eugenics as Viewed by the Physiologist," 93; Andrew Pearson, "Burbank Says Foreign Cultures Proves Us Half Civilized," *St. Petersburg Times*, 5 Apr. 1925 (this material may have first appeared in the *Dearborn Independent*, Henry Ford's national weekly publication, in 1923; see Scrapbook, 50-1, box 40, Luther Burbank Papers, Library of Congress); "Human Weeds," *Syracuse Herald*, 12 Feb. 1935; Margaret Sanger, "The Need of Birth Control in America," in *Birth Control: Facts and Responsibilities*, ed. Adolf Meyer (Baltimore: Williams and Wilkins, 1925), 47–48; *Deseret News*, 15 Feb. 1923. Annie Dewey, a librarian like her husband, Melvil Dewey, similarly alluded to and approved of Wells's thinking that two types of people—the creative and the trustworthy—were responsible for "weeding out the dull and the base." See Annie Dewey, "Standards of Living in the Home," 491. Wells was the son and grandson of English gardeners; in his utopian writings, freeing gardens of weeds is a recurring theme. Like John Dewey, Wells greatly admired Thomas Huxley; according to one biographer, Wells's career was devoted to applying Huxley's thinking to human life. See John Reed, *The Natural History of H. G. Wells* (Athens: Ohio University Press, 1982), 33–41; David Smith, *H. G. Wells: Desperately Mortal* (New Haven: Yale University Press, 1986), 5–19.

25. Sumner quoted in Russel Nye, *Midwestern Progressive Politics* (New York: Harper and Row, 1965), 29; Thorstein Veblen, *The Theory of the Leisure Class* (New York: Macmillan, 1899), 235; W. E. Burghardt Du Bois, "Relation of the Negroes to the Whites in the South," *America's Race Problems* (New York: McClure, Phillips, 1901), 132; Samuel Lindsay, "Child Labor a National Problem," *Annals of the American Academy of Political and Social Science* 27 (1906): 332; Athel Burnham, *The Community Health Problem* (New York: Macmillan, 1920), 103. On physicians in industrial environments, see Christopher Sellers, *Hazards of the Job: From Industrial Disease to Environmental Health Science* (Chapel Hill: University of North Carolina Press, 1997); Thomas Leonard, "'More Merciful and Not Less Effective': Eugenics and American Economics in the Progressive Era," *History of Political Economy* 35 (2003): 688–94, 707; Thomas Leonard, "Mistaking Eugenics for Social Darwinism: Why Eugenics Is Missing from the History of American Economics," *History of Political Economy* 37, supp. (2005): 225.

26. Herbert Davenport, *Outlines of Economic Theory* (New York: Macmillan, 1896), 326–29; Hessler, "Plants and Man," 54; Charles Davenport, *Heredity in Relation to Eugenics* (New York: Henry Holt, 1911), 212; Charles Davenport, "The Geography of Man in Relation to Eugenics," in *Heredity and Eugenics* (Chicago: University of Chicago Press, 1912), 299–300; Garland E. Allen, "The Eugenics Record Office at Cold Spring Harbor, 1910–1940: An Essay in Institutional History," *Osiris,* 2nd ser., 2 (1986): 225–64; Kevin T. Dann, *Across the Great Border Fault: The Naturalist Myth in America* (New Brunswick: Rutgers University Press, 2000), 114–38; Madison Grant, *The Passing of the Great Race,* 4th ed. (New York: Charles Scribner's Sons, 1921), 209. Two of the best essays that demonstrate that—despite the popularity of Darwin's metaphorical language and ideas to a variety of thinkers—the "survival of the fittest" should be associated primarily with Malthus and Spencer, rather than with Darwin (also suggesting the confusion perpetuated by many analyses that use the misnomer "social Darwinism"), are Gregory Claeys, "The 'Survival of the Fittest' and the Origins of Social Darwinism," *Journal of the History of Ideas* 61 (2000): 223–40, and James Allen Rogers, "Darwinism and Social Darwinism," *Journal of the History of Ideas* 33 (1972): 265–80.

27. Henry Williams, *Luther Burbank, His Life and Work* (New York: Hearst's, 1915), 323; Huntington and Whitney, *Builders of America,* 75, 81; Myre Iseman, *Race Suicide* (New York: Cosmopolitan Press, 1912), 131–32; Robie quoted in Linda Gordon, *Moral Property of Women,* 217; Andrew Pearson, "Burbank Says."

28. Granville Hall, *Educational Problems* (New York: Appleton, 1911), 185–86; Huntington and Whitney, *Builders of America,* 87; William Balch, "Is the Race Going Downhill?" *American Mercury* 8 (1926): 435.

29. Leonard, "'More Merciful and Not Less Effective,'" 704; Leonard, "Mistaking Eugenics for Social Darwinism," 204–5; Daniel Lichty, "Tobacco a Race Poison," in *Proceedings of the First National Conference on Race Betterment,* 230; Paul Popenoe and Roswell Johnson, *Applied Eugenics* (New York: Macmillan, 1920), 389; Burbank, *Harvest of the Years,* 161.

30. Popenoe and Johnson, *Applied Eugenics,* 413; Bernard Talmey, *Love: A Treatise on the Science of Sex-Attraction,* 3rd ed. (New York: Practitioners' Publishing Company, 1919), 344–46; "Aiding Nature by Sterilizing Human Weeds," *Sunset* 50 (Apr. 1923):

46; "Eugenics Wins in South," *Helena (MT) Independent*, 8 May 1927. Bisch's remarks appeared in papers such as the *Kokomo (IN) Tribune* and the *Newark (OH) Advocate* on 5 Mar. 1928, as well as the *Logansport (IN) Pharos-Tribune* of 22 Mar. 1928.

31. "Apostle of Birth Control Sees Cause Gaining Here," *NYT*, 8 Apr. 1923; Margaret Sanger, "Need of Birth Control in America," 47–48; Margaret Sanger to C. Smith, 7 May 1929, microfilm roll 131, frame 122, Sanger Papers, LOC; Huntington and Whitney, *Builders of America*, 145–46; Ged Martin, *Past Futures: The Impossible Necessity of History* (Toronto: University of Toronto Press, 2004), 111–48. The remarks—"value of child life," "breeding like weeds," and succoring "the dependent and the delinquent"—appeared in a draft article to be sent to Smith. Excellent discussions of Sanger and the contexts in which she operated include Esther Katz, "The Editor as Public Authority: Interpreting Margaret Sanger," *Public Historian* 17 (1995): 41–50, and Alexander Sanger, "Eugenics, Race, and Margaret Sanger Revisited: Reproductive Freedom for All?" *Hypatia* 22 (2007): 210–17. More than a decade before Sanger's "breeding like weeds" remark, Burbank told Illinois Department of Labor official Burt Bean that the insane were permitted to "breed like weeds." See Transcribed Interviews with Luther Burbank by Burt C. Bean, 17 Apr. 1913, 23, box 15, Burbank Papers.

32. Huntington and Whitney, *Builders of America*, 84–87. Albert Ashmead, "Asepsis—Prevention Better than Cure," *Science* 22 (11 Aug. 1893): 74–75.

33. Courtney Ryley Cooper, "Roots of Crime," *Saturday Evening Post*, 22 Dec. 1934; *Uniontown (PA) Daily News Standard*, 11 July 1923; Arthur Brisbane, "Human Weeds in Prison," in *Editorials from the Hearst Newspapers* (New York: Hearst's, 1914), 209–16; Benjamin Coombe, "Facts of Felony," *World To-Day* 20 (1911): 79; Urquhart, "Human Weeds"; "Weeds and Rattlesnakes," *Helena Independent*, 24 Dec. 1927.

34. John Wigmore, "Symposium of Comments," *Journal of the American Institute of Criminal Law and Criminology* 15 (1924): 400–405; "Weeds and Rattlesnakes"; "An Acceptable Substitute," *Helena Independent*, 28 Jan. 1928; "Eugenics Wins in South."

35. *CDFD*, 3 Nov. 1923; Felecia Ross and Joseph McKerns, "Depression in 'The Promised Land': The *Chicago Defender* Discourages Migration, 1929–1940," *American Journalism* 21 (2004): 55–73; James Barrett, "Unity and Fragmentation: Class, Race, and Ethnicity on Chicago's South Side, 1900–1922," *Journal of Social History* 18 (1984): 44–49; "We Can at Least Keep These Weeds Out of Our Garden," *CDFD*, 7 May 1921; "Hoodlums Insult Man with Fair Companion," *CDFD*, 25 June 1921; "Must Be Uprooted," *CDFD*, 25 May 1935, 12; "Wrong Impression," *CDFD*, 20 Oct. 1917.

36. Hall, *Educational Problems*, 103; Kevles, *In the Name of Eugenics*, 83; Henry Williams, "Can the Criminal Be Reclaimed?"; "Weeds," *Atlantic Monthly* 110 (1912): 719–20; Urquhart, "Human Weeds."

37. "Parent's Duty," *Centralia Daily Chronicle-Examiner*, 20 Sept. 1913; *Newark Advocate*, 22 July 1936; *Pampa Daily News*, 15 Oct. 1937; Joe Williams, "Stepfather Blues," in Michael Taft, *Blues Lyric Poetry: A Concordance* (New York: Garland Publishing, 1984), 308; "Crawford Talks on New Citizen," *Bluefield Telegraph*, 11 Aug. 1936.

38. Burbank, *Harvest of the Years*, 160–61; Robie quoted in Linda Gordon, *Moral Property of Women*, 217; Iseman, *Race Suicide*, 131–32; "The Meaning of Radio Birth Control," in *The Selected Papers of Margaret Sanger: The Woman Rebel, 1900–1928*, ed.

Esther Katz (Urbana: University of Illinois Press, 2007), 1: 386; Karl Pearson, ed., *The Life, Letters and Labours of Francis Galton* (London: Cambridge University Press, 1930), 3A: 220; Martha Van Rensselaer, "Household Bacteriology," *Cornell Reading-Courses* 2 (1913): 54.

39. Charlotte Gilman, "Is America Too Hospitable," *Forum* 70 (1923): 1985–86; Michael Guyer, *Being Well-Born: An Introduction to Eugenics* (Indianapolis: Bobbs-Merrill, 1916), 290–91.

40. Linda Nash, "The Agency of Nature or the Nature of Agency?" *Environmental History* 10 (2005): 67–69; A. Grout, "Some Successful Plants," *Harper's Monthly Magazine* 107 (1903): 546; Lester Ward, "Eugenics, Euthenics, and Eudemics," *American Journal of Sociology* 18 (1913): 737–54. Just as natural selection as the survival of fittest was misread, in Malthus's work, "competition was usually visualized as taking place between individuals with the same position in the economy, not between classes." See Peter J. Bowler, "Malthus, Darwin, and the Concept of Struggle," *Journal of the History of Ideas* 37 (1976): 643.

41. Gilmore, "Plant Vagrants in America," 71–72; Walter Rauschenbusch, "Some Moral Aspects of the 'Woman Movement,'" *Biblical World* 42 (1913): 198–99; Huntington and Whitney, *Builders of America,* 145; Wigmore, "Symposium of Comments," 402. On young women in this period, see Sara Evans, *Born for Liberty: A History of Women in America* (New York: Free Press, 1997), 175–86; Arnold Shaw, *The Jazz Age: Popular Music in the 1920's* (New York: Oxford University Press, 1987), 7–9; Paula Fass, *The Damned and the Beautiful: American Youth in the 1920's* (New York: Oxford University Press, 1977), 14–25. For the views of such a woman, see Dorothy Bromley, "Feminist—New Style," *Harper's* 155 (1927): 552–60.

42. While historians have demonstrated many similarities in how opiate regulation and marijuana criminalization traded on social fears of immigrants, black migrants, and other so-called urban degenerates, they have not explained what made marijuana an immediate, even exceptional threat, especially locally. Examining the biology, ecology, and phytochemistry of cannabis sharpens the historical perspective on why Americans wanted to control the plant and why their efforts proved unsuccessful. Key works include Henry Whiteside, *Menace in the West: Colorado and the American Experience with Drugs* (Denver: Colorado Historical Society, 1997); Michael Elsner, "The Sociology of Reefer Madness: The Criminalization of Marijuana in the U.S.A." (Ph.D. diss., American University, 1994); John C. McWilliams, *The Protectors: Harry J. Anslinger and the Federal Bureau of Narcotics, 1930–1962* (Newark: University of Delaware Press, 1990); Jerome Himmelstein, *The Strange Career of Marihuana* (Westport, CT: Greenwood Press, 1983); John Galliher and Allynn Walker, "The Puzzle of the Social Origins of the Marihuana Tax Act of 1937," *Social Problems* 24 (1977): 367–76; Richard Bonnie and Charles Whitebread, *The Marihuana Conviction* (Charlottesville: University Press of Virginia, 1974); David F. Musto, *The American Disease* (New Haven: Yale University Press, 1973). On the use of policy making to create criminalized space, see Joseph Spillane, "The Making of an Underground Market: Drug Selling in Chicago, 1900–1940," *Journal of Social History* 32 (1998): 37–43.

43. Saunders, "Botanizing on City Vacant Lots," 184–85; Arthur La Roe to Harry Anslinger, 14 Nov. 1938, Marihuana, American Narcotic Defense Association, box 46,

Bureau of Narcotics, Subject Files, Records of the Drug Enforcement Administration, RG 170, National Archives (hereafter BON); "City Called 'Big Reefer Farm,'" *Newark Ledger*, 17 Feb. 1938; Earle Rowell and Robert Rowell, *On the Trail of Marihuana: The Weed of Madness* (Mountain View: Pacific Publishing, 1939), 27; "League of Nations. Committee on Traffic in Opium and Other Dangerous Drugs . . . Indian Hemp. Memorandum Forwarded by the Representative of the United States of America," 10 Nov. 1934, 3, BON.

44. *State v. Bonoa*, 136 So. 15 (1931); A. E. Fossier, "Mariahuana Menace," *New Orleans Medical and Surgical Journal* 84 (1931–32): 249–50; "Arrests Bare Sale of Dope," *CT*, 19 Dec. 1927; "19 Arrested in Shack Colony Raid," *CT*, 2 Sept. 1933; "Officials Plan Death Drive on Narcotic Weed," *Minneapolis Star*, 8 July 1937. Identifying marijuana use as the cause of antisocial behavior may have been a calculated attempt by whites to shift blame by scapegoating blacks or Mexicans for their children's behavior, as native-born whites often disseminated racist attitudes and practices in Gary. See Neil Betten and Raymond A. Mohl, "The Evolution of Racism in an Industrial City, 1906–1940," *Journal of Negro History* 59 (1974): 51–64.

45. *American Practitioner and News* 80 (1900): 259–60; Robert Schauffler, *The Joyful Heart* (Boston: Houghton Mifflin, 1914), 13–14. The *Chicago Defender* reprinted portions of Schauffler's book as Robert Schauffler, "Discordant Thoughts," *CDFD*, 8 May 1915. See also James Putnam, *Human Motives* (New York: Little, Brown and Co., 1915), viii. Putnam also acknowledged that neurological differences were not necessarily useless: "It may be said that there is probably something constructive in a migraine, or in acroparaesthesia, or even an epilepsy. These processes are no doubt like weeds that spring up on the failure of the co-ordinations that favor nobler growth. But they spring up as something which has a significance of its own." See James Putnam, "On the Utilization of Psycho-Analytic Principles in the Study of the Neuroses," *Journal of Abnormal Psychology and Social Psychology* 11 (1916–17): 175; George James, *Quit Your Worrying!* (Boston: Page, 1916), 224–25; *Des Moines News*, 7 July 1918; "Your Mental Garden," *CDFD*, 14 June 1941; "Root of Evil," *CDFD*, 5 Sept. 1942; Donald B. Meyer, *The Positive Thinkers: Popular Religious Psychology*, rev. ed. (Middletown, CT: Wesleyan University Press, 1988), 42–45; Steven Ward, *Modernizing the Mind: Psychological Knowledge and the Remaking of Society* (Westport, CT: Praeger, 2002), 139–51.

46. M. Hayes and L. Bowery, "Marihuana," *Journal of Criminal Law and Criminology* 23 (1933): 1086–98; Rowell and Rowell, *On the Trail of Marihuana*, 12–15; "Traffic in Opium and Other Dangerous Drugs, for the Year Ending 1934," 41, Marihuana Seizures for 1936, box 110, BON.

47. "Traffic in Opium . . . 1935," 30; Musto, *American Disease*, 216–29; David F. Musto, "The Marihuana Tax Act of 1937," *Archives of General Psychiatry* 26 (1972): 101–8; "Weeds Cause Insanity," *LAT*, 1 July 1914, and "Insanity Weeds," *LAT*, 7 Feb. 1915 (republished, respectively, from the *New York Sun* and the *Baltimore American*); Albert Parry, "The Menace of Marihuana," *American Mercury* 36 (1935): 488–89; Bonnie and Whitebread, *Marihuana Conviction*, 45–90, 120–21, 145–64, 354; Lynn Zimmer, "The History of Cannabis Prohibition," in *Cannabis Science: From Prohibition to Human Right*, ed. Lorenz Bollinger (Frankfurt am Main: Peter Lang, 1997), 16; McWilliams, *Protectors*, 49; Himmelstein, *Strange Career*, 4.

48. Bonnie and Whitebread, *Marihuana Conviction*, 45–51, 145, 354; McWilliams, *Protectors*, 49; Himmelstein, *Strange Career*, 4. Egyptian and South African leaders supported Americans' condemnation of cannabis use in 1925 because of their concerns about cannabis use by lower-class and native peoples in these countries. See Zimmer, "History of Cannabis Prohibition," 16–21. For an indiscriminate lumping together of an array of struggling people—migrant workers, Mexicans, underworld addicts, and more—on the basis of a supposedly shared vocabulary, see Vernon Saul, "Vocabulary of Bums," *American Speech* 4 (1929): 337–46.

49. Samuel Bryan, "Mexican Immigrants in the United States," *Survey* 28 (1912): 729–30; Abraham Hoffman, *Unwanted Mexican Americans in the Great Depression* (Tucson: University of Arizona Press, 1996): 59–72; D. Bagley, "Report on Marihuana," 23 Mar. 1934, Marihuana, District 15 folder, BON; C. Goethe, "Quotas," *NYT*, 15 Sept. 1935; Clora Bryan, ed., *Central Avenue Sounds: Jazz in Los Angeles* (Berkeley: University of California Press, 1999), 32; *Dallas Daily Times Herald*, 29 June 1937; *San Antonio Light*, 25 Aug. 1929; "League of Nations . . . Indian Hemp. Memorandum," 1, BON; "Marijuana," *Minneapolis Star*, 9 July 1937; "Ban on Hashish Blocked," *CT*, 3 June 1929; Hayes and Bowery, "Marihuana," 1088; Bonnie and Whitebread, *Marihuana Conviction*, 32–42; Patricia Morgan, "The Political Uses of Moral Reform: California and Federal Drug Policy, 1910–1960" (Ph.D. diss., University of California Santa Barbara, 1978), 77–86. Deportation and repatriation of Mexicans did not eliminate the marijuana problem. Even as the number of Mexicans in American cities plummeted, marijuana use and fears about cannabis continued to increase.

50. James R. Grossman, *Land of Hope: Chicago, Black Southerners, and the Great Migration* (Chicago: University Of Chicago Press, 1991), 89–94, 138–75, 262; Bill Mullen, *Popular Fronts: Chicago African-American Politics, 1935–1946* (Urbana: University of Illinois Press, 1991), 2–3; Bonnie and Whitebread, *Marihuana Conviction*, 51, 92; Joshua Berrett, *Louis Armstrong and Paul Whiteman* (New Haven: Yale University Press, 2004), 112–13; Douglas Daniels, *Lester Leaps In: The Life and Times of Lester "Pres" Young* (Boston: Beacon Press, 2002), 160–63; Woody Herman and Stuart Troup, *Woodchopper's Ball: The Autobiography of Woody Herman* (New York: Dutton, 1990), 80–95; Charlie Barnet with Stanley Dance, *Those Swinging Years* (Baton Rouge: Louisiana State University Press, 1984), 14, 40; James Collier, *Louis Armstrong, an American Genius* (New York: Oxford University Press, 1983), 220–23; Anita O'Day with George Eells, *High Times, Hard Times* (New York: Putnam, 1981), 27–28, 54–62; Clora Bryan, *Central Avenue Sounds*, 32; Hoagy Carmichael with Stephen Longstreet, *Sometimes I Wonder* (New York: Farrar, Straus and Giroux, 1965), 101–2; Malcolm X with the assistance of Alex Haley, *The Autobiography of Malcolm X* (New York: Ballantine Books, 1992), 50–57; Elsner, "Sociology of Reefer Madness," 51–93; Malcolm Fulcher, "Reefers Give Don Redman [*sic*] an Idea," *CDFD*, 8 June 1935; James Donovan, "Jargon of Marihuana Addicts," *American Speech* 15 (1940): 337. On jazz and jazz-era cultural change, see Davarian L. Baldwin, *Chicago's New Negroes: Modernity, the Great Migration, and Black Urban Life* (Chapel Hill: University of North Carolina Press, 2007), 5–27; Mark R. Schneider, *African Americans in the Jazz Age* (Lanham: Rowman and Littlefield, 2006), 85–88; Joel Dinerstein, *Swinging the Machine: Modernity, Technology, and African American Culture*

between the World Wars (Amherst: University of Massachusetts Press, 2003), 130–36, 164–72; Lewis A. Erenberg, *Swingin' the Dream: Big Band Jazz and the Rebirth of American Culture* (Chicago: University of Chicago Press, 1998), xvii, 111–12; Burton Peretti, *The Creation of Jazz: Music, Race and Culture in Urban America* (Urbana: University of Illinois Press, 1992), 76–99; Shaw, *Jazz Age*, 14–22; Fass, *Damned and the Beautiful*, 300–310; Skip G. Gates, "Of Negroes Old and New," *Transition* 46 (1974): 44–58.

51. Blumenberg quoted in Kathy Ogren, *The Jazz Revolution* (New York: Oxford University Press, 1989), 15; *Etude* 42 (1924): 588; *Etude* 42 (1924): 515; Gerald Early, "The Lives of Jazz," *American Literary History* 5 (1993): 144–45; Chadwick Hansen, "Social Influences on Jazz Style," *American Quarterly* 12 (1960): 503–7; *Newark Ledger*, 17 Feb. 1938; "Marihuana Smoking," *Science News Letter* 34 (1938): 340; David Maurer, "Argot of the Underworld Narcotic Addict," *American Speech* 13 (1938): 184; Bonnie and Whitebread, *Marihuana Conviction*, 180–84; "Jive Jumps from Dumps," *NYT*, 25 Sept. 1944; "Single Weed Addict Can Wreck an Entire Band," *CDFD*, 27 Sept. 1941; Erenberg, *Swingin' the Dream*, 79–81. Jazz historians often perfunctorily discuss the role of marijuana use in jazz and jazz culture. In part, scholars wish to avoid glamorizing musicians' use of cannabis because narcotics use ravaged jazz artists in the 1950s and 1960s. Some assert that musicians who used cannabis before or during a performance played miserably. How such effects can be untangled from alcohol consumption is unstated. Reports about the prevalence of usage and its effects are unreliable. Peretti claims less talented white jazz musicians consumed more marijuana than blacks, and he seems to fault Mezz Mezzarow, a white delinquent and unremarkable jazz musician from Chicago's West Side, as a marijuana peddler par excellence who single-handedly created a romantic mythology around jazz musicians' marijuana use. A *Defender* columnist estimated "almost a third of the entire bronze profession" used marijuana. He also warned musicians that "agony, pain and other sufferings" came from regular consumption of "the deadly marihuana weed." These remarks do not clarify the relationship of the drug and the music but do hint at the strategies of social control attempted by moral entrepreneurs like Anslinger. W. E. B. Du Bois, "Winds of Time: Individual Reconversion," *CDFD*, 8 Sept. 1945; Lawrence Schenbeck, "Music, Gender, and 'Uplift' in the *Chicago Defender*, 1927–1937," *Musical Quarterly* 81 (1997): 344–58; Thomas Holt, "The Political Uses of Alienation: W. E. B. Du Bois on Politics, Race, and Culture, 1903–1940," *American Quarterly* 42 (1990): 303, 319–20; Gaines, *Uplifting the Race*, 153–63.

52. New York City Police Department, *Annual Report for the Year Ending 1921*, 281; District Supervisor to Commissioner of Narcotics, 13 Apr. 1934, District 14 files, box 110, BON; "Marihuana Smoking," 340; H. Anslinger with Courtney Cooper, "Marijuana: Assassin of Youth," *American Magazine* 124 (July 1937): 152; Bromberger quoted in House Committee on Ways and Means, *Taxation of Marihuana*, 75th Cong., 1st sess., 1937, *H. Report 6385*, 24; Hugh Pendexter Jr., "Don't Be a 'Mugglehead'!" *Worcester Telegram*, 11 Oct. 1936; "Rhode Island to End Weed as Drug Source," *NYT*, 20 Jan. 1935; "'Reefers' in Trumpet Case," *CDFD*, 5 Nov. 1938; "Nabs Reefer Man," *CDFD*, 23 July 1938; Caroline Acker, *Creating the American Junkie* (Baltimore: Johns Hopkins University Press, 2001), 17–32; H. Wayne Morgan, *Drugs in America: A Social History, 1800–1980* (Syracuse: Syracuse University Press, 1981), 127–43.

53. Allen Bernard, "Growth of Marijuana Habit," *New York Evening Journal,* 5 Nov. 1934; "Drug Flower Grown," *Oakland Tribune,* 28 Sept. 1928; Himmelstein, *Strange Career,* 6–22; Elsner, "Sociology of Reefer Madness," 34–66.

54. "Government to Aid State Agencies," *New Orleans Times Picayune,* 2 Sept. 1937; Fossier, "Mariahuana Menace," 249–50; "Finds Hashish Field," *NYT,* 13 Nov. 1925; "Narcotic 'Garden' Found," *NYT,* 18 Oct. 1934; "Traffic in Opium . . . 1935," 31–32; Assistant Sanitary Superintendent to Deputy Commissioner, 1 Aug. 1935, Records of the New York City Department of Health, box 07-026047, New York City Municipal Archives; Frank Igoe to H. Anslinger, 18 Mar. 1936, District 2 files, box 111, BON; "WPA Workers Assigned to Marijuana Eradication," *Journal of the American Medical Association* 107 (1936): 437; "$3,000,000 Bonfire Destroys Marijuana," *NYT,* 14 Aug. 1936, 12; "City Called 'Big Reefer Farm,'" BON; "Parents Urged to Guard against Killer Narcotic," *Patterson Morning Call,* 31 July 1940; "Police Raid 'Reefer' Nest," *CDFD,* 2 Sept. 1933; "U.S. Fight on Reefers," *CDFD,* 23 Jan. 1937. Figures regarding plant weight probably included soil, as the *Journal of the American Medical Association* warned that "to prevent the spread of the weed it is necessary to dig up the root completely, since it sometime multiplies ten times from one season to the next."

55. Acting Secretary Wilson to Josephine Roche, 30 Nov. 1935, Marihuana, vol. 4, BON; "Mississippi Marijuana Farms Raided," *CDFD,* 15 Oct. 1938; Anslinger, "Assassin of Youth," 19; *NYT,* 30 Apr. 1938. The nature of cannabis growth and marijuana use in cities cannot be fully recovered. Official records and journalistic accounts are fragmentary and full of exaggeration manufactured by marijuana prohibitionists to combat the problem. Without the "records" of peddlers, growers, and smugglers, it is unknown how much cannabis was grown and harvested inside cities, how much was supplied from crops produced in agricultural hinterlands, and how much was imported from abroad. Those who believed much of the supply was imported had to confront reports that migrant beat workers were transporting Midwestern cannabis into Texas, including San Antonio and small Crystal City, which was within fifty miles of the state's border with Mexico. See Harry Hornby to William Whalen, 25 Oct. 1938, Marihuana Eradication folder, box 2, BON.

56. Zachary J. S. Falck, "Controlling Urban Weeds: People, Plants and the Ecology of American Cities, 1888–2003" (Ph.D. diss., Carnegie Mellon University, 2004), 110–13; Lyster Dewey and Andrew Wright, "Hemp Fiber Production," 3–4, 12 Dec. 1933, Marihuana, vol. 3, box 46, BON; Parker Barnes, *Suburban Garden Guide* (New York: Suburban Press, 1911), 24.

57. McWilliams, *Protectors,* 48–56; Bonnie and Whitebread, *Marihuana Conviction,* 39; Fossier, "Mariahuana Menace," 247; Rowell and Rowell, *On the Trail of Marihuana,* 12; *CT,* 23 Feb. 1938; Anslinger, "Assassin of Youth," 152; "Traffic in Opium . . . 1933," 36; "Marihuana Crop Found," *NYT,* 18 Aug. 1938; Pendexter, "Don't Be a 'Mugglehead'!"; "Parents Urged to Guard against Killer Narcotic"; "Marijuana Weed Patch," *CT,* 16 Aug. 1935; "City Called 'Big Reefer Farm'"; William Frey and Otto Johnson to J. Biggins, 18 Oct. 1938, Crimes, vol. 1937–42, box 2, BON.

58. Jim Grant to Narcotic Division, 22 Aug. 1938, Marihuana, 1938–55, vol. 9, BON; "Marihuana Crop Found"; "$1,000,000 Marijuana Crop Seized," *Dallas Daily Times*

Herald, 29 June 1937; Pepoon, *Annotated Flora of the Chicago Region*, 281; Frey and Johnson to Biggins, 18 Oct. 1938, BON; Bradley Kirschberg to Harry Anslinger, 1 July 1938, Marihuana, 1937–38, folder 1, box 112, BON; "Rhode Island to End Weed as Drug Source"; "Marihuana Weed Grows Where Rope Factory Failed," *Science News Letter* 33 (15 Jan. 1938): 38; "Marihuana Conference Held December 5, 1938," 27–29, Marihuana, box 1, BON; Durham, *Your Hay Fever*, 145–46; Pendexter, "Don't Be a 'Mugglehead'!"

59. Hayes and Bowery, "Marihuana," 1087–88; Fossier, "Mariahuana Menace," 249, 251; H. Anslinger, "The Importance of Cooperation in Narcotic Work . . . 1 Nov. 1938," 5–7, box 69, BON; Kent Hunter, "Some Mexican Slayings," *CT*, 17 Sept. 1919; "Something like Hooch," *Kokomo Tribune*, 19 Oct. 1921; Young quoted in Fossier, "Mariahuana Menace," 251; Lyster Dewey and Wright, "Hemp Fiber Production," 3–4; Richard Thomas, *An American Glossary* (Philadelphia: Lippincott, 1912), 172; Committee on Ways and Means, *Taxation of Marihuana*, 81–83; H. Smith et al., "The Implications of Variable or Constant Expression Rates in Invasive Weed Infestations," *Weed Science* 47 (1999): 62. In contemporary ecological terms, invasive plants are plants that, once introduced and established in a new area, expand at rates exceeding those of long-established plants. *Cannabis sativa* is the scientific name for the plant that provides seed, oil, and medicinal compounds. *Hemp* is the English word for the plant, often when it is cultivated to produce fiber as well as seed and oil. *Marijuana* refers to parts of the plant used to produce a drug, typically for nonmedical uses that often lack legal sanction. The argument in this section is grounded in the view that cannabis is a monotypic genus and that there is only one species of the plant—a species capable of adaptation to a wide variety of conditions. However, Lester Grinspoon and James B. Bakalar claim that "most botanists agree that there are three species: *Cannabis sativa* . . . *Cannabis indica* . . . and *Cannabis ruderalis*," an interpretation based on Schultes's argument that cannabis is a polytypic genus, a revision that accounted for ongoing morphological research and long-neglected studies by Russian taxonomists. The most sophisticated study of this question is Ernest Small's 1979 two-volume work *The Species Problem in Cannabis: Science and Semantics*. Small asserts that the binomial species names of *sativa*, *indica*, and *ruderalis* seem to reflect phenotypic variation rather than an untraversable species divide. They are varieties of one plant, *Cannabis sativa*. Although cannabis may have first evolved in central Asia, it is a species capable of adaptation to a wide variety of conditions. The nativity of cannabis seems insignificant given the plant's remarkable adaptability. See Lester Grinspoon and James Bakalar, *Marihuana, the Forbidden Medicine* (New Haven: Yale University Press, 1993), 1–3; R. Schultes et al., "Cannabis: An Example of Taxonomic Neglect," in *Cannabis and Culture* (The Hague: Mouton Publishers, 1975), 21–38; Ernest Small, *The Species Problem in Cannabis: Science and Semantics* (Toronto: Corpus, 1979), 1: 71–75; R. E. Schultes, "Random Thoughts and Queries on the Botany of Cannabis," in *The Botany and Chemistry of Cannabis*, ed. C. R. B. Joyce and S. H. Curry (London: J. and A. Churchill, 1970), 11–26; Alan Haney and Fakhri A. Bazzaz, "Some Ecological Implications of the Distribution of Hemp (*Cannabis sativa L.*) in the United States of America," in Joyce and Curry, *Botany and Chemistry of Cannabis*, 40–46; Alan Haney and Benjamin B. Kutscheid, "An Ecological Study of Naturalized Hemp (*Cannabis sativa L.)* in East-Central Illinois," *American Midland Naturalist* 93 (1975): 1–24. An excellent study of the economic and

scientific production of knowledge of cannabis in the British Empire is James H. Mills, *Cannabis Britannica: Empire, Trade, and Prohibition 1800–1928* (Oxford: Oxford University Press, 2003).

60. Memorandum to H. Anslinger from H. Wollner, 14 Feb. 1938, Marihuana, 1937–38, folder 1, box 112, BON; Bureau of Plant Industry to W. Herwig, 15 Dec. 1926, Division of Drug and Related Plants, General Correspondence, Records of the Bureau of Plant Industry, RG 54, National Archives; Rowell and Rowell, *On the Trail of Marihuana,* 85–89 (League of Nations quotation on 82); H. Anslinger to Bradley Kirschberg, 5 July 1938, Marihuana, 1937–38, folder 1, box 112, BON; Memorandum for Mr. H. Anslinger from John Matchett, 12 July 1940, Marihuana, 1937–38, vol. 1, box 112, BON; Medical Officer to Surgeon General, 4 Sept. 1940; Marihuana, 1937–38, vol. 1, box 112, BON; "War on 'Reefers,'" *CDFD,* 11 Sept. 1937; Perry Duis, *Challenging Chicago: Coping with Everyday Life, 1837–1920* (Urbana: University of Illinois Press, 1998), 193–201. Tobacco was disparaged as a weed in earlier periods in ways the presaged the treatment of marijuana. John Burroughs wrote, "We have given Europe the vilest of all weeds, a parasite that sucks up human blood, tobacco." Given the fervor with which reformers attacked tobacco, tobacco crusaders may have invigorated attacks on cannabis, and cannabis crusaders may have simply recycled existing antitobacco rhetoric. Tobacco was a "race poison" that "stupefies and kills." A reformer's comparison of being "alcoholized" with being "tobacco narcotized" suggests that some well-intentioned people might even have confused these two "weeds." As late as 1944, newspaper reporters were calling marijuana a "narcotic tobacco." See Burroughs, *Pepacton,* 202–3; Lichty, "Tobacco," 222–32; "Jive Jumps from Dumps."

61. Rowell and Rowell, *On the Trail of Marihuana,* 23; "Marijuana Pickers Sentenced," *CT,* 16 Sept. 1947; "Marijuana Growing Wild," *CT,* 21 Sept. 1948; John Wagner, "Marijuana Eradication by the New York City Department of Sanitation," in *Proceedings of the Northeastern Weed Control Conference* (*NEWCC* hereafter) (1952): 119–21; *Brooklyn Eagle,* 19 Sept. 1951; *Brooklyn Eagle,* 18 July 1951; Falck, "Controlling Urban Weeds," 145–54; "City Blamed for Hill 'Reefer' Bed," *Pittsburgh Courier,* 28 July 1962.

62. "What Is a Weed?" *NYT,* 7 Nov. 1915.

63. "Idaho Weeds," *American Botanist* 29 (1923): 70.

64. Cresswell, *Tramp in America,* 147–56; John Culhane, *Walt Disney's Fantasia* (New York: Abrams, 1983), 30–57; Susan Willis, "Fantasia: Walt Disney's Los Angeles Suite," *Diacritics* 17 (Summer 1987): 84–95; Woody Guthrie, *Bound for Glory* (New York: Dutton, 1943), 33–114, 131–59, 182–83, 252–72, 293–328; Kusmer, *Down and Out,* 193–237. For one family's efforts to survive on Los Angeles's margins, see Becky M. Nicolaides, "The Quest for Independence: Workers in the Suburbs," in *Metropolis in the Making: Los Angeles in the 1920s,* ed. Tom Sitton and William Francis Deverell (Berkeley: University of California Press, 2001), 77–95.

65. Francis Sayre, "The Question of Self-Sufficiency," *Annals of the American Academy of Political and Social Science* 186 (1936): 129–30; *Helena Independent,* 8 Jan. 1937; Frank Thone, "Wealth from Weeds," *Science News Letter* 40 (1941): 166–67. Sayre estimated that conversion to "economic self-sufficiency" would involve "violent economic and social dislocations" and "intense human suffering" for "10,000,000" or more "human weeds."

CHAPTER THREE

1. Philip Gorlin, "Ragweed Control in New York City," *Hay Fever Bulletin* (Spring 1955): 4. Gorlin estimated that 230,000 of these people were allergic to ragweed pollen. On other wars against nature, see Robert J. Spear, *The Great Gypsy Moth War: The History of the First Campaign in Massachusetts to Eradicate the Gypsy Moth, 1890–1901* (Amherst: University of Massachusetts Press, 2005); Joshua Blu Buhs, *The Fire Ant Wars: Nature, Science, and Public Policy in Twentieth-Century America* (Chicago: University of Chicago Press, 2004); Clinton L. Evans, *The War on Weeds in the Prairie West: An Environmental History* (Calgary: University of Calgary Press, 2002); Edmund Russell, *War and Nature: Fighting Humans and Insects with Chemicals from World War I to Silent Spring* (New York: Cambridge University Press, 2001). Most historical studies of herbicides examine their agricultural and horticultural applications; see J. L. Anderson, "War on Weeds: Iowa Farmers and Growth-regulator Hormones," *Technology and Culture* 46 (2005): 719–44; Evans, *War on Weeds;* Virginia Scott Jenkins, *The Lawn: A History of an American Obsession* (Washington, DC: Smithsonian Institution Press, 1994); James Young, "Public Response to the Catastrophic Spread of Russian Thistle (1880) and Halogeton (1945)," *Agricultural History* 62 (1988): 122–30; Kass Green, *Forests, Herbicides and People: A Case Study of Phenoxy Herbicides in Western Oregon* (New York: Council on Economic Priorities, 1982). Throughout this chapter, NYOR is used in place of Operation Ragweed, an effort to which a number of departments in New York City's government contributed.

2. On the environmental history of New York City, see David Stradling, *Making Mountains: New York City and the Catskills* (Seattle: University of Washington Press, 2007); Matthew Gandy, *Concrete and Clay: Reworking Nature in New York City* (Cambridge: MIT Press, 2002); Max Page, *The Creative Destruction of Manhattan, 1900–1940* (Chicago: University of Chicago Press, 1999); Andrew Hurley, "Creating Ecological Wastelands: Oil Pollution in New York City, 1870–1900," *Journal of Urban History* 20 (1994): 340–64.

3. William Atwood, *Civic and Economic Biology* (Philadelphia: Blakiston's, 1922), 398; "Doctor Demands Laws," *CT,* 14 Jan. 1940. On the history of hay fever, see Gregg Mitman, *Breathing Space: How Allergies Shape Our Lives and Landscapes* (New Haven: Yale University Press, 2007), 10–96; Kathryn J. Waite, "Blackley and the Development of Hay Fever as a Disease of Civilization in the Nineteenth Century," *Medical History* 39 (1995): 190–96; M. Emanuel, "Hay Fever, a Post Industrial Revolution Epidemic," *Clinical Allergy* 18 (1988): 295–304.

4. *Sanitary Code of the Board of Health of the Department of Health of the City of New York,* sec. 221 (Brown 1929); "Curbing Hay Fever," *NYT,* 6 June 1932; "Preventing Hay Fever," *NYT,* 12 June 1932.

5. Zachary J. S. Falck, "Controlling Urban Weeds: People, Plants and the Ecology of American Cities, 1888–2003" (Ph.D. diss., Carnegie Mellon University, 2004), 304–10; "Report of the Pollen Survey Committee," *Journal of Allergy* 20 (1949): 148.

6. O. Durham, "The Contribution of Air Analysis to the Study of Allergy," *Journal of Laboratory and Clinical Medicine* 13 (1928): 969, 973; "Hay Fever Is Here," *NYT,* 9 Aug. 1938.

7. R. Wodehouse, "Weeds, Waste, and Hayfever," *Natural History* 43 (1939): 162;

Willard Payne, "The Source of Ragweed Pollen," *Journal of the Air Pollution Control Association* 17 (1967): 653–54; "Ragweed, Hayfever, and 2,4-D" (1946), folder 56, box 4898, Mayor O'Dwyer Papers (MODP hereafter), New York City Municipal Archives; Hay Fever Prevention Society, "Facts on Hay Fever," box 35, Frank E. Egler Papers (hereafter FEEP), Aton Forest, Norfolk, CT (emphasis in original); "Ragweed's Toll," *NYT*, 19 Aug. 1915; "Hay-Fever Plaint," *NYT*, 19 Aug. 1935; O. Durham, "Cooperative Studies in Ragweed Pollen Incidence," *Journal of Allergy* 1 (1929): 12, 20. On air pollution in this period, see David Stradling, *Smokestacks and Progressives: Environmentalists, Engineers and Air Quality in America, 1881–1951* (Baltimore: Johns Hopkins University Press, 1999); Joel A. Tarr, *The Search for the Ultimate Sink: Urban Pollution in Historical Perspective* (Akron: University of Akron Press, 1996); Christopher Sellers, "Factory as Environment: Industrial Hygiene, Professional Collaboration, and the Modern Sciences of Pollution," *Environmental History Review* 18 (1994): 55–83.

8. Thomas Kessner, *Fiorello H. La Guardia and the Making of Modern New York* (New York: McGraw-Hill, 1989), 202, 589–91; Robert A. Caro, *The Power Broker: Robert Moses and the Fall of New York* (New York: Knopf, 1974) 325, 463–67; Page, *Creative Destruction of Manhattan*, 1–5; Gandy, *Concrete and Clay*, 115–37; Eleanora Schoenebaum, "Emerging Neighborhoods: The Development of Brooklyn's Fringe Areas" (Ph.D. diss., Columbia University, 1976), 240–46, 301–5; Peter Derrick, *Tunneling to the Future: The Story of the Great Subway Expansion That Saved New York* (New York: New York University Press, 2001), 1–3, 231–32, 245–48; Kenneth T. Jackson, *Crabgrass Frontier: The Suburbanization of the United States* (New York: Oxford University Press, 1985), 6, 11; Robert Armstrong and Homer Hoyt, *Decentralization in New York* (Chicago: Urban Land Institute, 1941), 20–39; David A. Johnson, *Planning the Great Metropolis: The 1929 Regional Plan of New York and Its Environs* (London: Spoon, 1996), 22; Jameson W. Doig, "Joining New York City to the Greater Metropolis," in *The Landscape of Modernity: New York City, 1900–1940*, ed. David Ward and Olivier Zunz (Baltimore: Johns Hopkins University Press, 1992), 76–105.

9. John Kieran, *Natural History of New York City* (Boston: Houghton Mifflin, 1959), 181–241; *George v. Cypress Hills Cemetery* (1898), 32 A.D. 281; Lewis Gannett, "The Wilderness of New York," *Century Magazine* 110 (1925): 302–4; Melvin Gilmore, "Plant Vagrants in America," *Papers of the Michigan Academy of Science, Arts, and Letters* 15 (1931): 66–67; Alfred Kazin, *A Walker in the City* (New York: Harcourt, Brace, 1951), 10–11, 83–88; Harry Ahles, "Interesting Weeds in New York City," *Bulletin of the Torrey Botanical Club* 78 (1951): 266–70; Donald Peattie, *The Road of a Naturalist* (Boston: Houghton Mifflin, 1941), 138–45; Ethel Hausman, *Illustrated Encyclopedia of American Wild Flowers* (Garden City: Garden City Publishing, 1947), x–xi.

10. Payne, "Source of Ragweed Pollen," 653–54; Kristin Gremillion, "Early Agricultural Diets in Eastern North America," *American Antiquity* 61 (1996): 525–28; Wodehouse, "Weeds, Waste," 151–52, 160–62; John Ruskin, "Public Health Significance of Ragweed Control Demonstrated in Detroit," *NEWCC* (1954): 494; "Sensible Attack on Ragweed," *BDE*, 7 July 1947; Thomas McMahon, "Ragweed Control," *NEWCC* (1959): 287.

11. "Pittsburgh Really Not So Black," *Tycos-Rochester* 21 (Apr. 1931): 68; "Hay-Fever Season," *NYT*, 3 July 1944; *Bronx Home News*, 29 Aug. 1937; Caro, *Power Broker*, 331–34; "Hay Fever in New York," *NYT*, 5 Sept. 1938; W. J. Beecher, "The Lost Illinois Prairie," *Chicago History* 2 (1973): 166; "Ragweed Time," *NYT*, 21 Sept. 1938.

12. Falck, "Controlling Urban Weeds," 287–304; Edward Meierhof to David Goldstein, 29 June 1930, box 07-022818, Records of the Commissioner's Office, Department of Health, New York City Municipal Archives (NYDOH hereafter); J. Adrianzen to Shirley Wynne, 6 Sept. 1932, box 07-030354; NYDOH (emphasis in original); Paul DuBois to Health Commissioner, 6 Aug. 1931, box 07-023707, NYDOH; William Rappeport to Shirley Wynne, 9 Jan. 1931, box 07-023707, NYDOH.

13. Julia Ellsworth Ford, "A National Menace," unpublished manuscript, 1947, 1, 11, Julia Ellsworth Ford Papers (JEFP hereafter), Beinecke Manuscript Library, Yale University; Philip Cornick, *Premature Subdivision and Its Consequences* (New York: Institute of Public Administration, 1938), 65–69; *Rye (New York) Chronicle*, 31 Aug. 1945; Julia Ford to Shirley Wynne, 28 Aug. 1931, box 07-023707, NYDOH; Harold Tessler to Ernest Stebbins, 17 Aug. 1945, box 07-026101, NYDOH; *Hay Fever* 3 (Aug. 1952): 1; Hay Fever Prevention Society, "Resolution," 20 May 1947, box 07-031977, NYDOH; *Hay Fever* 7 (June 1956): 2.

14. John Barry to George Harvey, 28 Dec. 1929, box 07-023703, NYDOH; "Weeding Out Hay Fever," Sept. 1931, NYDOH; "Health and Weeds," *American City* 54 (Jan. 1939): 15.

15. "Preventing Hay Fever," *NYT*, 12 June 1932; "Hay Fever Season," *NYT*, 14 Aug. 1932; Sanitary Superintendent to Health Commissioner, 1 Oct. 1935, box 07-208375, NYDOH; John Barry to Sanitary Superintendent, 6 Sept. 1935, box 07-028375, NYDOH; Caro, *Power Broker*, 453, 465; Kessner, *Fiorello H. La Guardia*, 339; "Eradicating Ragweed," *NYT*, 22 Aug. 1938; "Chicago Employs 1,350 in Hay Fever Fight," *NYT*, 12 Aug. 1932; "Pollen-Laden Chicago," *NYT*, 3 Sept. 1933.

16. New York City WPA Public Information Section, press release, 20 May 1937, box 07-023728, NYDOH; "New 'Pistol' Kills Weeds," *CDFD*, 27 Apr. 1935; Brehon Somervell to John Rice, 16 Dec. 1937, box 07-023728, NYDOH; Commissioner to Robert Wagner, 5 Aug. 1939, box 1979, Records of the Works Progress Administration, RG 69, National Archives (WPA hereafter); Howard Hunter to James Fay, 10 Oct. 1940, box 1979, WPA; Henry Wallace to Morris Sheppard, 5 Nov. 1938, box 2912, General Correspondence of the Secretary of Agriculture, Records of the Department of Agriculture, RG 16, National Archives (GC-SOA hereafter); Henry Wallace to James Pope, 25 Mar. 1938, GC-SOA.

17. Barbara Blumberg, "The Works Progress Administration in New York City" (Ph.D. diss., Columbia University, 1974), 239–57, 574–75; Joseph Verdicchio, "New Deal Work Relief and New York City, 1933–1938" (Ph.D. diss., New York University, 1980), 240–45; Johnson, *Planning the Great Metropolis*, 244–51; Neil M. Maher, *Nature's New Deal: The Civilian Conservation Corps and the Roots of the American Environmental Movement* (New York: Oxford University Press, 2008), 43–113; Robert Cooke, "Hay Fever," *Neighborhood Health* 4 (July 1938): 2.

18. *Bronx Home News*, 7 Sept. 1941; "Dooms Ragweed Plant," *NYT*, 5 June 1942; John Duffy, *A History of Public Health in New York City, 1866–1966* (New York: Russell Sage Foundation, 1974), 623; Ernest Stebbins to Fiorello LaGuardia, 6 Oct. 1943, box 07-026080, NYDOH; "Hay-Fever Season," *NYT*, 3 July 1944.

19. James Bryan to Robert Moses, 9 July 1945, box 102688, Records of the Department of Parks, New York City Municipal Archives (NYDOP hereafter); W. Latham to A. Hodgkiss, 16 July 1945, box 102688, NYDOP; Alfred Fletcher to Katharine Morgan, 22

Apr. 1947, box 07-031977, NYDOH. This opinion, first published in the *Journal of the American Medical Association,* was printed in "Weinstein Defends Drive on Ragweed," *NYT,* 7 Sept. 1946; T. LeBlanc to Israel Weinstein, 14 Oct. 1947, box 07-031977, NYDOH; Israel Weinstein and Alfred Fletcher, "Essentials for the Control of Ragweed" (paper presented to the American Public Health Association), 7 Oct. 1953, box 07-031977, NYDOH. The authors claimed that Wodehouse's remark regarding "treating the environment" appeared in a letter from Wodehouse to a health official written in 1939.

20. Ada Georgia, *Manual of Weeds* (New York: Macmillan Company, 1919), 13–16; Albert Hansen, "Danger Lurks in City Weeds," *American City* 31 (July 1924): 17; W. Latham to G. Spargo, 8 June 1942, box 102602, NYDOP; New York State Department of Health Division of Public Health Education, *Eradicate Ragweed: A Cause of Hay Fever,* n.d.; E. Hildebrand, "War on Weeds," *Science* 103 (1946): 465–68; Alfred Fletcher and C. Velz, "Pollens: Sampling and Control," *Industrial Medicine and Surgery* 19 (1950): 130–31. On the development of herbicides, see Nicolas Rasmussen, "Plant Hormones in War and Peace: Science, Industry, and Government in the Development of Herbicides in 1940s America," *Isis* 92 (2001): 291–316; J. Troyer, "In the Beginning: The Multiple Discovery of the First Hormone Herbicides," *Weed Science* 49 (2001): 290–97; Gale Petersen, "The Discovery and Development of 2,4-D," *Agricultural History* 47 (1967): 243–51.

21. "Film Shows New Weed-Killing Techniques," *American City* 62 (Apr. 1947): 119; "2-4 Dow Weed Killer," box 102716, NYDOP; George Avery, letter dated 5 Apr. 1947, in *Eastern Regional Conference on the Control of Plants Harmful and Annoying to Man . . . April 16, 1947,* box 07-031977, NYDOH; Sherman Dryer Productions, *Hay Fever,* script, 31 July 1946, 23–24, box 07-031976, NYDOH; Frank Thone, "How about 2,4-D on that Weedy Lot?" *Brooklyn Eagle,* 18 Aug. 1947.

22. Alfred Fletcher to Ben Harris, 29 Jan. 1947, box 07-031977, NYDOH; "Drive on Ragweed," *NYT,* 27 July 1946; "New Methods of Weed Control," in *Eastern Regional Conference,* flyers, box 07-031977, NYDOH.

23. "Monthly Report for July, 1946," box 07-008451, NYDOH; press release, 4 Aug. 1946, folder 56, box 4898, MODP; press release, 7 Oct. 1946, folder 56, box 4898, MODP; Seneca Furman to Israel Weinstein, 25 May 1947, box 07-031977, NYDOH; Lillian Zimmerman to Israel Weinstein, 30 Apr. 1947, box 07-031977, NYDOH; Gorlin, "Ragweed Control," 5; "Hopes to Keep City Healthy," *Brooklyn Eagle,* 15 June 1946; "Israel Weinstein," *NYT,* 29 May 1975; Duffy, *History of Public Health,* 388–90.

24. Ruskin, "Public Health Significance," 495; Charles Gilmore to Robert Wagner, 5 June 1958, box 07-051737, NYDOH.

25. Israel Weinstein to Mayor, 29 Sept. 1947, box 07-031977, NYDOH; *Bronx Home News,* 8 July 1947.

26. Gorlin, "Ragweed Control," 4; Samuel Frant to Stephen Urban, 7 July 1947, box 07-031977, NYDOH.

27. "Ragweed Drive Mapped as Rain Benefits Enemy," *NYT,* 11 Apr. 1947; "It's Ragweed Time," *NYT,* 11 Aug. 1954; W. Latham to Hazel Tappan, 27 Sept. 1947, box 102824, NYDOP; Duane Everson to Israel Weinstein, 24 Aug. 1947, box 07-031977, NYDOH; Matthew Walzer and Bernard Siegel, "Effectiveness of the Ragweed Eradication Campaigns in New York City," *Journal of Allergy* 27 (1956): 113, 124; Alfred Salomon to John Mahoney, 7 Aug. 1952, box 08-015337, NYDOH.

28. J. Adrianzen to Shirley Wynne, 6 Sept. 1932, box 07-030354, NYDOH; H. Abul-Faith and F. A. Bazzaz, "The Biology of Ambrosia trifida L. II. Germination, Emergence, Growth and Survival," *New Phytologist* 83 (1979): 825–27; F. A. Bazzaz, *Plants in Changing Environments: Linking Physiological, Population, and Community Ecology* (New York: Cambridge University Press, 1996), 50–51; J. P. Grime, *Plant Strategies, Vegetation Processes, and Ecosystem Properties,* 2nd ed. (Chichester: Wiley, 2001), 249–56.

29. Keith D. Revell, *Building Gotham: Civic Culture and Public Policy in New York City, 1898–1938* (Baltimore: John Hopkins University Press, 2003), 217–23, 264–67; Richard A. Plunz, "Zoning and the New Horizontal City," in *Planning and Zoning New York City: Yesterday, Today, and Tomorrow,* ed. Todd W. Bressi (New Brunswick, NJ: Center for Urban Policy Research, 1993), 37–43; Norman Marcus, "Zoning from 1961 to 1991," in Bressi, *Planning and Zoning,* 71.

30. William Loucks, "Increments in Land Values in Philadelphia," *JLPUE* 1 (1925): 472–83; William Loucks, "Unearned Increment in Land Values and Its Social Implications," *Annals of the American Academy of Political and Social Science* 148 (1930): 72–78; Mason Gaffney, Fred Harrison, and Kris Feder, *The Corruption of Economics* (London: Shepeard-Walwyn, 1994), 100–103; Howard Shannon and H. Bodfish, "Increments in Subdivided Land Values in Twenty Chicago Properties," *JLPUE* 5 (1929): 29, 38–45; G. Arner, "Land Values in New York City," *Quarterly Journal of Economics* 36 (1922): 579–80.

31. Herbert Dorau and Albert Hinman, *Urban Land Economics* (New York: Macmillan Company, 1928), 138; Ernest Fisher, "Speculation in Suburban Lands," *American Economic Review* 23 (1933): 152–57; Cornick, *Premature Subdivision,* 86–89; Rosalind Tough, "Production Costs of Urban Land in Sunnyside," *JLPUE* 8 (1932): 43–54; Walter Firey, "Ecological Considerations in Planning for Rurban Fringes," *American Sociological Review* 11 (1946): 411–13; Eric H. Monkkonen, *America Becomes Urban: The Development of U.S. Cities and Towns, 1780–1980* (Berkeley: University of California Press), 7, 70, 222; Roy Rosenzweig and Elizabeth Blackmar, *The Park and the People: A History of Central Park* (Ithaca: Cornell University Press, 1992), 439–42; photographs of squatters' colonies, Percy Loomis Sperr Collection, New York Public Library.

32. Philip Cornick, "Land Prices in a Commodity Price System," *JLPUE* 10 (1934): 225; Fisher, "Speculation in Suburban Lands," 155; Cornick, *Premature Subdivision,* 299; Joseph Laronge, "The Subdivider of Today and Tomorrow," *JLPUE* 18 (1942): 423–27.

33. Shirley Wynne to Julia Ford, 25 Feb. 1931, box 07-028369, NYDOH; "Tax Remedy for Hay Fever," *NYT,* 27 July 1940; Joseph Kinsley to Ernest Stebbins, 22 July 1942, box 07-026073, NYDOH; Ernest Stebbins to Fiorello LaGuardia, 6 Oct. 1943, box 07-026080, NYDOH.

34. Revell, *Building Gotham,* 245–55; Joseph Kinsley to Ernest Stebbins, 22 July 1942, box 07-026073, NYDOH; Sol Pincus to George Palmer, 20 June 1942, box 07-026073, NYDOH; John Oberwager to Sol Pincus, 20 June 1942, box 07–026073, NYDOH; Frank Calderone to George Palmer, 29 June 1942, box 07-026073, NYDOH; "LaGuardia Signs Bill to Compel Destruction of Ragweed in City," *New York Herald Tribune,* 4 June 1942.

35. William Best to Benjamin Weiss, 10 Sept. 1935, box 07-026047, NYDOH; Francis Jones to Shirley Wynne, 8 Sept. 1932, box 07-028369, NYDOH; Edward Bowyer to John Rice, 11 Sept. 1935, box 07-028375, NYDOH; Joshua Lichtenstein to Leona Baumgartner, 16 Aug. 1955, box 08-045015, NYDOH.

36. Bernard Masket to Robert Moses, 11 Aug. 1944, folder 26, box 102675, NYDOP. For critiques of the neoclassical economic model's failure to account for environmental costs, see Thomas Prugh, *Natural Capital and Human Economic Survival* (Solomons, MD: ISEE Press, 1995), 129–30; Herman E. Daly and John B. Cobb Jr., *For the Common Good: Redirecting the Economy toward Community, the Environment, and a Sustainable Future* (Boston: Beacon Press, 1994), 51–58.

37. "WPA Hay Fever Squad Begins," *NYT*, 21 May 1937; Brehon Somervell, *Administration of Work Relief in the City of New York, August 1936 to December 1937* (New York: 1938), 14; Philip Gorlin to Alfred Fletcher, 8 Sept. 1947, box 07-031977, NYDOH; press release, 20 Sept. 1950, box 07-030199, NYDOH; John Mahoney to Rudolph Halley, 3 Oct. 1952, box 08-015337, NYDOH; Gorlin, "Ragweed Control," 4; Roscoe Kandle to Louis Fucci, 18 June 1954, box 08-010292, NYDOH.

38. Sol Pincus to William Best, 9 Sept. 1935, box 07-026047, NYDOH; "Sunflower Found Enemy of Ragweed," *NYT*, 7 Sept. 1937; Frances Cormier to James Dawson, 8 Mar. 1946, box 102716, NYDOP; William Latham to James Dawson, 3 Apr. 1946, box 102716, NYDOP. During the Depression, botanist Paul Sears advocated using Russian thistle to protect the drought-ravaged West; see Frank Thone, "The Worth of Weeds," *Science News Letter* 28 (1935): 382. For ground cover run amok, see Mart Allen Stewart, "Cultivating Kudzu: The Soil Conservation Service and the Kudzu Distribution Program," *Georgia Historical Quarterly* 81 (1997): 151–67.

39. Weinstein and Fletcher, "Essentials for the Control of Ragweed," 5; Wodehouse, "Weeds, Waste," 178.

40. Frederick Aschman, "Dead Land: Chronically Tax Delinquent Lands in Cook County, Illinois," *Land Economics* 25 (1949): 240–42; J. Sienold to Mrs. Frothingham Wagstaff, 24 Apr. 1947, box 2, Outdoor Cleanliness Association Records, New York Public Library; Edward Vogel to John Mahoney, 9 Oct. 1951, box 08-003343, NYDOH; "Cashmore Acts to End Weed Lair Menace," *Brooklyn Eagle*, 27 July 1946; Harold Connolly, *A Ghetto Grows in Brooklyn* (New York: New York University Press, 1977), 117–33; Wendell E. Pritchett, *Brownsville, Brooklyn: Blacks, Jews and the Changing Face of the Ghetto* (Chicago: University of Chicago Press, 2002), 105–74; Walther Thabit, *How East New York Became a Ghetto* (New York: New York University Press, 2003), 7–39; Firey, "Ecological Considerations," 411.

41. Allen Solomon and Murray Buell, "Effects of Suburbanization upon Airborne Pollen," *Bulletin of the Torrey Botanical Club* 96 (1969): 435–45.

42. Alfred Fletcher to Israel Weinstein, "Monthly Report for September, 1946," box 07-008451, NYDOH; "Sneezes Greet Weinstein's Plea," *NYT*, 30 July 1946; John Mahoney to Robert Wagner, 27 Apr. 1953, box 08-015341, NYDOH; Richard Wiseman et al., "Studies on Factors Influencing Ragweed Pollen Counts in the New York Metropolitan District," *Journal of Allergy* 25 (1954): 10; Herman Aaronoff to George James, 12 June 1964, box 09-015821, NYDOH.

43. Benton MacKaye, *The New Exploration: A Philosophy of Regional Planning* (New York: Harcourt, Brace and Company, 1928); Homer Hoyt, "The Influence of Highways and Transportation on the Structure and Growth of Cities and Urban Land Values," in *Highways in Our National Life*, ed. Jean Labatut and Joshua Lane (Princeton: Princeton University Press, 1950), 205; Royce Hanson, ed., *Perspectives on Urban Infrastructure*

(Washington, DC: National Academy Press, 1984); Joel A. Tarr and Josef W. Konvitz, "Patterns in the Development of the Urban Infrastructure," in *American Urbanism: A Historiographical Review*, ed. Howard Gillette and Zane L. Miller (New York: Greenwood Press, 1987), 195; Joel A. Tarr and Gabriel Dupuy, *Technology and the Rise of the Networked City in Europe and America* (Philadelphia: Temple University Press, 1988), xiii; John Stilgoe, *Outside Lies Magic: Regaining History and Awareness in Everyday Life* (New York: Walker and Company, 1998), 54–66, 96–97. On postwar suburbanization and changing environmental values, see Adam Rome, *The Bulldozer in the Countryside: Suburban Sprawl and the Rise of American Environmentalism* (New York: Cambridge University Press, 2001), 6–10, 119–53; Robert D. Lewis and Richard Walker, "Beyond the Crabgrass Frontier: Industry and the Spread of North American Cities, 1850–1950," *Journal of Historical Geography* 27 (2001): 3–19; Richard N. L. Andrews, *Managing the Environment, Managing Ourselves: A History of American Environmental Policy* (New Haven: Yale University Press, 1999), 201–54; Samuel P. Hays, *Beauty, Health, and Permanence: Environmental Politics in the United States, 1955–1985* (Cambridge: Cambridge University Press, 1987), 21–29, 329–62.

44. *Railway Engineering and Maintenance Encyclopedia*, 2nd ed. (New York: Simmons-Boardman Publishing, 1926), 140–44; *Railway Engineering and Maintenance Encyclopedia*, 7th ed. (New York: Simmons-Boardman Publishing, 1948), 530–42; *Garden and Forest* 8 (1895): 310; R. Gunderson, "Vegetation Control on Western Maryland Railway," *NEWCC* (1962): 384–89; R/W Maintenance Corporation Forester's Manual, 910.1.1, Dec. 1952, box 36, FEEP; "What's the Best Way to Keep Right-of-Way Cleared?" *Power Engineering* 68 (Jan. 1964): 46; "For Watershed Brush Control," *Water Works Engineering* 112 (1959): 286–87, 314–15; William Boyd, "Chemical Control of Vegetation on Areas Bordering Public Water Supplies," *NEWCC* (1966): 400; Wilfred Robbins, Alden Crafts, and Richard Raynor, *Weed Control: A Textbook and Manual* (New York: McGraw-Hill, 1942), 429–36; Wilbur Garmhausen, "Chemicals Bring a Revolution to Roadside Weed Control," *Public Works* 90 (Feb. 1957): 115–16; Dale Rake, "Use of Herbicides under Asphalt," *Public Works* 95 (May 1964): 110. Chemicals used as soil sterilants were not hormone herbicides, but borate compounds that microorganisms could not break down. They lasted longer than organic compounds to kill slow-developing plant life. See Robert McMahon, "Roadside Spraying for Weed Control," *NEWCC* (1954): 525. Many of the best histories of highway development—including Clay McShane, *Down the Asphalt Path: The Automobile and the American City* (New York: Columbia University Press, 1994); Bruce E. Seely, *Building the American Highway System: Engineers as Policy Makers* (Philadelphia: Temple University Press, 1987); and Mark H. Rose, *Interstate: Express Highway Politics, 1941–1956* (Lawrence: Regents Press of Kansas, 1979)—say little about the environmental impacts of transportation arteries.

45. Frank E. Egler, "Vegetation Management for Rights-of-Way and Roadsides," in *The Smithsonian Report for 1953* (Washington, DC: Smithsonian Institution, 1954), 299–322; Frank E. Egler, "Science, Industry, and the Abuse of Rights of Way," *Science* 127 (1958): 573–76.

46. Gregg Mitman, "Hay Fever Holiday: Health, Leisure, and Place in Gilded-Age America," *Bulletin of the History of Medicine* 77 (2003): 600–635; "Weeding Out Hay Fever"; "Hayfeverless Maine," *NYT*, 7 Sept. 1930; Falck, "Controlling Urban Weeds,"

304–10; S. Sack, "How Far Can Wind-Borne Pollen Be Disseminated," *Journal of Allergy* 20 (1949): 459.

47. "Highway Men Aid War on Hay Fever," *NYT,* 1 Sept. 1935; "Refuges from Hay Fever," *NYT,* 22 Aug. 1937; "Little Ragweed Found Up-State," *NYT,* 9 Sept. 1943; New York State Ragweed Pollen Survey Results (1959), New York State Library, Albany.

48. Stradling, *Making Mountains,* 8–19, 84–108, 139, 179–209; Alf Evers, *The Catskills: From Wilderness to Woodstock* (Woodstock, NY: Overlook Press, 1972), 327–568, 659–83; Manville Wakefield, *To the Mountains by Rail* (Fleischmanns, NY: Purple Mountain Press, 1989), 244–45; Phil Brown, *Catskill Culture: A Mountain Rat's Memories of the Great Jewish Resort Area* (Philadelphia: Temple University Press, 1998), 11, 40–42. In her memoir, Esterita Blumberg writes that the construction and proliferation of large resorts in the 1930s through the 1950s led one Catskills vacationer to remark, "When I go to the *country,* it's only the Concord [a large resort]. . . . It's a regular *city*" (emphasis in original). See Esterita Blumberg, *Remember the Catskills: Tales by a Recovering Hotelkeeper* (Fleischmanns, NY: Purple Mountain Press, 1996), 222; *The People of Sullivan County New York* (Ithaca: Cornell University Press, 1963), 1–4. Gandy, *Concrete and Clay,* 43–67. Sullivan County is at one edge of the New York City metropolitan area. Although the Regional Plan Association did not consider it part of the tristate metropolitan area in 1960, the association does today. While the U.S. Census Bureau does not consider Sullivan County part of the New York City primary metropolitan statistical area, the environmental organization the Wildlife Trust includes Sullivan County in the metropolitan area's "bioscape." See William Niering, *Nature in the Metropolis: Conservation in the Tri-State New York Metropolitan Region* (New York: Regional Plan Association, 1960); http://www.rpa.org/tristateregion.html, http://quickfacts.census.gov/qfd/states/36/36105.html, and http://www.thewildones.org/Bioscape/nybioscape.html. For the urbanization of neighboring Ulster County, see James LaGro and Stephen DeGloria, "Land Use Dynamics in an Urbanizing Non-Metropolitan County in New York State (USA)," *Landscape Ecology* 7 (1992): 275–89.

49. Peter Hugill, *The Elite, the Automobile, and the Good Roads Movement in New York* (Syracuse: Syracuse University Press, 1981), 12; Evers, *Catskills,* 684–92; Wakefield, *To the Mountains,* 60–61; John Conway, *Retrospect: An Anecdotal History of Sullivan County, New York* (Fleischmanns, NY: Purple Mountain Press, 1996), 58–62.

50. *Roadside Vegetation Cover Project* (1953): 11–22; D. Curtis et al., "Herbicide Work on New York State Highways," *NEWCC* (1955): 463–70; New York State Department of Health, press release, 31 Mar. 1947, box 07-031977, NYDOH; Thomas McMahon, "Ragweed Control," 286.

51. David Funk and Bertram Loeb, "Ragweed Eradication in Pine Hill (Ulster Co.) New York, 1946–1950," *NEWCC* (1951): 287–94; Stradling, *Making Mountains,* 209–26; Roland Miller, "No Ambrosia for Sullivan County," *New York State Conservationist* (Oct.–Nov. 1954): 9; "Official Survey Finds Ragweed Spray Success," *Liberty (NY) Register,* 26 Aug. 1954.

52. *Annual Proceedings of the Board of Supervisors of the County of Sullivan, State of New York* (1954), 101; Raymond McMahon, "Chemical Control of Roadside Vegetation" (paper presented at Convention of the American Association of Highway Officials, Pittsburgh, 11

Nov. 1953), 3, 7; Central Archives, 1233, American Museum of Natural History, New York (AMNH hereafter); Thomas McMahon, "Control of Roadside Weeds and Brush," n.d., 1, box 35, FEEP.

53. *Hay Fever Bulletin* 5 (Summer 1954): 7; "Official Survey Finds Ragweed Spray Success," 1; Louis Fucci to Frank Egler, 4 Feb. 1955, box 35, FEEP; *NYT,* 26 Aug. 1954.

54. New York State Health Department, Statement on Ragweed Control in Sullivan County, 25 Apr. 1955, box 34, FEEP; "Mass Spray to Kill Ragweed Criticized," *NYT,* 15 Sept. 1954; "He Has Better, Cheaper Plan," *Liberty (NY) Register,* 16 Sept. 1954; Frank E. Egler, "Roadside Brush Control: An Application of Plant-Community Management," *National Shade Tree Conference Proceedings* (1952): 66–70; American Museum of Natural History Committee for Brush Control, press release, 20 Aug. 1954, FEEP.

55. Excerpt from Memorandum from Frank E. Egler to R. Pough, 26 Feb. 1952, AMNH; Frank E. Egler, "Chemical Sprays and Rightofways: Dow's Death to Game Habitat," 1953, AMNH; Frank E. Egler, "Chemical Sprays, Rightofways and Game: The Du Pont Desert," preliminary manuscript, 1952, FEEP.

56. Frank Egler to R. P. Wodehouse, 30 Dec. 1955, box 34, FEEP; Frank Egler to Richard Pough, 18 Jan. 1955, box 34, FEEP.

57. Oren Durham, "Contribution of Aerobiology to Weed Control," *NEWCC* (1955): 482; Israel Weinstein, "Administrative Aspects of a Municipal Ragweed Control Program," *NEWCC* (1959): 272–73.

58. Garmhausen, "Chemicals Bring a Revolution," 115–61; Raymond McMahon, "Public Acceptance of Chemical Control of Roadside Vegetation," *NEWCC* (1955): 56; Raymond McMahon, "Chemical Control of Roadside Vegetation," 3.

59. E. P. Sylwester, "Better Weed Control: If We Want It," *Down to Earth* (Spring 1954); Raymond McMahon, "Control of Roadside Weeds and Brush," 1–2; Raymond McMahon, "Public Acceptance," 54.

60. Frank Egler to Rachel Carson, 1 Feb. 1962, box 43, folder 781, Rachel Carson Papers, Yale Collection of American Literature, Beinecke Rare Book and Manuscript Library (RC-YC hereafter); Frank Egler to Rachel Carson, 25 Jan. 1962, RC-YC; Frank E. Egler, "Roadside Ragweed Control Knowledge, and Its 'Communication' between Science, Industry, and Society," *Proceedings of the IX International Botanical Congress* (1959): 1432–43. In both letters, underlined text is Egler's emphasis. On Carson's collaboration and friendship with Egler, see Linda J. Lear, *Rachel Carson: Witness for Nature* (New York: Holt, 1997), 400–401, 434, 441, 460.

61. Frank E. Egler, "Right of Way Maintenance by Plant-Community Management" (Norfolk, CT, 1949), 2; Warren G. Kenfield, *The Wild Gardener in the Wild Landscape* (New York: Hafner, 1966), 32.

62. Frank E. Egler, "Human Ecology, and Connecticut's Two Roadside Bulletins," *Ecology* 41 (1960): 820; Gabrielle R. Barnett, "Drive-By Viewing: Visual Consciousness and Forest Preservation in the Automobile Age," *Technology and Culture* 45 (2004): 38–47.

63. Egler, "Roadside Brush Control," 66; Richard Pough to Albert Parr, 18 Jan. 1955, AMNH. On automobiles and environmentalism, see David Louter, *Windshield Wilderness: Cars, Roads, and Nature in Washington's National Parks* (Seattle: University of Washington Press, 2006), 19–40, 116–22; Paul S. Sutter, *Driven Wild: How the Fight*

against Automobiles Launched the Modern Wilderness Movement (Seattle: University of Washington Press, 2006). While roadsides lined with wildflowers would once have been ubiquitous, ordinary, and unremarkable, infrastructure development made them increasingly extraordinary. In this regard, Egler, Pough, and like minds were not unlike preservationists who worked to protect redwood trees in California. See Barnett, "Drive-By Viewing," 41.

64. Robert McMahon to Rachel Carson, 3 June 1963, box 101, FEEP; Whitney Jacobs, "Spray Man Challenges 'Silent Spring' Author," *Hartford Times*, 6 July 1963; Thomas McMahon to Mrs. Jonathan Williams, 23 Sept. 1963, box 101, FEEP.

65. May Watts, *Reading the Landscape of America* (New York: Macmillan, 1975), 216–17; Mark McDonnell, Steward T. A. Pickett, and Richard Pouyat, "Application of the Ecological Gradient Paradigm to the Study of Urban Effects," in *Humans as Components of Ecosystems: The Ecology of Subtle Human Effects and Populated Areas*, ed. Mark McDonnell and Steward T. A. Pickett (New York: Springer-Verlag, 1993), 175–80.

66. Edward Cohart and Roscoe Kandle, "The Effects of a Ragweed Program on Ragweed Pollen Counts," *Journal of Allergy* 30 (1959): 287–98; Edward Cohart and Roscoe Kandle, "The Effect of a Ragweed Pollen Control Program on Hay Fever," *Journal of Allergy* 30 (1959): 299–310; Falck, "Controlling Urban Weeds," 209–16.

67. Marianne Hahn to Leona Baumgartner, 27 July 1959, box 07-034498, NYDOH; Louis Fucci to Arthur Bushel, 6 June 1966; box 09-001752, NYDOH; Muriel Stromberg, "Bruckner-Soundview," *Bronx Press-Review*, 19 Sept. 1963; Leona Baumgartner to Louis Fucci, 6 Aug. 1959, box 07-034498, NYDOH.

68. Israel Weinstein to George Denison, draft letter, 13 June 1947, box 07-031997, NYDOH.

69. "Frank Egler to Pierre Dansereau, 16 January 1963," *Sarracenia: Conversations on Ecology I–XIII, Vegetation Management, 1963–1964* (Apr. 1965): 13–14. Egler's view resembled Lewis Mumford's belief that urban America extended beyond the city's boundaries. Mumford wrote, "The city, by its incontinent and uncontrolled growth, not merely sterilizes the land it immediately needs but vastly increases the total area of sterilization far beyond its boundaries." See Lewis Mumford, "Natural History of Urbanization," in *Man's Role in Changing the Face of the Earth*, ed. W. Thomas (Chicago: University of Chicago Press, 1956), 393.

CHAPTER FOUR

1. Aldo Leopold, *A Sand County Almanac*, 1987 ed. (New York: Oxford University Press, 1949), 48, 173–74; Benton Mackaye, *The New Exploration: A Philosophy of Regional Planning* (New York: Harcourt, Brace and Company, 1928), 66, 160–61; Thomas R. Dunlap, *Faith in Nature: Environmentalism as Religious Quest* (Seattle: University of Washington Press, 2004), 61–67; Paul S. Sutter, *Driven Wild: How the Fight against Automobiles Launched the Modern Wilderness Movement* (Seattle: University of Washington Press, 2002), 54–99.

2. Thomas J. Sugrue, *The Origins of the Urban Crisis* (Princeton: Princeton University Press, 1996), 33–55; Neil Wynn, *The Afro-American and the Second World War*, rev. ed. (New York: Holmes and Meier, 1993), 60–73; Ronald Takaki, *Double Victory: A*

Multicultural History of America in World War II (Boston: Little, Brown, 2000), 23–38; "Weeds in Uncle Sam's Garden," *CDFD*, 10 July 1943; "Ruins of Empire," *Pittsburgh Courier*, 17 July 1943 (emphasis in original).

3. Sam Bass Warner, *The Urban Wilderness: A History of the American City* (New York: Harper & Row 1972), 3.

4. Mark Gelfand, *A Nation of Cities: The Federal Government and Urban America, 1933–1965* (New York: Cambridge University Press, 1975), 277; John A. Jakle and David Wilson, *Derelict Landscapes* (Savage, MD: Rowman and Littlefield, 1992), xv; Samuel Wood and Alfred Heller, *California Going, Going: Our State's Struggle to Remain Beautiful and Productive* (Sacramento: California Tomorrow, 1962), 9–13; "Weeds Growing Faster Than Ever," *St. Louis Globe-Democrat*, 21–22 May 1966. Despite municipal boundaries, social differences, and the capital invested in landscape production and maintenance, cities, suburbs, and exurbs were part of metropolitan entities. By 1928, land economists recognized the fallacy of limiting "urban land" to the political boundaries of cities in general, much less to a central city in particular. Suburbs were urban land, regardless of the politics and culture of those who lived in them. See Herbert Dorau and Albert Hinman, *Urban Land Economics* (New York: Macmillan, 1928), 151–52; Jon C. Teaford, *The Metropolitan Revolution: The Rise of Post-Urban America* (New York: Columbia University Press, 2006), 1–7; Wayne Zipperer, "The Application of Ecological Principles to Urban and Urbanizing Landscapes," *Ecological Applications* 10 (2000): 685–88; Eric Sandweiss, *St. Louis: The Evolution of an American Urban Landscape* (Philadelphia: Temple University Press, 2001), 7–9; Joseph Wood, "Suburbanization of Center City," *Geographical Review* 78 (1988): 325–29; Melvin Webber, "Urban Growth: What Are Its Sources," in *Cities: The Forces that Shape Them*, ed. Lisa Taylor (New York: Cooper Hewitt Museum, 1982), 148–49; Laurence Wolf, "The Metropolitan Tidal Wave in Ohio, 1900–2000," *Economic Geography* 45 (1969): 142–45. On the making of metropolitan American, see Robert Beauregard, *When America Became Suburban* (Minneapolis: University of Minnesota Press, 2006); Teaford, *Metropolitan Revolution;* Kevin Fox Gotham, *Race, Real Estate, and Uneven Development: The Kansas City Experience, 1900–2000* (Albany: State University of New York Press, 2002); Peter Dreier, "America's Urban Crisis: Symptoms, Causes, and Solutions," in *Race, Poverty, and American Cities*, ed. John Boger and Judith Welch Wegner (Chapel Hill: University of North Carolina Press, 1996): 79–141; Sugrue, *Origins of the Urban Crisis;* Carol Abbott, *The Metropolitan Frontier: Cities in the Modern American West* (Tucson: University of Arizona Press, 1993); Susan Greenbaum, "Housing Abandonment in Inner-City Black Neighborhoods," in *The Cultural Meaning of Urban Space*, ed. Robert Rotenberg and Gary W. McDonogh (Westport: Bergin and Garvey, 1993), 139–56; Mark Gottdiener, ed., *Cities in Stress: A New Look at the Urban Crisis* (Beverly Hills: Sage Publications, 1986); Michael Ebner, "Re-Reading Suburban America: Urban Population Deconcentration, 1810–1980," *American Quarterly* 37 (1985): 366–81.

5. Richard Kurtz and Joanne Eicher, "Fringe and Suburb: A Confusion of Concepts," *Social Forces* 37 (1958): 36; Ray Northam, "Vacant Urban Land in the American City," *Land Economics* 47 (1971): 351; George Wehrwein, "The Rural-Urban Fringe," in *Readings in Urban Geography*, ed. Harold Mayer and Clyde Kohn (Chicago: University of Chicago Press, 1959), 538; William H. Whyte Jr., "A Plan to Save Vanishing U.S. Countryside,"

Life, 17 Aug. 1959, 88, 94; Wolf von Eckardt, *A Place to Live: The Crisis of the Cities* (New York: Delacourte Press, 1967), 15; National Commission on Urban Problems, *Zoning Controversies in the Suburbs* (Washington, DC: Government Printing Office, 1968), 53; "Growing a New Atlanta," *Atlanta Journal-Constitution*, 10 June 1997; Robert Weaver, "HUD's First Secretary Looks Ahead," in *The Housing Yearbook* (New York: Abco Press, 1966), 8. On the problem of defining the shifting fringe, see Zachary J. S. Falck, "Controlling Urban Weeds: People, Plants and the Ecology of American Cities, 1888–2003" (Ph.D. diss., Carnegie Mellon University, 2004), 361–62.

6. Carl Abbott, *The New Urban America: Growth and Politics in Sunbelt Cities* (Chapel Hill: University of North Carolina Press, 1987), 1–59, 123–45; "Weeds," *Santa Fe New Mexican* (*SFNM* hereafter), 6 Aug. 1923; "City to Force Cutting of Weeds," *San Antonio Light* (*SANLI* hereafter), 28 July 1929; "Think" (column), *San Antonio Express* (*SANE* hereafter), 11 Nov. 1935; "City to Fight Sneeze Weeds," *SANE*, 24 Sept. 1939; *City of Houston v. Quinones*, 177 S.W.2d 259 (1944).

7. "Five Charges Filed," *SANLI*, 7 July 1949; "Lot Owners Get Warning," *SANLI*, 27 May 1958; "Where's the Street," *SANE*, 20 June 1965; Sigman Byrd, *Sig Byrd's Houston* (New York: Viking, 1955), 8–14, 110, 173–74, 214; "Owners Told to Cut Weeds," *Arizona Republic* (*AZREP* hereafter), 19 Jan. 1961; John Harlow, "Vacant Lot Villain," *Tucson Daily Citizen*, 12 Nov. 1966; "Weed Capital of the World," *SFNM*, 5 Oct. 1959; "Eradicate Weeds Now," *SFNM*, 21 June 1961.

8. Herman Berkmann, "Decentralization and Blighted Vacant Land," in Mayer and Kohn, *Readings in Urban Geography*, 596; *1963 Survey of Vacant Land in the City of Chicago* (Chicago Department of City Planning, 1964), 11; Ernestine Cofield, "Battle of Woodlawn," *CDFD*, 19 Nov. 1962; Amanda I. Seligman, *Block by Block: Neighborhoods and Public Policy on Chicago's West Side* (Chicago: University of Chicago Press, 2005), 81–89; "City Blamed for Hill 'Reefer' Bed," *Pittsburgh Courier*, 28 July 1962; Mayor Ripley quoted in Mitchell Gordon, *Sick Cities* (New York: Macmillan, 1963), 318–19; Jon C. Teaford, *Rough Road to Renaissance: Urban Revitalization in America, 1940–1985* (Baltimore: Johns Hopkins University Press, 1986), 206–7; Teaford, *Twentieth-Century American City*, 123; James Little et al., *The Contemporary Neighborhood Succession Process: Lessons in the Dynamic of Decay from the St. Louis Experience* (St. Louis: Washington University Institute for Urban and Regional Studies, 1975), 16–17; Ralph Blumenthal, "Parks Deteriorating for Lack of Upkeep," *NYT*, 22 May 1972.

9. "Stairs Leading under the Detroit-Superior Bridge" (1946), http://images.ulib.csuohio.edu/cdm4/item_viewer.php?CISOROOT=/urbanohio&CISOPTR=345&CISOBOX=1&REC=1, http://www.clevelandmemory.org/press/; "Old Jefferson Street Bridge in Cleveland" (1963), http://images.ulib.csuohio.edu/cdm4/item_viewer.php?CISOROOT=/urbanohio&CISOPTR=281&CISOBOX=1&REC=1, also in Cleveland Press Collection, Cleveland State University Library; Susan Sargoy, "Brooklyn's Miracle Gardeners," *NYT*, 16 Aug. 1964; "7100 Block of Upland St. Tackles a Problem," *Pittsburgh Courier*, 26 May 1962; "NYC Scrubs Chicago's Face," *CDFD*, 30 Aug. 1965; Ralph Blumenthal, "Parks Deteriorating for Lack of Upkeep," *NYT*, 22 May 1972; T. M. Crowe, "Lots of Weeds: Insular Phytogeography of Vacant Urban Lots," *Journal of Biogeography* 6 (1979): 169–81.

10. Charles Gaines, "Whites Make Slums of Negro Areas," *CDFD*, 28 Sept. 1963;

Bruce Weber, "Riot Report," *Merchandising Week* 99 (14 Aug. 1967): 12, reprinted in Frederick Sturdivant, ed., *The Ghetto Marketplace* (New York: Free Press, 1969), xiii–xvii, 8; Sterling Tucker, *Beyond the Burning: Life and Death of the Ghetto* (New York: Association Press, 1968), 14; Peter A. Coclanis, "Dandelion Greens," *Callaloo* 24 (2001): 46.

11. J. Fuerst, "Class, Family, and Housing," *Society* 12, no. 1 (1974): 50–51; Richard Scobie, "Public Housing Woes," *Society* 12, no. 4 (1975): 9–10; Eric Monkkonen, *America Becomes Urban: The Development of U.S. Cities and Towns, 1780–1980* (Berkeley: University of California Press, 1988), 15; Roger Lewis, "In Boston, a 'Point' for Public Housing," *WP*, 13 March 1993. As with pre–World War II concerns over so-called human weeds, postwar economic developments permitted the adaptation and application of this metaphor to blacks who migrated from the rural South to Northern cities. One factor that many analysts included in explanations of their migration was advances in weed control, especially in the cotton fields where they once worked as sharecroppers and temporary laborers. The 1967 federal minimum-wage law mandating $1 per hour for farm workers formerly paid $3.50 for a twelve-hour day made "chemicals far more economical than human weed pickers." See Joseph Vandiver, "The Changing Realm of King Cotton," *Trans-Action: Social Science and Modern Society* 4 (Nov. 1966): 28; Alfred McClung, "Race Riots Are Symptoms," in *Race Riot, Detroit 1943* (New York: Octagon Books, 1968), xi; "Negroes Face Grim Future," *Des Moines Register*, 27 Feb. 1968; Eugene Methvin, *The Riot Makers: The Technology of Social Demolition* (New Rochelle: Arlington House 1970), 22–23.

12. "Auntie Litter," *SANLI*, 14 June 1970; Lewis Fisher, *Saving San Antonio: The Precarious Preservation of a Heritage* (Lubbock: Texas Tech University Press, 1996), 261–62; "Huebner, COPS," *SANLI*, 5 Feb. 1977; A. Lloyd Ryall, letter to the editor, *Las Cruces Sun-News*, 24 Sept. 1972; "Weeds Flourish as City Waits," *Albuquerque Tribune*, 5 Aug. 1965; "Fix-It," *Albuquerque Tribune*, 21 Oct. 1974; "Owners Cleanup Citysores," *Arizona Republic*, 30 June 1968; "978 Citysores Are Reported," *AZREP*, 10 Nov. 1968; "Citysore Score," *AZREP*, 29 Nov. 1968; "West Side Needs War on Weeds," *Tucson Daily Citizen*, 18 Nov. 1970; Valerie Alvord, "Meet the Wild Side of Leucadia," *San Diego Union-Tribune*, 8 June 1985. For an anthropologist's investigation of the homeless living amid ordinary vegetation under highway overpasses in San Jose in the early 1990s, see Talmadge Wright, *Out of Place: Homeless Mobilizations, Subcities, and Contested Landscapes* (Albany: State University of New York Press, 1997), 272–77.

13. H. Hester, "Vacant Lot Clearance," *American City* 71 (Oct. 1956): 27; "One Man's Opinion," *Lubbock Avalanche-Journal*, 11 June 1973; "Pocatello Unhappy with Weed Title," *Spokane Spokesman-Review*, 16 Aug. 1985.

14. Hugh S. Gorman, "Manufacturing Brownfields: The Case of Neville Township, Pennsylvania, 1899–1989," *Technology and Culture* 38 (1997): 539–74; "New Grant Program Deals with Brownfields Sites," *Environmental Progress* 23, no. 4 (1999), http://www.epa.state.il.us/environmental-progress/v23/n4/brownfields-grant-program.html; Kimberly A. Youngblood, "Voices of Discord: The Effects of a Grassroots Environmental Movement at the Brio Superfund Site," in *Energy Metropolis: An Environmental History of Houston and the Gulf Coast*, ed. Martin V. Melosi and Joseph A. Pratt (Pittsburgh: University of Pittsburgh Press, 2007), 273; Paul Stanton Kibel, "Los Angeles' Cornfield: An Old Blueprint for New Greenspace," *Stanford Environmental Law Journal* 23 (2004):

310; Carol M. Browner, "Brownfields '98 Speech," Los Angeles, CA, 16 Nov. 1998, http://yosemite.epa.gov/opa/admpress.nsf/a162fa4bfc0fd2ef8525701a004f20d7/9cce4c ecca908b278525701a0052e41a!OpenDocument; Keith Schneider, "How a Rebellion over Environmental Rules Grew from a Patch of Weeds," *NYT*, 24 Mar. 1993; EPA Outreach and Special Projects Staff, "Mustard Plants Helping to Clean Up Site," EPA 500-F-98-xxx, May 1998, http://web.archive.org/web/20041031062324/http://www.epa.gov/swerosps/bf/pdf/ss_trnt1.pdf.

15. "Group Weeds Out 'Slum' Image," *CST*, 24 July 1987; Alf Siewers, "Lawndale Tries to Weed Out Eyesores," *CST*, 30 Apr. 1990; Ann Scott Tyson, "A Garden and Some Hope Grow in Englewood," *Christian Science Monitor*, 3 Sept. 1996; Larry Bennett, *Fragments of Cities: The New American Downtowns and Neighborhoods* (Columbus: Ohio State University Press, 1990), 68–71; Jakle and Wilson, *Derelict Landscapes*, 180; James Howard Kunstler, *The Geography of Nowhere: The Rise and Decline of America's Man-Made Landscape* (New York: Simon and Schuster, 1993), 160; Camilo J. Vergara, *American Ruins* (New York: Monacelli Press, 1999), 11–15, 82–83; Mindy Thompson Fullilove, "Psychology of Place," President's Lecture Series, Carnegie Mellon University, 12 Oct. 1999; Mindy Thompson Fullilove and Robert Fullilove, "Place Matters," in *Reclaiming the Environmental Debate: The Politics of Health in a Toxic Culture*, ed. Richard Hofrichter (Cambridge: MIT Press, 2000), 81. On modes of writing about "wasteland" landscapes, see Matthew Potteiger and Jamie Purington, *Landscape Narratives: Design Practices for Telling Stories* (New York: Wiley, 1998), 213–22.

16. Edgar Anderson, *Plants, Man, and Life* (Boston: Little, Brown and Company, 1952), 16–17; G. Ledyard Stebbins, "Edgar Anderson," *Biographical Memoirs* (Washington, DC: National Academy of Sciences, 1978), 49: 3–23; Kim Kleinman, "His Own Synthesis: Corn, Edgar Anderson, and Evolutionary Theory in the 1940s," *Journal of the History of Biology* 32 (1999): 293–320.

17. Herbert Dorau and Albert Hinman, *Urban Land Economics* (New York: Macmillan, 1928), 138–39; James Neal Primm, *The Lion of the Valley: St. Louis, Missouri, 1764–1980*, 3rd ed. (St. Louis: Missouri Historical Society Press, 1998), 445–59; Sandweiss, *St. Louis*, 231–35; Joseph Heathcott and Marie Agnes Murphy, "Corridors of Flight, Zones of Renewal: Industry, Planning, and Policy in the Making of Metropolitan St. Louis, 1940–1980," *Journal of Urban History* 31 (2005): 155; Lewis Thomas, "The Sequence of Areal Occupance in a Section of St. Louis," *Annals of the Association of American Geographers* 21, no. 2 (1931): 85.

18. Edgar Anderson, "A Classification of Weeds and Weed-Like Plants," *Science* 89 (1939): 364–65; Stebbins, "Edgar Anderson," 3–23; Edgar Anderson, *Landscape Papers*, ed. Bob Callahan (Berkeley: Turtle Island Foundation, 1976), 28–32; Charles B. Heiser, "Student Days with Edgar Anderson; or, How I Came to Study Sunflowers," *Annals of the Missouri Botanical Garden* 59 (1972): 363.

19. 1941–42 Class of Field Botany, "Sunflowers of St. Louis and Their Relation to the Compositae," 12–14, Research—Helianthus, box 18, RG 3, Edgar Anderson Papers, Missouri Botanical Garden (EA-MBG hereafter); "The Sunflower in St. Louis: A General Examination," 25–26, EA-MBG; Jean Mitchell, "Distribution of Helianthus in St. Louis," 1–4, EA-MBG.

20. Eloise Fay, "Botany 325: Sunflower Distribution in St. Louis," 1948, EA-MBG.

21. "Sunflower in St. Louis," 26.

22. Heiser, "Student Days with Edgar Anderson," 363; Anderson, *Landscape Papers*, 62–63, 83–84, 93. Anderson's landscape studies and insights were impressionistic; he did not use them to produce the rigorous scientific knowledge that European ecologists attempted to produce from bombed-out, rubble-filled land in the same period. See Jens Lachmund, "Exploring the City of Rubble: Botanical Fieldwork in Bombed Cities in Germany after World War II," *Osiris*, 2nd ser., 18 (2003): 234–54. On postwar St. Louis, see Colin E. Gordon, *Mapping Decline: St. Louis and the Fate of the American City* (Philadelphia: University of Pennsylvania Press, 2008), 22–38, 153–87; Andrew Hurley, "Environmental Hazards since World War II," in *Common Fields: An Environmental History of St. Louis*, ed. Andrew Hurley (St. Louis: Missouri Historical Society Press, 1997), 249–50; Heathcott and Murphy, "Corridors of Flight," 157–70.

23. William Brady, "Personal Health Service," *Oakland Tribune*, 10 Apr. 1946; Hal Boyle, "Garden of Life," *Hayward (CA) Daily Review*, 13 Nov. 1959; the column also appeared with the titles "Opportunities Are Endless in Planting Love Gardens," "How to Plant Your Own Garden of Love, Hate," and "Gardens of Hate and Love." See *Tuscon Daily Citizen*, 3 Apr. 1959; "Comic Dictionary," *Bridgeport Telegram*, 4 May 1967.

24. J. Ralph Audy, "Man and the Land," in *Tomorrow's Wilderness*, ed. Francois Leydet (San Francisco: Sierra Club, 1963), 101–6; J. de Wet, "The Origin of Weediness in Plants," *Proceedings of the Oklahoma Academy of Sciences* 47 (1966): 14.

25. Malcolm X quoted in Louis A. DeCaro, *Malcolm and the Cross: The Nation of Islam, Malcolm X, and Christianity* (New York: New York University Press, 1998), 138–39; "Violence in the City," *NYT*, 1 June 1964; "Wilkins Denounces Negro 'Hoodlums,'" *NYT*, 24 June 1964; "Violence in New York," *CDFD*, 20 June 1964; *The Autobiography of Martin Luther King, Jr.*, ed. Clayborne Carson (New York: Warner Books, 1998), 305. In Spike Lee's 1992 film *Malcolm X*, during the minister's infamous "chickens come home to roost" remarks after the assassination of President John F. Kennedy, Denzel Washington's Malcolm X adds, "In the soil of America the white man planted the seeds of hate. He allowed the weeds that sprang up to choke the life out of thousands of black men." The remark appears in neither Haley's *Autobiography* nor *One Day, When I Was Lost*, written by James Baldwin in 1973, on which Lee's film is based.

26. Mitchell Gordon, *Sick Cities*, 322–24; Arthur Goldberg, "The Rule of Law in an Unruly World," *Department of State Bulletin* 54 (13 June 1966): 938; House Committee on Un-American Activities, *Subversive Influences in Riots, Looting and Burning, Part 1*, 90th Cong., 1st sess., 1967, 735–38; Franklin Parker, "Human Relations and the School," *Journal of Human Relations* 16 (1968): 28.

27. Ludwig Hilberseimer, *The Nature of Cities: Origin, Growth, and Decline* (Chicago: Theobald, 1955), 111–12; Berkmann, "Decentralization and Blighted Vacant Land," 593; Edward Vogel to John Mahoney, 9 Oct. 1951, box 08-003343, Records of the Commissioner's Office, Department of Health, New York City Municipal Archives; "Schoolgirl, 12, Ambushed," *Pittsburgh Courier*, 18 Nov. 1961; Garth Jones, "Survey Crew Finds Two Coeds' Bodies," *Galveston News*, 31 July 1965; Charles Matthias Goethe, *Garden Philosopher* (Sacramento: Keystone Press, 1955), 237–45; "Scourge of a Nation," *CDFD*, 7 July 1951. Demographic change in Sacramento County was likely particularly unsettling to Goethe. Mexicans had accounted for less than 6 percent of the foreign-born white

population between 1900 and 1920 but made up more than 14 percent by 1950. See Warren Thompson et al., *Growth and Changes in California's Population* (Los Angeles: Haynes Foundation, 1955), 83–84, 355–57. Latin America's population explosion prompted Goethe to insist that "human weeds" were "social inadequates." C. M. Goethe, "Of Human Weeds," *Salt Lake City Tribune*, 25 Dec. 1965. Goethe was under the impression that Barbados was some sort of tropical paradise, as it was "the only area in the entire world that is absolutely free from weeds." He wrote this at a time when the island was still a British colony dominated by sugar plantations; he was perhaps unaware of democratization and independence movements afoot. See *Garden Philosopher*, 236. On Goethe's activism, see Alexandra Minna Stern, *Eugenic Nation: Faults and Frontiers of Better Breeding in Modern America* (Berkeley: University of California Press, 2005), 134–47.

28. Isaac Rehert, "More American than Apple Pie," *Baltimore Sun*, 5 July 1969; Anderson, *Plants, Man and Life*, 132, 196; Charles B. Heiser, *The Sunflower* (Norman: University of Oklahoma Press, 1976), 92–93. Rehert may have just read *Plants, Man and Life*, which was originally published in 1952 but reprinted again in 1967 and 1969.

29. Loren Eiseley, "And as for Man," in *The Innocent Assassins* (New York: Scribner, 1973), 74; Gale E. Christianson, *Fox at the Wood's Edge: A Biography of Loren Eiseley* (New York: Henry Holt, 1990), 29–31; Ruth Howell, *A Crack in the Pavement* (New York: Atheneum, 1970), n.p.; Anne Ophelia Dowden, *Wild Green Things in the City: A Book of Weeds* (New York: Crowell, 1972); Anne Ophelia Dowden, "Weed Hunting in Manhattan," *Garden Journal* 22 (1972): 135–38; Leonard Marcus, "Nature into Art: An Interview with Anne Ophelia Dowden," *Lion and the Unicorn* 6 (1982): 30–32; Maida Silverman, *A City Herbal: A Guide to the Lore, Legend, and Usefulness of Thirty-four Plants that Grow Wild in the City* (New York: Knopf, 1977), 3; Nancy M. Page and Richard E. Weaver Jr., *Wild Plants in the City* (New York: Quadrangle, 1975), x, 1–7; Amy Clampitt, "Vacant Lot with Pokeweed," *Christian Science Monitor*, 18 Oct. 1984, 35. Field guides for plants could limit their subjects to weeds or simply subsume weeds under the term *wildflowers*. See Roger Tory Peterson and Margaret McKenny, *A Field Guide to Wildflowers: Northeastern and North-Central North America* (Boston: Houghton Mifflin, 1968); Alexander Martin, *Weeds* (New York: Golden Press, 1972); Pamela Forey and Cecilia Fitzsimons, *An Instant Guide to Wildflowers* (New York: Crescent Books, 1986). For an interpretation of the diversity of bird guides, see Tom Dunlap, "Early Bird Guides," *Environmental History* 10 (2005): 110–18. Not all books celebrating nature, including children's books, praised these plants. Illa Podendorf, a teacher at the University of Chicago's Laboratory School, conveyed to children the conventional thinking that people should decide where every plant belonged. Podendorf's values were those of Americans for whom neat lawns and flower gardens were the only acceptable urban landscapes. See Illa Podendorf, *The True Book of Weeds and Wild Flowers* (Chicago: Children's Press, 1955). Dowden's tremendous botanical illustrations frequently appeared in books by popular naturalists such as Hal Borland, a man who developed a strong disgust for the idea of beauty in the city. For a discussion of city dwellers' reception of nature writers such as Hal Borland and Edwin Teale, see Michael G. Kammen, *A Time to Every Purpose: The Four Seasons in American Culture* (Chapel Hill: University of North Carolina Press, 2004), 199–204.

30. "Crowded Streets and Empty Lots," *NYT*, 1 Sept. 1983; "Notes and Comment,"

New Yorker, 23 Aug. 1982, 23; Tessa Huxley, "Urban Wildflowers," *NYT*, 20 June 1985; Tyson, "A Garden and Some Hope Grow in Englewood."

31. Steven Clemants and Gerry Moore, "Patterns of Species Richness in Eight Northeastern United States Cities," *Urban Habitats* 1 (2003): 4–16; Yeqiao Wang and Debra Moskovits, "Tracking Fragmentation of Natural Communities and Changes in Land Cover," *Conservation Biology* 15 (2001): 838, 840, 841; Ann Durkin Keating, "Annexations and Additions to the City of Chicago," in *Electronic Encyclopedia of Chicago*, http://www.encyclopedia.chicagohistory.org/pages/3716.html (expansions after 1920 included the Mount Greenwood area in 1927 and land for O'Hare Airport in 1956); Gordon Whitney and Stanley Adams, "Man as a Maker of New Plant Communities," *Journal of Applied Ecology* 17 (1980): 431–48; Ralph Sanders, "Estimating Satisfaction Levels for a City's Vegetation," *Urban Ecology* 8 (1994): 269–83; Paul Robbins, *Lawn People: How Grasses, Weeds, and Chemicals Make Us Who We Are* (Philadelphia: Temple University Press, 2007), 30. There are 58 million home lawns in the United States, which are part of the 25–40 million acres of land covered with turf in the country. See Ted Steinberg, *American Green: The Obsessive Quest for the Perfect Lawn* (New York: Norton, 2007), 4. The two decades of lawn studies have emphasized the behaviors and systems that perpetuate and expand this vegetation cover. See Virginia Jenkins, *The Lawn: A History of an American Obsession* (Washington, DC: Smithsonian Institution Press, 1994), 91–115; Georges Teyssot, ed., *The American Lawn* (New York: Princeton Architectural Press, 1999). For a significant analysis of the environmental costs of lawns and the environmental benefits of natural landscaping, a short history of the natural landscaping movement, and the policy basis for abandoning existing weed laws, see Bret Rappaport, "As Natural Landscaping Takes Root We Must Weed Out the Bad Laws—How Natural Landscaping and Leopold's Land Ethic Collide with Unenlightened Weed Laws and What Must Be Done About It," *John Marshall Law Review* 26 (1993): 865–940.

32. Ecologists alarmed by the destructive impact of plants from afar—variously labeled "alien," "exotic," "invasive," and "nonnative" species—have expressed their wariness with language that has alarmed historians sensitive to the belligerent antihumanism of nativists' political activities. An important exchange between ecologist Daniel Simberloff and historian Philip J. Pauly illustrates the divergent approaches; see "Letters to the Editor," *Isis* 87 (1996): 676–78. Criticism of the "invasive" concept has led ecologists to examine the history of the terminology. One recent review article claims, "The term 'invasion' was first used in an ecological context by Goeze (1882:109) in his book 'Pflanzengeographie,' in connection with the spread of non-native species. On the same page, he presented the invasion of mango in Jamaica as an example of a beneficial invasion. Therefore, the term 'invasion' by itself was used without any necessary connection with negative or positive impacts. This is also how 'invasion' was understood by Clements (1904 et seq.) and other ecologists in the first half of the last century." See Marcel Rejmanek et al., "Commentary: Biological Invasions: Politics and the Discontinuity of Ecological Terminology," *Bulletin of the Ecological Society of America* (Apr. 2002): 131–33. The *Oxford English Dictionary* indicates that *invade* is not necessarily a militaristic word. It can be legalistic. One definition of *invade* is, "To intrude upon, infringe, encroach on, violate (property, rights, liberties, etc.)." Interestingly, this definition leads to other pos-

sible terms for the behavior in question, such as *encroach,* which means, "To make gradual inroads on, extend (its) boundaries at the expense of, something else," and "To advance, intrude beyond natural or conventional limits." Thus, invasive species might have instead been or be called "encroaching species" or "intruding species." In terms of implications for the environment and ecological systems, what matters more than the word chosen to describe such species is the effect of their spread. For this reason, ecologists are working to limit when potentially objectionable terminology is used by distinguishing which species potentially can cause or have caused serious environmental problems. A good example is David M. Richardson et al., "Naturalization and Invasion of Alien Plants: Concepts and Definitions," *Diversity and Distributions* 6 (2000): 93–107. Historians Peter Coates and Philip Pauly have also studied how plants came to be perceived, recognized, and scrutinized as being "alien" and "invasive." In quite different ways, they explain the possible and the exaggerated threats that "invasive" plants pose to existing ecological dynamics, especially in the United States. Coates's review of terms for these organisms leads him to recommend *newcomer,* although *newcomer* was not quite the neologism that Coates suggested. David Fairchild proposes the term for the organisms gathered and imported by the federal government's Office of Foreign Seed and Plant Introduction. See David Fairchild, *The Annual Catalogue of Plant Immigrants* 8 (1917): 500; Peter Coates, *American Perceptions of Immigrant and Invasive Species: Strangers on the Land* (Berkeley: University of California Press, 2006), 172; Philip J. Pauly, *Fruits and Plains: The Horticultural Transformation of America* (Cambridge: Harvard University Press, 2007), 126–29, 147–49, 246. Since assessments of nativity and invasiveness have become subject to historical scrutiny, perhaps more accurate assessments of problematic organisms may be developed that do not facilely rely on their geographic origins as an explanation for their out-of-placeness. In this section, the term *transformer species* (as discussed by Richardson et al., "Naturalization and Invasion," 102) emphasizes that some plants—from any part of the world, in any part of the world—can alter environments in ways other than people desire them to be altered. *Transformer species* avoids geographical or chronological jingoism as it focuses on environmental effects that could be judged harmful, or simply powerful, in particular economic, political, social, and cultural contexts. Many "transformer species" may, in fact, be from afar, but they are not necessarily so, as summer grape vines and poison ivy show in the United States. Such a term creates no dissonance when native species create problematic environments

33. Randy Westbrooks, *Invasive Plants: Changing the Landscape of America* (Washington, DC: Federal Interagency Committee for the Management of Noxious and Exotic Weeds, 1998), 11; "Pest Alert—Japanese Knotweed," USDA Forest Service, Northeastern Area, NA-PR-04-99 (July 1999). An important essay on the arrival and dispersal of these plants written by a scientist is Richard N. Mack, "The Commercial Seed Trade: An Early Dispenser of Weeds in the United States," *Economic Botany* 45 (1991): 257–73.

34. Art Kozelka, "Lythrums for Color, Hardiness," *CST,* 7 July 1957; Alfred Putz, "Successful City Garden Requires Tough Plants," *WP,* 17 July 1955; Coates, *American Perceptions,* 69; Frank Thone, "Nature Ramblings: Get after the Weeds!" *Science News Letter* 35 (1939): 159; Thomas McMahon, "Ragweed Control," *NEWCC* (1959): 288; P. Kaufman and A. Pridham, "Results of Preliminary Tests in Highway Weed Control

Work in Tompkins County," *NEWCC* (1950): 252. In another attempt to misrepresent the environmental sensitivity of their work, McMahon Brothers carried out demonstrations of their work as the organization the National Council for the Control of Pesticides. See Frank E. Egler, "Pesticides in Our Ecosystem: Communication II," *Bioscience* 14 (1964): 33. Environmental activists concerned with "invasive" plants have a variety of names and are affiliated with a wide range of organizations. "Urban weed warriors," for example, were at work in city parks in Baltimore, as well as around the metropolitan region. See http://maryland.sierraclub.org/action/Weed-Warriors-050605.asp.

35. Warren G. Kenfield, "Borne Free: The Case for Immigrant Plants," *Landscape Architecture* 65 (1974): 404–9; Viktor Muhlenbach, "Contributions to the Synanthropic (Adventive) Flora of the Railroads in St. Louis, Missouri," *Annals of the Missouri Botanical Garden* 66 (1979): 1, 67, 83, 87. On fieldwork that is similar, but with more significance for the management of public space, see Lachmund, "Exploring the City of Rubble," 234–54. On the ecology of exotic plants in cities, see Ingo Kowarik, "On the Role of Alien Species in Urban Flora and Vegetation," in *Plant Invasions: General Aspects and Special Problems* (Amsterdam: SPB Academic Publishing, 1995), 85–103.

36. Kay Fanning and Maureen Joseph, "Discovering Dumbarton Oaks Park: Restoring a Masterwork for Modern Needs," *APT Bulletin* 30, nos. 2–3 (1999): 65; Ron Avery, "Weeding Out the Enemy along Wissahickon," *Philadelphia Daily News*, 7 Aug. 1995. On Philadelphia's parks, see Alec Brownlow, "Inherited Fragmentations and Narratives of Environmental Control in Entrepreneurial Philadelphia," in *In the Nature of Cities: Urban Political Ecology and the Politics of Urban Metabolism*, ed. Nikolas C. Heynen, Maria Kaika, and Erik Swyngedouw (London: Taylor and Francis, 2006), 208–25; Susan Davis, "Cape Weed Caper," *California Waterfront Age* 6, no. 1 (1990): 22–24; Westbrooks, *Invasive Plants*, 19.

37. Alf Siewers, "Gardeners Sue over City Ban on Wild Things," *CST*, 7 June 1991.

38. William Cronon, "The Trouble with Wilderness; or, Getting Back to the Wrong Nature," in *Uncommon Ground: Rethinking the Human Place in Nature*, ed. William Cronon (New York: Norton, 1996), 69–90; Wang and Moskovits, "Tracking Fragmentation of Natural Communities," 842; Carol Kaesuk Yoon, "Plants Hang On in Concrete Jungle," *NYT*, 6 Aug. 2002; Clemants and Moore, "Patterns of Species Richness," 8–9.

39. *Revised Code of the City of Saint Louis, Missouri, 1948* (Michie, 1949), chap. 27, sections 35–38; *Revised Code of the City of Saint Louis, Missouri* (Board of Aldermen, 1960), chap. 245; *Revised Ordinances of St. Louis County, Missouri, 1974* (St. Louis County Counselor, 1974), chap. 619.

40. Neale Copple, *Tower on the Plains: Lincoln's Centennial History, 1859–1959* (Lincoln: Lincoln Sunday Journal and Star, 1959), 25–46, 103–5; Roscoe Pound and Frederic Clements, *The Phytogeography of Nebraska*, 2nd ed. (Lincoln, 1900), 400–415; Frederic Clements, *Research Methods in Ecology* (Lincoln: University Publishing Company, 1905), 312–13; Louisa Spalding Millspaugh, "Women as a Factor in Civic Improvement," *Chautauquan* 43 (1906): 315–16; *Lincoln: Its Suburbs and Pleasure Resorts* (Lincoln: Lincoln Commercial Club, n.d.), 2–10; Loren Eiseley, *The Star Thrower* (New York: Times Books, 1978), 161; Christianson, *Fox at the Wood's Edge*, 29–31. J. Sterling Morton, a leading Nebraska citizen, would begin a campaign to plant trees across the state and later

founded Arbor Day. On the ecological implications of Morton's tree planting, see Lisa Knopp, *The Nature of Home: A Lexicon and Essays* (Lincoln: University of Nebraska Press, 2002), 16–27

41. *Susan M. Anthony v. City of Lincoln,* Case File No. 32673, Records of Nebraska Supreme Court, RG 069, Nebraska State Historical Society, Lincoln (NSC-NSHS hereafter); *Anthony v. Lincoln,* 41 N.W. 2d 149 (1950); Lincoln City Council, *Proceedings* 101 (27 Sept. 1948): 424; *Ben A. Greenwood v. City of Lincoln,* Case File No. 33175, NSC-NSHS; *Greenwood et al. v. City of Lincoln,* 55 N.W. 2d 343–47 (1952).

42. *Lincoln City Directory* (1950), 792; Lincoln City Council, *Proceedings* 104 (26 Sept. 1949): 247; Lincoln City Council, *Proceedings* 107 (2 Oct. 1950): 279–80; Lincoln City Council, *Proceedings* 109 (1 Oct. 1949): 532; *Lincoln Sunday Journal and Star,* 28 Feb. 1937.

43. *Greenwood et al. v. City of Lincoln,* 55 N.W. 2d 345-6 (1952); City of Lincoln, Ordinance No. 3780, in Lincoln City Council, *Proceedings* 88 (18 Nov. 1940): 8–9.

44. *Samuels et al. v. City of Beaver Falls,* 5 Pa. D. & C.2d 503, 509–10 (1955); *Flesch v. Metropolitan Dade County,* 240 So. 2d 506 (1970). The ordinance turned to *Webster's* to define *improve* as raising land values with infrastructure or buildings. While Dade County's law seemed to give ample space to wild plants beyond a one-hundred-foot buffer, as more improved properties appeared each year, space for such plants diminished. *Galt* was again cited in Pennsylvania when a property owner contended that he had not violated Williamsport's ordinance "because the term 'weeds' is defined as 'any plant which grows where not wanted,' and since he wanted the plants on his property, they [were] not weeds, offensive or otherwise." The court dismissed "this disingenuous argument because the vegetal preferences of Appellant are hardly relevant to the interpretation and enforcement of the ordinance." See *Sobocinski v. City of Williamsport,* 319 A.2d 699 (1974).

45. Stephen Kenney, "Home Grown: The Native American Spirituality of Henry David Thoreau" (Ph.D. diss., SUNY-Buffalo, 1989), epigraph, 17–18; Stephen Kenney to Daniel Martin, 28 May 1984, *Village of Kenmore v. Stephen A. Kenney, David Tritchler,* Case File, Village of Kenmore, New York (*VK-SK* hereafter).

46. "Man to Defend His Unmown Lawn in Court," *NYT,* 16 Sept. 1984; Trial Transcript, 18, 19 Sept. 1984, Village of Kenmore Justice Court, 14, 46, 67, 82, *VK-SK;* "Not Everyone Is Wild about Wildflowers," *Christian Science Monitor,* 1 July 1985; Section 25-201, *Kenmore Municipal Code* (Municipal Code Corporation, 1982), 1507; Village of Kenmore, Inspection Report and Violation Order, Issued to David Tritchler, 20 June 1983, *VK-SK;* Village of Kenmore, Inspection Report and Violation Order, Issued to David Tritchler, 16 May 1984, *VK-SK.* Coincidentally, after Lyster Dewey retired from the USDA, he moved to Kenmore in the early 1940s to live with his daughter, Grace, and her husband, Carl Frost. This house on Knowlton Avenue was a little less than a mile to the east of the house Kenney rented on Victoria Boulevard in the mid-1980s.

47. "Judge Rules Out Overgrown Yard," *NYT,* 21 Sept. 1984; Thomas Viksjo, "Brief for Respondent," on Appeal from the Village of Kenmore Justice Court, State of New York County Court: County of Erie, 26–27, *VK-SK;* "Kenmore 'Rebel' Gains International Fame for His Homegrown Suburban Meadow," *Ken-Ton Bee* (*KTB* hereafter), 29 Aug. 1984; "It's Been No Bed of Roses Living Near Wildflower Lawn," *KTB,* 19 Sept. 1984; "Man to Defend His Unmown Lawn in Court." On the diverse groups advocating

environmental reform that developed in the 1960s, see Adam Rome, "'Give Earth a Chance': The Environmental Movement and the Sixties," *Journal of American History* 90 (2003): 525–54.

48. Judge Hawthorne quoted in "Vegetation Litigation," *National Law Journal*, 8 Oct. 1984, 43; Judge Hawthorne quoted in John Galeziowski, "Brief for Defendants-Appellants," on Appeal from the Village of Kenmore Justice Court, State of New York County Court: County of Erie, 39, *VK-SK;* H. Walker Hawthorne, Order, Village Court of the Village of Kenmore, County of Erie, 20 Sept. 1984, 6–7, *VK-SK;* David Robinson, "Dismissal Sought for Two in Mowing of Lawn," *Buffalo News,* 17 Sept. 1985; "The 'Lawn' Case," *Buffalo News,* 25 Sept. 1984.

49. Office Penke, Police Report, 14 Sept. 1984, 186 Victoria Blvd., *VK-SK;* Margaret Sullivan, "Village, Defendant Mow Down Good Neighbor Policy," *Buffalo News,* 23 Sept. 1984; Walker Hawthorne, Order, 3, *VK-SK;* H. Walker Hawthorne to Agnes Hess, 27 June 1984, *VK-SK;* Thomas Dolan, "Kenney Leaving Natural Lawn for Mountaintop with Deer, Bear," *Buffalo News,* 25 Aug., 1985; David Halle, *America's Working Man: Work, Home, and Politics among Blue-Collar Property Owners* (Chicago: University of Chicago Press, 1984), 11–12. Hawthorne's letter to Hess may have been unsent; she likely received a 3 July 1984 letter that concluded, less angrily, "In the not-too-distant-future, he will leave Kenmore, leaving behind a lawn which is out of character with the community." On the tangled politics of hair in the years before Kenney lived in Kenmore, see Timothy Hodgdon, *Manhood in the Age of Aquarius: Masculinity in Two Countercultural Communities* (New York: Columbia University Press, 2008), 41–42; Gael Graham, "Flaunting the Freak Flag: *Karr v. Schmidt* and the Great Hair Debate in American High Schools, 1965–1975," *Journal of American History* 91 (2004): 522–43; Timothy Miller, *The Hippies and American Values* (Knoxville: University of Tennessee Press, 1991), 116–17. On the relationship of rights and community, see Samuel Walker, *The Rights Revolution: Rights and Community in Modern America* (New York: Oxford University Press, 1998), 61–114. An important article on Americans and rights is Gordon Wood, "The Origins of Vested Rights in the Early Republic," *Virginia Law Review* 85 (1999): 1421–45. Originally delivered as a lecture, Wood's article expresses dismay that people have become "serious about the rights of plants. Trees may have rights today, but as far as I know weeds do not, at least not yet." I thank Jim Longhurst for (long ago) bringing this article to my attention.

50. John Bonfatti, "Thoreau Disciple Is Fined by Court for Not Cutting His Lawn," 20 Sept. 1984, Associated Press Files, LexisNexis; Brian Meyer, "WNY's Dubious Achievement Awards," *Buffalo News,* 1 Jan. 2000; Tom Buckham, "'Vigilantes' Mow Kenney's Lawn," *Buffalo News,* 13 July 1985; Joseph Forma, Memorandum and Order, State of New York, County of Erie, 14 Nov. 1985, 6, *VK-SK;* Marc Lacey, "Kenney Puts Down Roots in the Wild," *Buffalo News,* 5 June 1988; Karen Brady, "Kenmore Party Recalls a 'Mow Down,'" *Buffalo News,* 25 July 1988. A concurrent development in New York State, not referenced in the Kenney trial, was a case that reasserted the principles of *Galt.* In *People v. Resnick,* a state judge ruled that "long grass and weeds attract litter and debris, and by providing a refuge for garbage and waste generally despoil the character and aesthetics of the City. . . . It is the opinion of this court that grass and weeds 2 to 2 1/2 feet high would be considered long to any person of ordinary intelligence. Grass and weeds of that length

could also be considered unsightly, provide an area for vermin to nest, thus possibly causing a health problem, and could possibly attract litter." See *People of the State of New York v. Jack Resnick & Sons, Inc.*, 487 N.Y.S. 2d 988, 991–92 (1985). For a case of bulldozer-aided vigilantism against plants and wildlife in a suburb southwest of Chicago, see Eric Zorn, "Weed Warrior in a Legal Thicket," *CST,* 18 June 1991.

51. Howard Campbell, "Weed Control in Suburbia," *NEWCC* (1961): 17; "'Lawn' Case"; Ben De Forest, "Battle over Wildflowers," 2 June 1985, Lexis-Nexis Associated Press Files; "Pursuit of Happiness," *KTB,* 19 Sept. 1984; *Frederick H. Goldbecker v. Fairfax County Board of Supervisors,* 38 Va. Cir. 264 (1995). The Goldbeckers lived on a nearly half-acre lot on a street near country clubs and George Mason University. The creative construction of the law was indicative of the measures wealthier municipalities used to fight vegetation growth that resembled the vegetation in urban areas perceived to be devoid of value. As what constituted an acceptable lawn changed over time, gradations of lawn care in a metropolis sometimes distinguished some communities from others. For example, one 1957 guide for lawn aficionados stated that clover was no longer "socially acceptable" as a lawn plant "except for in a few die-hard communities. . . . Within the past ten years, clover has been relegated to a place where it is considered a pesky weed, to be destroyed as rapidly as possible." See R. Milton Carleton, *New Way to Kill Weeds in Your Lawn and Garden* (New York: Arco Publishing Co., 1957), 90; Samuel P. Hays, *Beauty, Health, and Permanence: Environmental Politics in the United States, 1955–1985* (Cambridge: Cambridge University Press, 1987), 21–39; Jenkins, *Lawn,* 185. On lawn regulations by covenants that were stricter than municipal laws, see Evan Mckenzie, *Privatopia: Homeowner Associations and the Rise of Residential Private Government* (New Haven: Yale University Press, 1994), 14–18. *Goldbecker* may have been the most ambitious test of such laws, as the plaintiffs appealed to the Supreme Court, which denied their petition for writ of certiorari. See *Goldbecker v. Fairfax County,* 510 U.S. 1074 (1994).

52. *The Empty Lot: Obstacle or Opportunity: A Study of Vacant Land in Two Chicago Neighborhoods* (Chicago: University of Illinois at Chicago, 1982), 1–14; City of Chicago, Complaint Number CC-743-128, Order for Court Appearance, issued to Marie Wojciechowski, 25 July 1990, *City of Chicago v. Marie Wojciechowski,* Case File 90 MC1 322079, Circuit Court of Cook County Municipal Department, First Division, Chicago (*CCMW*-322079 hereafter); affidavit of John Felter, *CCMW*-322079.

53. Municipal Code of Chicago, Section 99-9, Journal of the City Council of Chicago, 20 Dec. 1989, 10124; Municipal Code of Chicago, Section 7-28-120, . . . , 810; Report of Proceedings, *City of Chicago v. Marie Wojciechowski,* 12 Oct. 1989, 27–28, 34, 51–52, *CCMW*-322079.

54. Reply Brief in Support of Motion for Summary Judgment, 22 Jan. 1992, 2, *CCMW*-322079; Patricia K. Armstrong, Summary of 30 October 1990 Botanical Assessment of 1917–1923 North Honore, Exhibit 2 of Motion for Summary Judgment, 2 July 1992, *CCMW*-322079; City of Chicago's Memorandum in Opposition to Defendant's Motion for Summary Judgment, 22 Jan. 1992, 2–3; *CCMW*-322079; Report of Proceedings, *Chicago v. Wojciechowski,* 36–41, *CCMW*-322079.

55. Report of Proceedings, *Chicago v. Wojciechowski,* 50, *CCMW*-322079; Defendant's First Set of Interrogatories and Request for Production of Documents, 11 Mar. 1993, 5–6,

CCMW-322079; Plaintiff's Response to Defendant's First Set of Interrogatories, 10 June 1993, 2, *CCMW*-322079.

56. Report of Proceedings, *Chicago v. Wojciechowski*, 43, 58–59, 61, 66–68, *CCMW*-322079.

57. Motion for Summary Judgment, 2 July 1992, 3, *CCMW*-322079.

58. Report of Proceedings, *Chicago v. Wojciechowski*, 62–63, 66, *CCMW*-322079; Agreed Order, drafted Dec. 1993 and entered 21 Jan. 1994, *CCMW*-322079.

59. *Schmidling et al. v. Chicago*, 1 F.3d 501 (1993); Agreed Order; *CCMW*-322079. Alf Siewers, "Judge Won't Mow Down City's Weed Ordinance," *CST*, 10 Mar. 1992; Rappaport, "As Natural Landscaping Takes Root," 887, 893–95, 916. The *Schmidling* case began in June 1991 in a U.S. District Court before being argued in the U.S. Court of Appeals, 7th District, in Nov. 1992 and being decided in 1993. For a profile of Schmidling, see Alf Siewers, "Gardeners Sue over City Ban on Wild Things," *CST*, 7 June 1991. Despite the defeats of gardeners like Kenney and Wojciechowski, Rappaport celebrated Lorrie Otto as the "High Priestess of Natural Landscaping Movement" with her victory over her local government to protect her native plant prairie in Bayside, WI; the proliferation of such spaces in her village; and the emergence of the Wild Ones network to support natural landscapers. Although Otto's home in the village of Bayside was only about seven miles from Milwaukee's north border, on the east side of the city, it was in a somewhat exclusive tip of the village, with many lots about an acre in size. Otto's property was more than 1.5 acres. The yard that she cultivated as a natural landscape was considerably larger than most Americans' yards, particularly when she began this work in the 1950s and 1960s. For a profile of Otto, see Steve Lerner, *Eco-Pioneers: Practical Visionaries Solving Today's Environmental Problems* (Cambridge: MIT Press, 1998), 41–45.

60. *Robert L. Howard v. City of Lincoln*, 497 N.W.2d 58-9 (1993). Not only did *Howard* resemble *Greenwood*, but one judge who wrote the opinion in the *Greenwood* case was Justice Paul Boslaugh, who sat on the court from 1949 to 1961 and who swore in his son to the court in 1961. Justice Leslie Boslaugh served from 1961 to 1994. See James W. Hewitt, *Slipping Backward: A History of the Nebraska Supreme Court* (Lincoln: University of Nebraska Press, 2007), 84, 196–97.

61. *Lundquist v. City of Milwaukee*, 643 F.Supp. 776-7 (1986); *Schmidling v. Chicago*, 1 F.3d 502; *Howard v. Lincoln*, 56–57; *City of Newark v. Garfield Development Corporation*, 495 N.E.2d 481, 482, 484 (1986).

62. *Operation Weed and Seed Implementation Manual* (Washington, DC, 1992), 1–2; Robert Samuels, "Letter from the Director," *Insites* 10, no. 3 (2002): 3; *The Weed and Seed Strategy* (U.S. Department of Justice, 1992); Janice Roehl et al., *National Process Evaluation of Operation Weed and Seed*, National Institute of Justice Research in Brief (Washington, DC: National Institute of Justice, 1996), 1–6. A related form of law enforcement was "fixing broken windows" policing, a method articulated by Harvard professor James Q. Wilson and promoted by law-enforcement official William Bratton. When Los Angeles hired Bratton, who had previously headed the police departments in Boston and New York, as police chief in 2002 to combat escalating gang violence, Bratton's determination to punish even the most minor violations was described as a strategy to prevent the neglect that led to "more serious crime, the way weeds take over an untended lot." In such a

formulation, illegal actions are equated with weeds; since people, not acts, were ultimately punished, people, not the crimes, were weeds. See Ana Figueroa, "A City in Need of an Angel," *Newsweek*, 9 Dec. 2002, 48; Bernard Harcourt, "Broken-Windows Policing," in *The Encyclopedia of Police Science*, 3rd ed., ed. Jack Greene (New York: Routledge, 2007), 112–18. For a critique, see Bernard Harcourt, *Illusion of Order: The False Promise of Broken Windows Policing* (Cambridge: Harvard University Press, 2005), esp. 215–16.

63. Roehl et al., *National Process Evaluation*, 4–6, 9–11; *Operation Weed and Seed Implementation Manual*, 3; *Report to the Attorney General: Seeding America's Neighborhoods* (Washington, DC, 1992), 38; Terence Dunworth et al., *National Evaluation of Weed and Seed: Cross-Site Analysis*, National Institute of Justice Research Report (Washington, DC: National Institute of Justice, 1999), xv.

64. Dunworth et al., *National Evaluation of Weed and Seed: Cross-Site Analysis*, xiii–xviii, 57–58; Terence Dunworth et al., *National Evaluation of Weed and Seed: Salt Lake City Case Study* (Washington, DC: National Institute of Justice, 1999), 3–4; Roehl et al., *National Process Evaluation*, 12; Dick Lilly, "Attack Crime or Harass Teens?—Central Area Groups Protest 'Weed and Seed' Grant," *Seattle Times*, 24 Mar. 1992; Urban Strategies Group, *A Call to Reject the Federal Weed and Seed Program in Los Angeles* (Los Angeles, 1992), 6–9. For a geographer's analysis of OWS, see Tim Cresswell, "Weeds, Plagues, and Bodily Secretions: A Geographical Interpretation of Metaphors of Displacement," *Annals of the American Association of Geographers* 87 (1997): 335–36.

65. Jim Casey and Phillip O'Connor, "Raid on CHA's Wells," *CST*, 7 Apr. 1992; Mary Johnson, "Lost Lives, Lost Hopes," *CST*, 23 Oct. 1994; Mary Mitchell, "Weedy Lots Cause Thicket of Problems," *CST*, 4 Aug. 1995. In 2000, federal investigators scrutinized Chicago's OWS program after police in two municipalities just west of Chicago, Bellwood and Maywood, were allegedly "shaking down drug dealers." See Michael Sneed, "Feds Dig in to Weed Out Police Corruption," *CST*, 14 May 2000.

CONCLUSION

1. Chris Rose, "Weeds Shall Overcome," *New Orleans Times-Picayune*, 29 July 2007, http://blog.nola.com/chrisrose/2007/07/weeds_shall_overcome.html; Malcolm Jones and Cathleen McGuigan, "Toward a New New Orleans," *Newsweek*, 5 May 2008, 53. On New Orleans's pre-Katrina environmental history, see Craig Colten, *An Unnatural Metropolis: Wresting New Orleans from Nature* (Baton Rouge: Louisiana State University Press, 2005); Ari Kelman, *A River and Its City: The Nature of Landscape in New Orleans* (Berkeley: University of California Press, 2003); Craig Colten, ed., *Transforming New Orleans and Its Environs: Centuries of Change* (Pittsburgh: University of Pittsburgh Press, 2000).

2. Jim Wasserman, "Growing like Weeds," *Sacramento Bee*, 9 Sept. 2007; J. Elphinstone, "Neighborhoods Endure that Empty Feeling," *Chicago Tribune*, 18 Nov. 2007; David Streitfeld, "In the Ruins of the Housing Bust," *NYT*, 24 Aug. 2008; "Weeds Choking Bellevue's Green Space," *Merced Sun-Star*, 23 Apr. 2009, http://www.mercedsunstar.com/167/story/806497.html, and comment by monkeyinwrench, http://www.mercedsunstar.com/167/story/806497.html#Comments_Container; Victor Patton, "Dense Weeds Could Lead to Fines," *Merced Sun-Star*, 22 Mar. 2010, http://

www.mercedsunstar.com/2010/03/22/1358434/dense-weeds-could-lead-to-fines.html; Gretchen Wenner, "Weeds Grow at Greg Norman Golf Course," *Bakersfield Californian*, 26 June 2008, http://www.bakersfield.com/news/business/local/x361189707/Weeds-grow-at-Greg--Norman-golf-course.

3. Chelsea Schneider and Astrid Galvan, "Tumbleweeds Taking over Neighborhoods," *AZREP*, 31 July 2008, http://www.azcentral.com/realestate/articles/2008/07/31/20080731gr-tumbleweeds0801-CP.html; Edythe Jensen, "Chandler Ramps Up War on Weeds," *AZREP*, 5 Aug 2010, http://www.azcentral.com/community/chandlers/articles/2010/08/05/20100805chandler-targets-weeds0806.html; Diana Balazs and Weldon Johnson, "Cities Struggle to Tame Weeds Near Empty Homes," *AZREP*, 22 Mar. 2008, and comment posted on 22 Mar. 2008 by TheTruth9251, http://www.azcentral.com/community/chandler/articles/0321cr-weeds0321.html; Chelsea Schneider, "Foreclosures Create Problems for Neighbors," *AZREP*, 17 July 2008, http://www.azcentral.com/community/gilbert/articles/2008/07/17/20080717gr-thegardensporter0717.html; Edythe Jensen, "Phoenix-Metro Areas Fight Weed Problem," *AZREP*, 7 Aug. 2010, http://www.azcentral.com/news/articles/2010/08/07/20100807phoenix-cities-fight-weed-problem.html; Weldon Johnson, "Direct Action by Neighbors Seen as Effective Tool against Blight," *AZREP*, 19 June 2010, http://www.azcentral.com/news/articles/2010/06/18/20100618chandler-neighborhood-blight.html.

4. Lesterhead, "Clinton Hill's Urban Prairie," Clinton Hill Blog, comment posted on 17 July 2006, http://www.clintonhillblog.com/?p=1759; Cathy Woodruff, "Weeds Mar Traffic Islands," *Albany Times Union*, 17 Aug. 2007, http://timesunion.com/AspStories/storyprint.asp?StoryID=614844; "Volunteering Introduces College Students to Area," *Washington Times*, 24 Aug. 2005; "Volunteers Pitch in to Spruce Up Area," *Washington Times*, 23 May 2004; "Mountains of Weeds," *Florida Times-Union*, 20 Aug. 2007, http://www.jacksonville.com/tu-online/stories/082007/opi_192369650.shtml; Fran Spielman, "On the Blight Track," *Chicago Sun-Times*, 13 Sept. 2007, http://www.suntimes.com/business/555047,CST-FIN-home13.article#; Fran Spielman, "Crackdown on Vacant Lots," *Chicago Sun-Times*, 21 July 2009, http://www.suntimes.com/news/cityhall/1676165,CST-NWS-vacant21web.article#; Daniel Chacon, "City Leaders to Parks: Whack Those Weeds," *Rocky Mountain News*, 17 Aug. 2007, http://www.rockymountainnews.com/drmn/local/article/0,1299,DRMN_15_5675281,00.html; Genevieve Judge, "War on Weeds Campaign Kicks Off in Pocatello," KIFI-TV, 1 Aug. 2007, http://www.localnews8.com/Global/story.asp?S=6867563; Paul Rolly, "Weeding Out a Sidewalk Nuisance," *Salt Lake Tribune*, 5 Sept. 2007; Gary Richards, "Roadshow," *San Jose Mercury News*, 17 Sept. 2007, http://www.mercurynews.com/mrroadshow.

5. Baltimore Recreation and Parks, "Become an Urban Weed Warrior," http://web.archive.org/web/20071020002616/, http://www.baltimorecity.gov/government/recnparks/images/docs/WeedWarriorBrochure.pdf; "Urban Weed Warriors," http://web.archive.org/web/20080130145641/, http://www.ci.baltimore.md.us/government/recnparks/conservation.php; Cameron Barr, "The Prickly Task of Clipping a Habit," *Washington Post*, 3 June 2005; "Story Hillers Beat Back the Evil Weed, for Now," Milwaukee Rising Blog, posted 10 May 2009, http://wwwmilwaukeerising.net/wordpress/2009/05/10/story-hillers-beat-back-the-evil-weed-for-now/; Lynn Marshall, "Seattle Faces a Foe That's Mean, Green and Growing," *Los Angeles Times*, 18 Dec. 2005,

http://www.earthcorps.org/news.php?articleId=117); Home Depot Foundation, Case Study, 2008 Awards of Excellence, Community Trees, the Green Seattle Partnership Cascade Land Conservancy and City of Seattle, http://www.homedepotfoundation.org/pdfs/green_seattlepartnership.pdf; Lisa Stiffler, "Volunteers Spend a Soggy, Muddy Holiday Helping a Park," *Seattle Post-Intelligencer,* 17 Jan. 2006, http://www.seattlepi.com/local/255922_reforest17.html; Lisa Stiffler, "City, Conservation Groups Are Battling Invasive Plants," *Seattle Post-Intelligencer,* 16 June 2004, http://www.seattlepi.com/local/178085_ivy16.html. Some of these plants may have been in Seattle for more than a century; see Matthew W. Klingle, *Emerald City: An Environmental History of Seattle* (New Haven: Yale University Press, 2007), 126; Paul Tolme, "Itch and Sneeze," *Newsweek,* 11 Aug. 2008, 52.

6. F. A. Bazzaz, *Plants in Changing Environments: Linking Physiological, Population, and Community Ecology* (New York: Cambridge University Press, 1996), 264–79; Andrew Revkin, "U.S. Report, in Shift, Turns Focus to Greenhouse Gases," *NYT,* 26 Aug. 2004; J. Hatfield et al., "Agriculture," in *The Effects of Climate Change on Agriculture, Land Resources, Water Resources, and Biodiversity in the United States* (Washington, DC: U.S. Department of Agriculture, 2008), 59; Tom Christopher, "Can Weeds Help Solve the Climate Crisis?" *New York Times Magazine,* 29 June 2008, http://www.nytimes.com/2008/06/29/magazine/29weeds-t.html; Sharon Begley, "Get Out Your Handkerchiefs," *Newsweek,* 4 June 2007, http://www.msnbc.msn.com/id/18881800/site/newsweek/; Diego Cupolo, "A Wasteland Thrives in Jersey," *Newark Star-Ledger,* 14 Jan. 2009.

7. Alan Weisman, *The World without Us* (New York: St. Martin's Press, 2007), 3, 26–28; *I Am Legend,* DVD, dir. Francis Lawrence (Warner Bros. Pictures, 2007); Molly Wolf, *Hiding in Plain Sight: Sabbath Blessings* (Collegeville, MN: Liturgical Press, 1998), 3–6; Neal Mayerson, "Weeding for Character," *Psychology Today,* 13 April 2010, http://www.psychologytoday.com/node/41044; Donald Perry and Sylvia Merschel, "As Cities Crumble, Plants May Be at the Root of It," *Smithsonian* (Jan. 1987): 72–79. Such imagery was also used to promote the History Channel's *Life after People* series.

8. J. R. McNeill, *Something New under the Sun: An Environmental History of the Twentieth-Century World* (New York: Norton, 2000), 262–64; Richard Stalter and Steven Scotto, "The Vascular Flora of Ellis Island," *Journal of the Torrey Botanical Society* 126 (1999): 367–75; Will Turner, Toshihiko Nakamura, and Marco Dinetti, "Global Urbanization and the Separation of Humans from Nature," *BioScience* 54 (2004): 585–90.

9. Patti Smith, "Ain't It Strange?" *NYT,* 12 Mar. 2007, http://www.nytimes.com/2007/03/12/opinion/12smith.html; Sherry, "Cracks in the Pavement," Stay of Execution Blog, comment posted on 18 July 2006 by anonymous, comment posted on 22 July 2006 by Nancy Drew, http://civpro.blogs.com/civil_procedure/2006/07/cracks_in_the_p.html; Bill O' Driscoll, "Wild Times Ahead: Waiting for the End of Civilization with Anarcho-Primitivist Kevin Tucker," *Pittsburgh City Paper,* 13 July 2006, http://www.pittsburghcitypaper.ws/gyrobase/Content?oid=oid%3A28655; Linda Galindo, *Way to Grow! Cultivating the Weeds, Daisies, and Orchids in Your Organization* (Dallas: Walk the Talk Company, 2004), 19, 34, 77; Robert E. Kohler, *Landscapes and Labscapes: Exploring the Lab-Field Border in Biology* (Chicago: University of Chicago

Press, 2002), 8–9; Chet Raymo, "Science Musings," *Boston Globe*, 12 Feb. 1996; David Quammen, "Planet of Weeds," *Harper's Magazine*, Oct. 1998, 66–68.

10. Tim, "A Thistle Grows in New Orleans" comment posted on 2 Apr. 1007, ~ Tim's ~ Nameless ~ Blog, http://timsnamelessblog.blogspot.com/2007/04/thistle-grows-in-new-orleans.html; Tim, "Correction and Update," comment posted on 17 Apr. 2007, ~ Tim's ~ Nameless ~ Blog, http://timsnamelessblog.blogspot.com/2007/04/correction-and-update.html; Chris Rose, "The 60-Second Interview: Matt Perrine," *New Orleans Times-Picayune*, 25 Jan. 2008, http://blog.nola.com/chrisrose/2008/01/the_60second_interview_matt_pe.html; "St. Louis Ranked over Kansas in Sunflowers," *Berkeley Daily Gazette*, 27 July 1944. Tim later learned from his mother and his wife that some of the plants that had inspired him were cleomes rather than thistles. However, like the bull thistles (*Cirsium vulgare*) that Tim likely did see, the cleomes (*Cleome hassleriana*) were also plants growing without cultivation. Neither, native plant enthusiasts would note, is indigenous to New Orleans.

11. Edgar Anderson, *Plants, Man, and Life* (Boston: Little, Brown and Company, 1952), 205; "E. O. Wilson, interview with Scott Simon," *Weekend Edition Saturday*, NPR, 30 June 2007, http://www.npr.org/templates/story/story.php?storyId=11626223; *Encyclopedia of Global Environmental Change*, vol. 2, *The Earth System*, ed. Ted Munn (Chichester: Wiley, 2002), s.v. "weed"; William Cronon, "The Uses of Environmental History," *Environmental History Review* 17 (1993): 1–22; Lewis Mumford, "Natural History of Urbanization," in *Man's Role in Changing the Face of the Earth*, ed. W. Thomas (Chicago: University of Chicago Press, 1956), 388; William H. McNeill, "Passing Strange: The Convergence of Evolutionary Science with Scientific History," *History and Theory* 40 (2001): 14.

INDEX

Note: Page numbers in italic type indicate illustrations.